Cadets on Campus

★ ★ ★

Number 155
Williams-Ford Texas A&M University
Military History Series

Cadets on Campus

History of Military Schools of the United States

★ ★ ★

John Alfred Coulter II

TEXAS A&M UNIVERSITY PRESS
COLLEGE STATION

This paper meets the requirements of ANSI/NISO Z39.48–1992
(Permanence of Paper).
Binding materials have been chosen for durability.
Manufactured in the United States of America

LIBRARY OF CONGRESS CATALOGING-IN-PUBLICATION DATA

Names: Coulter, John A., 1953– author.
Title: Cadets on campus: history of military schools of the United States /
    John Alfred Coulter II.
Other titles: Williams-Ford Texas A&M University military history series;
    no. 155.
Description: First edition. | College Station: Texas A&M University Press,
    [2017] | Series: Williams-Ford Texas A&M University military history
    series; no. 155 | Includes bibliographical references and index.
Identifiers: LCCN 2016057975 (print) | LCCN 2016059789 (ebook) | ISBN
    9781623495213 (printed case: alk. paper) | ISBN 9781623495220 (ebook)
Subjects: LCSH: Military education—United States—History.
Classification: LCC U408 .C68 2017 (print) | LCC U408 (ebook) | DDC
    355.0071/073—dc23
LC record available at https://lccn.loc.gov/2016057975

In memory of

Col. Harry S. Temple (1911–2004)

A true Virginia gentleman, soldier, Virginia Tech historian, veteran of World War II and Korea, designer of the Presidential Medal of Freedom, chief of the US Army's Institute of Heraldry, and author of *The Bugle's Echo: The Chronology of Cadet Life at the Military College at Blacksburg, Virginia*, volumes I–VI.

Colonel Temple started me on the study of the unique history and culture of the military school.

Dr. Richard Lee Henderson (1944–2011)

Professor and Coordinator for Organizational Leadership in the PhD program at University of the Incarnate Word, San Antonio, Texas; holder of the Sister Theophane Power Chair, 2000–2005; and campus administrator, China Incarnate Word campus, February–July 2010.

Without Dr. Henderson's inspiration I could never have completed this educational journey, which included three states, three countries, and the war in Afghanistan.

Contents

Illustrations

Preface

With the establishment of the Military Academy at West Point, a new educational format that included military training was introduced in the United States. From 1802 to 2014, 842 military schools have operated in the United States (appendix A). These schools, both privately and publicly funded, provided education at the elementary, secondary, and college levels. The importance of understanding this subject has grown as the number of military charter and public secondary schools has increased tenfold since 1999.

This volume's overarching thesis is that the number of military schools, enrollment in them, and their locality have been heavily influenced by political, cultural, and economic factors ever since the military school concept was launched in the United States in 1802. This study required the assembly of the most extensive list of military schools and best available information on their dates of operation. Explanations for openings, closings, and trends in enrollment are presented and explained in their relationships to political, cultural, and economic trends regionally or nationally.

Lester Webb provided a nineteenth-century study that served as a starting point for this book. It was Webb's hope that others would "dig deeper into the history of military education, a field hitherto ignored and neglected by educational historians." Likewise, Alvan Hadley identified the need for future research to address the historical "social, economic, cultural, or political circumstances" that impact military schools and to pull together meaningful statistics to provide greater understanding of military school failures. His work proved to be the first comprehensive list of military schools and was a starting point for the formation of the list of 842 military schools and their years of operation addressed here.[1]

Historians traditionally have attributed the popularity of military schools to the Southern culture and regional orientation. On the contrary, military schools have had a wide regional representation throughout the United States. The popularity of the military school, rather than being a product of Southern culture, should be attributed to historical, economic, and cultural events that stretch far across the nation.

The influence of political, cultural, and economic factors upon the American military school does not overshadow individual educational leadership. Four men stand out in their impact on military schools in the United States. The expansion of the military format and development of the military school culture were facilitated by four educators and associated alumni of three institutions: Alden B. Partridge (1785–1854), founder of Norwich University; Sylvanus Thayer (1785–1872), superintendent of West Point; Francis H. Smith (1812–90), superintendent of Virginia Military Institute; and Stephen B. Luce (1827–1917), champion of maritime academy education.

The military school format is characterized by students as cadets or midshipmen in uniform and under military discipline. The format began in the United States when Congress created the United States Military Academy at West Point, New York, but the opening of West Point in 1802 did not originate that educational environment; rather, it was imported from Europe. Since then the military format has experienced periods of varying popularity in the educational history of the United States.

The number of military schools grew steadily after the formation of West Point in 1802, with 171 military schools operating between 1855 and 1866. During that time the line between secondary- and higher-education military institutions was ill-defined, with terms such as *institute, academy,* and even *college* used freely for both levels.[2] But clearly, military schools were dominated by secondary schools, with approximately 128 (75 percent) at that level.

The growth of military schools peaked between 1903 and 1926, when at least 278 to 280 military schools operated in the United States. By that time the line between secondary and higher education was well defined. Secondary military schools numbered approximately 257, or about 91 percent of the military schools in the country. Public education was firmly established by then; almost all secondary military schools were private schools.

After the start of America's involvement in the Vietnam War, the twentieth century saw the number of military schools in the United States decline dramatically. In the early 1960s approximately 203 military schools were operating in the United States. Antimilitary sentiment "engendered by the Vietnam War was reflected in much public anathema toward military education." As schools closed or dropped their military requirements, the number of military schools declined by 63 percent. In 1998 only 75 military schools remained in the United States. Of those, 42 were secondary schools, 4 of which were public military schools.[3]

The dramatic decline in the number of military schools in the United States began its reversal with the opening of the Chicago Military Academy in 1999, twenty-four years after the fall of Saigon and the end of the Vietnam War. Since 1999, forty-four military schools have opened in the United States; seventeen were public military high schools, twenty-one were charter military high schools, and six were private military schools. Among the newly established military schools, 86 percent were in public education rather than private schools, as most military schools traditionally had been. In 2014 there were ninety-seven military schools, seventy-four of which were secondary schools, an increase of more than 57 percent since 1998. The trend of establishing new military institutions does not appear to have slowed; the New Orleans Military Maritime Academy opened in the fall of 2011 in Louisiana, Riverside County Education Academy in California in 2011, Hollywood Hills Military Academy in Florida in 2012, Paul R. Brown Leadership Academy in North Carolina in 2013, North Valley Military Institute in California in 2013, and the Utah Military Academy in 2014.

In 2014 the United States had twenty-two military schools operating on the college level. This number included the five federal service academies, twelve military colleges or universities, and five military junior colleges. Forty private military schools operated on primary and secondary levels. Two of the private schools were primary schools, and thirty-seven schools were primarily or totally at the secondary level.[4] There were also public and charter schools; these included nineteen public secondary military schools and seventeen charter military schools. The total number of schools, private and public (charter included), totaled ninety-eight institutions, with approximately 52,575 cadets and midshipmen (see appendix B).

Organization

The opening chapter examines the military school culture and those factors that explain its strength in military schools. Chapter 2 addresses the establishment of the US Military Academy, and chapters 3–5, organized chronologically as much as possible, address three of the four influential figures in the military school movement: Alden Partridge, Sylvanus Thayer, and Francis Smith. Chapter 6 deals with the impact of the Civil War on military schools, and chapter 7 focuses on the establishment of maritime military schools, the nautical events that influenced their creation, and the fourth influential figure in military school movement: Stephen B. Luce.

Chapter 8 addresses the social, economic, and political events that have impacted military schools following the Civil War, the history of African American military schools, and the integration of military schools subsequently. Next, chapter 9 deals with the adoption of the military school format by religious denominations starting in 1838. The next two chapters, 10 and 11, take the reader from 1898 through 1998—through World War I, the Great Depression, and World War II. Chapter 12 focuses on the impact of the Vietnam War and the young counterculture, with their drastic impact on the closing of military schools and lowering enrollment, are addressed.

Next, chapter 13 addresses military schools from 1999 to 2014 and gives the reader a look at the proponents and opponents of the military education format and their arguments. It tracks the political and cultural influences and the individuals who influenced a dramatic resurgence in the number of military schools, and it provides a quantitative study of socioeconomically similar neighborhood, magnet, and charter secondary schools compared to their counterpart public and charter military schools. Finally, chapter 14 provides conclusions associated with the study's overarching thesis associated with political, cultural, and economic factors and illustrates them graphically against the number of military schools historically and geographically.

Definitions

Military school. An institution of education that organizes all or a portion of the student body) into a corps of cadets under military discipline. It requires all members of the corps to be habitually in military uniform when on campus. It has as its objective the development of the cadet's character through military training and regulation of conduct according to principles of military discipline.[5] The program is not limited to technical skills such as pilot or nautical navigation, nor is it described as a drug treatment or behavioral modification camp. The school may provide elementary, secondary, or college education.

Charter school. An institution of education not state or federally owned but receiving its charter and most of its funding from the local, state, or federal government. It is usually owned by an individual or a corporation. Charter schools are similar to private schools in that they are independent, and their students attend by choice. Unlike private schools, charter schools are accountable to governmental authorities for results.[6]

Neighborhood school. A public school with attendance boundaries, generally open to all students who live in the school's designated area. In some cases students from other areas may be offered enrollment, but students living within that school's boundaries have priority.[7]

Magnet school. A public school specializing in a specific subject area, such as mathematics and science, fine arts, world languages, or humanities. These schools may require application or lottery selection for attendance.[8]

Cadet. A cadet is a military student at a military school. This does not infer an association with the armed forces of the United States or a military obligation, but rather a lifestyle. Some schools use the term *midshipman* or *midshipwoman.* The term *cadet* is used for military students regardless of gender or academic class. Some military schools have both civilian students and military cadets in their student bodies.

Service academies. College-level institutions operated under the authority of the US federal government and associated with an armed service or the merchant marine. The cadets or midshipmen are organized into a corps of cadets, brigade of midshipmen, or similar formation. These institutions grant baccalaureate degrees with the objective of granting military commissions. Service academies develop cadet character through military training and regulation of conduct according to the principles of military discipline. They require all members of the corps to be habitually in military uniform when on campus.

Senior military college. A college or university that grants baccalaureate degrees and organizes all or a portion of the student body into a corps of cadets. It requires all members of the corps to be habitually in military uniform when on campus. Its mission is the development of the cadet's character through military training and regulation of conduct according to principles of military discipline similar to those maintained at the service academies.[9] Unlike service academies, postgraduate military service is not a requirement of its graduates.

Junior military college. A military school that does not grant baccalaureate degrees but provides high school and junior college education and organizes all or a portion of the student body into a corps of cadets. It requires all members of the corps to be habitually in military uniform when on campus. It has as objectives the development of the cadet's character through military training and regulation of conduct according to principles of military discipline similar to those maintained at the service academies.[10]

Maritime military academy. Institutions conducted as military colleges whose purpose is to prepare officers of the merchant marine rather than the US Navy. The vessels of the merchant marine are civilian owned and crewed. In time of war these vessels become an auxiliary force to the US Navy for a logistics mission associated with the transport of troops and supplies.

Lost Cause. The literary, intelligential, and social conscience of the post–Civil War South. The Lost Cause encompassed the beliefs held by the white population of the South that helped them deal with defeat and change. It presented the "wartime sacrifice and shattering defeat in the best light."[11] The objective was to justify the actions of the Southern states and find positive historical interpretation of the war. It focused on the military rather than the less positive political and social aspects of the war. Gen. Jubal A. Early is cited as the former Confederate who was most influential in the shaping of the Lost Cause with a reverential view particularly of Gen. Robert E. Lee.

Regional designations. Regional designations are based on the US Census Bureau's division of the United States for the census. The North (Northeast is used by the Census Bureau) includes Connecticut, Maine, Massachusetts, New Hampshire, New Jersey, New York, Pennsylvania, Rhode Island, and Vermont. The Midwest (or North Central region) contains Illinois, Indiana, Iowa, Kansas, Minnesota, Michigan, Missouri, Nebraska, North Dakota, Ohio, South Dakota, and Wisconsin. The South has Alabama, Arkansas, Delaware, the District of Columbia, Florida, Georgia, Kentucky, Louisiana, Maryland, Mississippi, North Carolina, Oklahoma, South Carolina, Tennessee, Texas, Virginia, and West Virginia. Because of the number of military schools located in the territory of Puerto Rico and its geographic location, I also include it in the South. The West has the remaining states of Alaska, Arizona, California, Colorado, Hawaii, Idaho, Montana, Nevada, New Mexico, Oregon, Utah, Washington, and Wyoming.

Political regional designations. The division of the United States in the Civil War is addressed in three groups. The Union States, which firmly remained with the United States, are California, Connecticut, Illinois, Indiana, Iowa, Maine, Massachusetts, Michigan, Minnesota, Nevada, New Hampshire, New Jersey, New York, Pennsylvania, Rhode Island, Vermont, and Wisconsin. The Border States, which were divided and had factions on both sides, are Delaware, Kansas, Kentucky, Maryland, and Missouri. The Confederate States are Alabama, Arkansas, Florida, Georgia, Louisiana, Mississippi, North Carolina, South Carolina, Tennessee, Texas, and Virginia.

Limitations

This study uses documentation of the 842 US military schools over the period 1802–2014 to reach conclusions of a historical nature based on the schools' years of operation, their founders' backgrounds and motives, and other contributing factors. In many cases individual schools have very limited documentation.[12] For example, the record of Daniel Boone Military Institute in Kentucky is limited to its act of incorporation, and that of Elm City Military Institute (likely also known as Stowe Military Academy) in Connecticut is limited to its act of incorporation and a mention in a biography.

Prior to the advent of public education, many military schools were basically one-room schoolhouses with fifteen to twenty-five cadets. The limited historical footprint of schools, especially those with short life spans, does not always provide the details required to confirm the exact year of establishment, year of closure, year of transition to or from military school format, or founder. The confirmed dates of operation have been used to estimate years of operation. Because analysis, particularly prior to 1900, is based on twelve-year time spans, accuracy of the number of schools operating in any particular year is minimal.

The school policies regarding uniform and discipline requirements are largely unavailable. If schools historically used the term *military* in their title, and no evidence was found that uniforms and military discipline were not required, they are assumed to have been military schools and are included in the statistical examination of the number of operational military schools in the United States. Another limitation includes difficulty in determining whether early military schools had elementary, secondary, junior college, or college orientation. Many of the schools listed actually included multiple levels of education. Before the twentieth century the lines between secondary and higher education were often blurred, and there is often little attempt to identify schools according to the level of education they offered.

The list of military schools presented in appendix C presents 842 military schools in operation between 1802 and 2014. This appendix displays

changes in names, many of which occurred when schools changed ownership. Where research indicated that the same faculty, facilities, or cadets
remained, I do not designate the institutional change as a new school but
track it as an evolution of the same institution. In some cases this does not
match the historical interpretation of particular schools, but my view is that
the list better reflects historical reality.

Finally, chapter 13, "Resurgence of an Old Educational Tradition,"
includes quantitative study of socioeconomically similar neighborhood,
magnet, and charter secondary schools versus their counterpart public and
charter military schools. The sample of only nineteen public or charter
military schools is such a small sample size that statistical conclusions are
limited. Further, performance indicators in public and charter military
schools are limited to dropout rates, graduation rates, attendance, and
standardized test scores.

Acknowledgments

I would like to thank my wife, Laura, for her encouragement over the past seven years. I also thank Judith E. Beauford, chair of my committee, who has guided me with great patience, and the members of the committee: Dorothy Ettling, for providing critical support in my final months of classwork; Kevin Vichcales, for agreeing to join the committee from the History Department; and Paul Davis, who volunteered his time from outside the university community. Others who supported my work are the staff and faculty of University of the Incarnate Word, especially Dreeben School of Education, including Absael Antelo, and the Center for Veteran Affairs. Bruce Allardice, professor of history at South Suburban College, Illinois, kindly loaned me his research files, and Jennifer Speelman of The Citadel provided her hard-to-find master's thesis. Kim Hays provided valuable insights, Linda Hanser Powell made editorial suggestions, and Rev. Walter Prehn, head of Trinity School of Midland-Odessa, gave valuable insights on Episcopal military schools. Bobbie St. Clair labored long hours and provided excellent editorial work.

Others among hundreds who assisted in my research included Paula Allen, Texas Military Institute; Ruth Atkins, Lake Elsinore Historical Society, California; L. Augustowski, Peekskill Military Association, New York; Philip Brach, vice president, St. John's College High School Alumni, Washington, DC; H. Clark Burkett, Historic Jefferson College, Mississippi; Charlie Cahn, headmaster, Suffield Academy, Connecticut, Lt. Col. Steven L. Dahlgren, Cretin-Derham Hall High School, Saint Paul, Minnesota; Peter Dans, La Salle alumnus; Jane Dieffenbacher, historian, Fairfield, New York; Kevin Dolan and John Downs, Kemper Military School alumni; Bob Duncan and David Bishop, Skillman Library, Lafayette College, Pennsylvania; and Tersina DiPietro, Bordentown Military Institute Alumni Association; Sean Flynt, Office of Communication, Samford University, Alabama; and Mando Garcia, archivist, St. Edward's University, Texas.

Others who helped me were Mike Gibson, archivist, Loras College, Iowa; Jennifer R. Green, Central Michigan University; Winn Herrschaft, Washington Country Museum, Oregon; Rob Hosier, Castle Heights Military Association Alumni Association; Andy Jackson and Les Marsh, Sons of Confederate Veterans, Shelbyville, Tennessee; and Randy Jennings, Carolina Military Academy Alumni Association; Kelly C. Jordan, Sharon Kellam, and Mary Kludy, VMI Archives; Christin Loehr, Julia Tutwiler Library, University of West Alabama; Doug Halbert, Brown Military Academy Alumni Association, California; and Hunter Leake and Al McCormick, Wentworth Military Academy; Henry Mitchell, Mitchells Publications; Linda Moody, Educational Directories Inc.; Ruth Ann Montgomery and Frayda Salkin, archivist, McDonogh School, Maryland; and Patricia Solin, Pennsylvania Schoolship Association.

Still others who provided assistance were Raymond F. Shea, Union Landing Historical Society, New York; and Judee Showalter, McGraw-Page Library, Randolph-Macon College; Eugene G. Stackhouse, Germantown Historical Society, Pennsylvania; Beth Shutts, Hamden Historical Society; Gary Toppings, Bernhard Thuersam, and Kathleen Urbanic, archivists, Sisters of St. Joseph, Rochester, New York; staff of John Varner Library, Auburn University; Carl Van Ness, archivist, University of Florida; Kenneth Veronda, headmaster, Southwestern Academy, California; Laura Vetter, archivist, Episcopal High School, Virginia; Mike Vlieger, chief executive officer, Peacock Military Academy Alumni Association, Texas; Carol Weiss, Nyack, New York, village historian; Richard Weaver, Spring Hill College, Alabama; and Gail Wiese, archivist, Norwich University, Vermont.

Cadets on Campus

★ 1 ★

Military School Culture

The literature on school culture is built upon organizational behavior. According to Edgar Schein, organizational culture is "the pattern of assumptions" that members have developed or learned to cope with through the experiences of the group. These cultures are passed to new members and considered the acceptable manner to address the challenges of everyday organizational life. These cultural understandings are found within three levels: artifacts, values, and tacit assumptions. The strongest of the three are those tacit assumptions that members are not conscience of and act on without forethought. The strength of organizational culture is based on the homogeneity of the group, socialization of the group, and length and intensity of the shared experiences.[1] Fig. 1.1 illustrates the military school organizational culture in terms of examples of artifacts, values, and tacit assumptions.

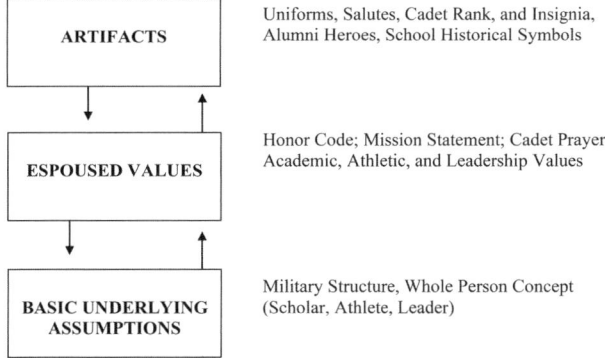

Fig. 1.1. Military School Organizational Culture. Adapted with permission from E. H. Schein, *The Corporate Culture Survival Guide*, rev. ed. (San Francisco: Jossey-Bass Publishers, 2009).

Texas Military Institute Corps of Cadets march-on for pass-in-review, 2005.
(Courtesy Allan Rupe/TMI—The Episcopal School of Texas)

Military school discipline, uniforms, daily ceremonies, and shared experiences provide a framework through which homogeneity, socialization,
and the degree of shared experiences are amplified. According to Kim
Hays, in a study of three military boarding schools, the military school
experience is full of dilemmas and impassioned debates: "Yet there is an
order within it, an order that comes from the feeling that one belongs
to something important to which one can contribute." That feeling of
belonging was reflected by Remi Hajjar in his study of a public military
school in which the academy "promoted a heightened sense of agency and
efficacy, fosters a concern for classmates and others."[2]

Artifacts of military school culture include uniforms, insignia, and intensity of the emotions associated with team efforts and unit competition,
as well as the manner in which cadets address their peers, upperclassmen,
faculty, and military staff. These artifacts play a significant role in the socialization process in the private secondary military school. A cadet recently
described wearing uniforms as an effect that made "everyone equal so that
you could not tell the rich from the poor." That theme is echoed by Harry
Temple in his experiences at a military college in the 1930s and also by
Leigh R. Gignilliat, the superintendent of Culver Military Academy from
1910 to 1939, who described military uniforms, room inspections, and

military discipline as creating a "spirit of democracy" in which all boys had equal opportunity. This is confirmed by Hajjar in his study of a public secondary military school in which he concluded that civility and sense of concern for others were traits enhanced by the school environment.[3]

Values espoused by military schools are generally associated with two kinds of codes: one positive in character and the other negative. The best known of the positive codes is the motto "Duty, Honor, Country" of the first American military school, the US Military Academy at West Point. This motto was designated in 1898 and described in the academy's first cadet handbook in 1907 (twenty-eight years prior to establishment of the academy's honor code): "The principles of sentiment and of conduct embodied in these three words penetrate the whole life of the institute and have been exemplified in the lives and deaths of her sons for over a hundred years." Other examples of positive codes include Texas Military Academy's core values: spiritual maturity, moral integrity, well-roundedness, and scholastic aptitude.[4] Fork Union Military Academy's principles include dedication, integrity, and consistency. Norwich University's guiding values include eight elements: "We are men and women of honor and integrity. We shall not tolerate those who lie, cheat, or steal. We are dedicated to learning, emphasizing teamwork, leadership, creativity, and critical thinking. We respect the right to diverse points of view as a cornerstone of our democracy. We encourage service to nation and others before self. We stress being physically fit and drug free. To live the Norwich motto—I will try!—meaning perseverance in the face of adversity. We stress self-discipline, personal responsibility, and respect for law. We hold in highest esteem our people and reputation."[5]

The most common negative code is that which a cadet must never do: break the military school's honor code. "The fundamental principle governing a cadet's conduct is the honor code, which is fundamentally the same at all schools . . . but with variant phrasing: A Cadet will not lie, cheat or steal nor tolerate those who do." That code became a written standard in 1935 at West Point, when the principal drafter, Cadet Lawrence Lincoln of the Honor Committee, typed it while he was in the field during summer training: "The code was not my product nor even the product of the entire Honor Committee; it was the product of years in the Corps developing the Honor System and hundreds of Cadets passing it on, in part by the Honor Book; but mainly by word of mouth and practice."[6]

The honor system or code was not set down by Sylvanus Thayer, West Point's superintendent, 1817–33, as some suggest. Thayer's source of

Virginia Tech freshmen cadets take the Honor Code oath, 2014. (Courtesy Virginia Tech)

honorable conduct was the code of ethical behavior that was expected of all commissioned officers in the army in the 1800s. Among the communicated behaviors was that an officer's or a cadet's word was his bond. This behavioral expectation was inherited from the British army, where an officer was expected to be a gentleman and "a gentleman's word is his bond."[7]

According to Edgar Schein, the final and most powerful element of organizational culture is shared tacit assumptions, which result in perceptions, feelings, and behaviors that are learned and taken for granted and are not debatable. Military schools' regulations, codes, rewards, and punishments create an environment in which there are very clear expectations and outcomes resulting from either failure to act in a certain way or rewards for positive performance. Organizational rewards in military schools are not associated with salary increases but rather with rewards that are related to self-esteem, including "reputation, or prestige, recognition, attention, importance or appreciation." Behaviors of cadets become automatic, including wearing a uniform cap when outside, removing it when inside, saluting, performing military greetings, good posture, and good behavior in formation and drill.[8]

The most common source of an organizational culture, according to Schein, comes from the founder of the organization. These founders shape the culture and select the first employees, who they believe will maintain their vision. They often define and influence the socialization process for new hires. Using the model of the American military schools as an educational movement, the founders, Alden B. Partridge, Sylvanus Thayer, Francis H. Smith, and Stephen B. Luce, shaped the culture and influenced the initial schools' leaders.

Missouri Military Academy is a good example of how the influence of the founders of the military school movement has extended to the present. The academy was established in 1889 by A. F. Fleet, who was followed by A. K. Yancey, who reopened the school after a fire. These men attended Fleetwood Academy and Howard College, respectively. Both military schools were run under the supervision of men educated by Francis Smith of VMI when Fleet and Yancey were cadets. Smith was educated under Sylvanus Thayer at West Point, and that institution was influenced by Alden B. Partridge. The influences of Missouri Military Academy's founders and the influences of the military schools that impacted their lives previously were evident in the school the author visited in December 2011. It was clear that it was a military school patterned on the model of discipline, moral behavior, and character development established by the aforementioned men who have had the greatest impact on military schools in the United States.

Stephen B. Luce garnered great influence at New York Maritime Academy and had direct influence on the establishment of Massachusetts Maritime Academy in 1891; the US Merchant Marine Academy, established in 1943; and, indirectly, the California Maritime Academy in 1929, the Maine Merchant Marine Academy in 1941, and the Texas Merchant Marine Academy in 1962.

Graduates from Partridge's Norwich University and Thayer's West Point, as well as some associated with Smith's Virginia Military Institute, Texas A&M College, and Virginia Tech, played a major role or held significant positions of responsibility in the newly founded Air Force Academy in 1954. Many of the military schools in operation in 2014 were not founded by graduates of those institutions; however, the military school model established by the forefathers has clearly been used as the guide for their operation. This is exemplified at the Army and Navy Academy, which was founded in California in 1910:

All notable schools have a grand history and great sense of tradition, and the Army & Navy Academies are no exception. Since their foundings in 1910 our Cadets have felt loyalty and kinship with our nation's institutions of military leadership and training. Our country is proud of the five service academies and the contributions they have made in the development of our nation. From the founding of the first military training school, the United States Military Academy—also known as West Point—in 1802, to the establishment of the United States Air Force Academy in 1954, the Army & Navy has inculcated their values, traditions and sense of duty to one's community and nation. Tradition plays an important part in our Cadets' educational endeavors and accomplishments.[9]

Although Schein is highly influential, other scholars define organizational culture from a different perspective. William Ouchi has stated that "organizational culture consists of a set of symbols, ceremonies, and myths that communicate the underlying values and beliefs of that organization to its employees."[10] The military school culture is heavy in its use of symbols, whether they are uniforms, flags, or salutes, symbolizing respect and relationship between leader and follower. Ceremonies are occasional occurrences in most nonmilitary schools, whereas in military schools they occur multiple times a day with formations, parades, and often reveille, with

Virginia Military Institute annual New Market ceremony, 2015, commemorates the bravery and sacrifice of the cadets called to serve at the Battle of New Market. (Courtesy Virginia Military Institute)

the raising of the national flag, and retreat, with its lowering. Myths and stories are often associated with the school's legendary alumni, who tell of wartime exploits and athletic victories.

At Texas Military Institute, Gen. Douglas MacArthur is not only remembered as a hero of World War I, World War II, and the Korean War, but also as the school's quarterback and football team captain. These legends enhance the school's rich environment and encourage conversations among cadets that often focus on military relationships, responsibilities, and interactions. At Virginia Military Institute the Battle of New Market is observed with Roll Call formation, in which ten cadets respond for those who in 1864 "died on the field of honor." The ceremony includes the reading of the New Market Prayer and playing of "Amazing Grace," and the cadets march past the New Market Monument.[11] Such practices link the present with the past and set a standard of leadership and devotion to duty.

Cohesion and Esprit de Corps

As previously mentioned, the strength of organizational culture is a product of socialization and the length and intensity of the shared experiences. James Burke saw military culture as having four elements: discipline, professional ethos, ceremonies and etiquette, and cohesion and esprit de corps. Schein addresses discipline, professional ethos, and ceremonies and etiquette within his concepts of artifacts, values, and tacit assumptions. But to further understand the unique degree of culture that develops in military schools, an examination of cohesion and esprit de corps is required.[12]

Cohesion is "the emotional bond of shared identity and camaraderie among [cadets] within their . . . unit in sociological terms, horizontal or primary group integration." Gen. Edward Meyer, West Point class of 1951, said cohesion is "the bonding together of [cadets] in such a way as to sustain their will and commitment to each other, the unit and mission accomplishment despite . . . stress." Rod Andrew's study of the Southern military school tradition expanded further: "This sense of organic unity pervaded cadet life. Cadets of the same class and same company worked, studied, played, ate, worshiped, and lived together, usually seven days a week."[13] Cohesion also is a process of vertical or hierarchical bonding, in which cadet leaders using a positive leadership style will influence the incorporation of the leader's goals and interpretation of mission.

Esprit de corps, on the other hand, is "the commitment and pride [cadets] take in the larger military establishment to which their immediate unit belongs; it is the outgrowth of secondary group relations or again formally

United States Naval Academy plebes climb the Herndon Monument, 2014, marking the end of their first year and a step toward becoming third-classmen. The tradition began about 1957. (Creative Commons)

vertical integration."[14] This is where a cadet's loyalties extend beyond his or her immediate classmates in his company and to the entire class, the company, and the corps of cadets.

The degree of both cohesion and esprit de corps in military schools varies. At the upper end of the scale, strong cohesion and esprit de corps will commonly appear at military colleges or federal military academies, where cadets experience a four-year bond and common stressful experiences. The degree of esprit de corps can be impacted by a shorter program, as in two-year military junior colleges or military secondary boarding schools, with their variation in retention and degree of demand on the cadets. Finally, military schools that are not boarding schools can achieve a level of cohesion and esprit de corps if the program facilitates retention and sufficient challenge for shared experience.

A part of the classic military culture from the first years of West Point, and central to all the military colleges and federal military academies and to a lesser degree most other military schools, is the class system. "Under the class system, a cadet attained more privileges vis-a-vis underclassmen in each year as he rose to the sophomore [third class], then to the junior [second class] and finally to the Senior Class [first class] rank. Class distinction bonded the advanced groupings as progressive forces on the privilege scale,

Virginia Military Institute rats meet their cadre for first time, 2015. (Courtesy Virginia Military Institute)

while at the same time bonding the underprivileged beginning group in a common cause defense."[15]

The least privileged class, of course, is the newly arrived freshmen (fourth classmen), who are given such traditional designations as rats, plebes, knobs, rooks, fish, and doolies. The experience of Cadet Harry Temple, Virginia Tech class of 1934, can in varying degrees apply to those new cadet trainees throughout the American military school's two hundred years: "Instead of making robots, the system strangely enough engendered creative, imaginative, and innovative individuals. A sense of self-discipline, self-reliance, loyalty, and responsibility grew from that sharing of traditions, standards and ideals. It forged solid and lasting bonds among the classmates who stuck it out."[16]

Allegiance Conflict

One aspect of cadet culture that emerges through cohesion and esprit de corps is the development of a strong sense of allegiance to one's classmates, company, traditions, and corps of cadets. Unfortunately this allegiance does not always mean the cadets are in synch with the regulations and desires of the administration or military department. There will often grow to a varied degree a notion that "the corps runs the corps." In a healthy manner this attitude means the cadets will use the good leadership they have been taught and address challenges and ethical dilemmas in a manner in accordance with school regulations. But this also can place cadets and their traditions and unwritten mores in conflict with the school's administration.

There are many historical examples of the conflicts between the cadets of military schools and the schools' administrations. Some clashed over policy enforcement or changes. One of the earliest was the result of a cadet's discontent with Sylvanus Thayer's order against alcohol on the West Point campus. On Christmas Eve, 1826, Cadet Jefferson Davis, the future president of the Confederacy, and other cadets participated in a celebration interrupted by the officer of the day. Cadet Davis obeyed orders to go to his room under military arrest; however, 19 cadets of the 250-man corps continued celebrating, ending the night with the barracks in a shambles and themselves dismissed. Another example was in 1912 at Mississippi Agricultural and Mechanical College, where cadets conducted a strike over restrictions associated with socializing with female civilian students. The strike was broken with 325 of 808 cadets of the college dismissed.[17]

Cadet perception of the duty to uphold the traditions of the corps regardless of institutional regulation brought about the dismissal of 64 of the 139 cadets at the Citadel in 1898. Cadets were caught physically removing from campus a cadet who had reported a violation of a regulation. Tradition held that reporting a deviation from policy was acceptable only when performing official duties, such as those as officer of the guard. Among the cadets dismissed for failure to terminate their actions upon the arrival of academy officials was the cadet adjutant and two of the three cadet company commanders. The senior class was reduced from 30 to 5 cadets. Once again, cadets as a group committed what they saw as their obligation to the traditions of the corps, disobeyed orders, and were dismissed.[18]

Dissatisfaction with school administration at Texas A&M in 1908 brought the majority of the senior, sophomore, and freshmen classes to leave campus in sympathy with faculty complaints of dismissals without cause, mess hall mismanagement, and other grievances related to presidential ethical failures. Cadets have historically reacted to "acts by school administrators that insulted their sense of 'honor' or manly independence." In the case of Texas A&M, unity of efforts by parents, the majority of faculty, and alumni resulted in the departure of the college president, but dismissals of cadets reduced the college from to 580 to 375.[19]

In 1924 the Clemson Corps of Cadets conducted a walkout in protest of worm-infested, moldy, and spoiled food. With no satisfaction from the administrative, and the dismissal of the most popular cadet on campus, approximately 500 cadets left campus. The result was that 23 cadets were dismissed, 112 were suspended, 356 cadets withdrew, and 63 were allowed to disenroll honorably. During World War II, Capt. Harry Temple, Virginia Tech class of 1934, was serving in Europe when he heard on a German propaganda broadcast that his alma mater's cadets were in revolt. In reality they, too, had conducted a protest over poor food. The wartime commandant took quick and immediate action and reduced every cadet officer involved to private.[20]

Military School Culture
vs. Culture of the Armed Forces

The organizational culture of a military school and that of the armed forces are not equivalent. In 1935 the president of Fork Union Military Academy summarized the difference: "Militarism is intensive preparation for war [whereas] military training in a Christian school is using the magnificent discipline of military life without any of the evils or spirit of militarism." The culture of the armed forces is characterized by combat and the masculine warrior. The armed forces' organizations, infantry divisions, fighter wings, and aircraft carriers, unlike military schools, are organized and function for their service in combat. The primary role of the armed forces is preparation for and conduct of combat, and their culture is synonymous with that image. This is stated in the second line of the Soldier's Creed: "I am a Warrior and a member of a team."[21]

This ethos of combat has very limited application to today's military schools. During the Civil War, six military schools participated in combat (Georgia Military Institute, Virginia Military Institute, the Citadel, the University of Alabama, West Florida Seminary, and the Confederate Naval Academy). During World War II, only one, the US Merchant Marine Academy, provided cadets as crew members of vessels in combat. Presently, there are only four military schools within the federal military academies that require postgraduate military service: the US Military Academy, the US Naval Academy, the US Air Force Academy, and the US Coast Guard Academy. The remaining school, the US Merchant Marine Academy, does not require military service; instead, it requires maritime service, which could be as a maritime lawyer or working in one of the nation's ports, but in turn, the combat ethos does not apply to it.

After World War II the amount of tactical military training decreased to a point that by 2000 it was absent at the secondary level. Military schools are not organized with a focus on combat. This applies to military colleges and universities as well as secondary military schools. Between 2006 and 2009, only 35 percent to 52 percent of the cadets who graduated from the Citadel, Texas A&M, and Virginia Military Institute accepted a service commission upon graduation. Military high schools that maintain Junior Reserve Officer Training Corps (JROTC) programs are federally mandated not to conduct any training related to combat. The violence is instead limited to the athletic fields, with emphasis on developing well-rounded cadet athletes. Military colleges and universities do not require military service upon graduation; however, several military colleges require attendance in Reserve Officer Training Corps (ROTC) college classes.

The Citadel, as early as 1920, sought to "prepare men for civil pursuits by giving them a sound education reinforced by the best features of military training." From its earliest years, Virginia Military Institute used military discipline "as a means of forming and developing individual character." Although the cultures of the armed forces and the military schools are different, there are some commonalities. Uniforms, rank insignia, saluting, military common language, and protocols are similar among the armed forces and military schools. In the armed forces, the rifle is a weapon to kill the enemy; in the military school, it is an instrument used to teach drill. In modern military schools drill with arms becomes an exercise in leadership for the cadets, a means of developing attention to detail, standards, and interpersonal skills and not a preparation for service in the armed forces.[22]

According to Fork Union, a school not affiliated with any JROTC program, "the goal of the military school is not to create 'soldiers' but to build 'solid citizens' who embody the values of integrity, honor, duty, self-discipline, and service to others." A public military school, Chicago Military Academy, states that it employs "a military structure to concentrate on academic achievement and individual responsibility." According to William Trousdale, in a study of ten military boarding schools, "military schools have justly and correctly shown that militarism and military style discipline in an educational environment involve wholly different concepts."[23]

A strong organizational culture will determine the means of dealing with external problems, and consensus is a critical part of that strength. According to Schein, this process culturally requires consensus in strategy development based on the core mission, goals, and means to accomplish those goals, as well as criteria for measuring success and adjustments to remediation or repair of strategy.[24] The external threat for a military high school is the less desirable influences brought into the school by its cadets from their homes and neighborhoods. The military school is countercultural, and a strong socialization process for both cadets and new faculty is critical to help guide and strengthen methods of dealing with the external problems.

Pillars of the Military School

Sylvanus Thayer, the superintendent of West Point, 1817–33, redefined the academy into what would become the model for the American military school. At West Point his efforts were focused on two pillars: intellect and character. In general terms, the intellect pillar provided for a demanding curriculum supported by a regime of regulated study periods to further augment instruction. A system of frequent comprehensive assessment was managed in such a way as to provide motivation and standing among cadet classmates.[25]

The second pillar centered on character, and key to its successful assimilation by cadets was the school's commandant of cadets and his staff, who acted as role models for character, discipline, and the appropriate deportment. The pillar of character demanded moral behavior, church attendance, outstanding uniform appearance, and conduct as gentlemen. Regulations and associated penalties supported these expectations.[26]

The United States Military Academy at West Point, 2008. (US Army photo)

Character's central importance to the military profession was explained by Secretary of War Newton Baker in 1920: "The exact or untruthful soldier trifles with the lives of his fellowmen, and the honor of his government; and it is, therefore, no matter of idle pride but rather of stern discipline necessity that makes West Point require of her students a character of trustworthiness which knows no evasions. In the final analysis of the West Point product, character is the most precious component."[27]

In 1817 Thayer assigned the academy's first commandant of cadets with oversight of the military aspect of the academy. Thayer had inherited the Master of the Sword position, created in 1814, which loosely equated to the position of an athletic director. Thayer thus became the first superintendent to manage these two interests successfully. The integration of these competing departments under a military school culture provides the administrator with several challenges, the first of which is the task of educating the faculty so they understand the military school ethos. The second challenge is the integration of departments that are competing for the cadets' efforts to maximize academic, athletic, and character development. The understanding of the military school ethos by faculty impacts each department, including the military department. Fig. 1.2 illustrates the alignment of structure and the model cadet.

Fig. 1.2. Military School Structure.

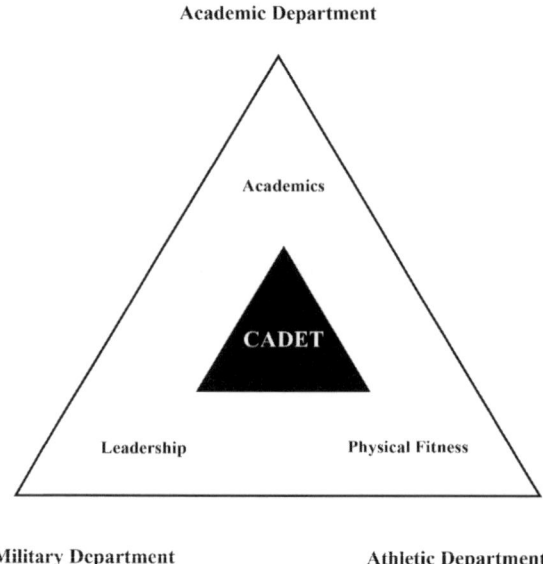

Over time Thayer's two pillars have matured within military schools, particularly in secondary military schools, as the three facets of academics, athletics, and character, which at various schools may be identified as the development of body, mind, and spirit or the development of the whole person. In support of these three facets, a military school has three departments: academic, military, and athletic. The challenge is to integrate these components within the concept of the overarching primary goal of the military school: character development.

Howe Military Academy, one of the oldest secondary military schools, established in 1884 and still in operation in 2014, and Oakland Military Institute College Preparatory Academy, a military charter school established in 2001, provide excellent examples of development of the whole person. Howe Military Academy's operating philosophy based on the three facets of academics, athletics, and leadership—body, mind, and spirit— is a common theme throughout military schools. The school is affiliated with the Episcopal Church and seeks development of the whole cadet through "rigorous academics and spiritual formation [which produce] self-discipline, physical well-being, leadership, logical thinking, and communication skills." Oakland Military Institute has a similar foundation of academic excellence, leadership development, and physical fitness. The school, through its "military frame work . . . inspires honor and pride within its cadets, cultivating life-long respect, confidence, physical fitness, and wellness, and appreciation for others."[28]

Culver Military Academy, a secondary military boarding school in Indiana, provides a good example of how the ethos of a military school has evolved over the years and has remained true to the school's core values:

> From its opening in 1894, Culver Military Academy, or CMA as it is commonly referred to, has remained committed to the education of the whole person. The traditions and rich history of Culver continue to influence how the leadership and education in the classical virtues are taught. The leadership system, based on a military model, is in place to prepare the young men of Culver to serve their country. The system is effective in providing skills for those few graduates who wish to pursue a career in the military, however, it more aptly provides essential and valuable leadership lessons for students who will go on to more traditional careers within our increasingly global community. Responsibility, accountability, service, and teamwork are all bedrocks

of a Culver education that will benefit each graduate in everyday life, no matter what path he chooses.[29]

An important step in maximizing the impact of the military school culture is having the faculty cross-check each cadet on their performance in each area. Faculty should report behavioral or academic issues to each other so that together they can address these issues in all areas and leverage character development. For example, if a football player's conduct or performance in English class is not acceptable, forums facilitating early identification of the issue can bring the invaluable influence of the coach or military department to assist the English teacher to inspire better performance. When faculty is well integrated and engaged, departments do not stand alone; they work as a unified team. Whether on the football field, on the drill field, or in the classroom, everyone is committed to the development of the cadet and helps develop a better student, athlete, and leader.

This cooperation among faculty affords a cadet the opportunity to achieve self-esteem through any one of these avenues, as well as achieve at least a minimal satisfactory performance in the other two components. The earlier motivational theories of Abraham Maslow apply here in a very important way. Cadets are no exception to the "desire for a stable, firmly based, (usually) high evaluation of themselves, for self-respect, or self-esteem, and for the esteem of others." In the military school environment, achievement can be met in any of the three components—athletics, academics, or military—and almost all cadets will find a component in which to excel. Achievement in one area cannot meet all self-esteem needs without satisfactory performance in the other two areas. This process is described by Trousdale as a measure of success of the secondary military school: "If they [the school] have given a boy a sense of self-worth, self-esteem, pride, and a degree of confidence, they have fulfilled their mission."[30]

For example, a cadet who is an outstanding athlete cannot achieve his desired goals without meeting requirements both academically and militarily. He will not be rewarded with military rank or responsibility, and the positive feedback from his coaches becomes more of a corrective counseling tool to impress on him the need for academic success. Without the meeting of expectations in the areas where the cadet does not excel, that cadet cannot gain the self-esteem needed.

The empowerment of the faculty to address the question of character within the military school culture is not the only critical empowerment required. Perhaps the most noteworthy aspect of the military school culture is the degree of empowerment provided to cadets. The military school's robust leadership challenges provide the role models and safe places to fail for cadet leaders to form their own leadership styles.

The results of cadet empowerment afford the cadet the opportunity to fail or succeed and be held accountable. Cadet units compete against each other in drill and uniform inspection and for designation as the best company of the year, which often considers grade-point averages as well. This competition gives cadets the opportunity to fail in an environment where the results do not damage their grades, career, or salary. Basically, the military department is a practice field for leadership skills not often taught in nonmilitary schools. Furthermore, the subject of leadership is not only addressed in an academic manner; there are daily events supporting the practice and perfection of leadership style.

An important part of a military school is the empowerment of cadets to become a part of a system, and an honor committee of cadets is the height of that empowerment. Cadets on the honor committee sit in judgment on violations such as lying, cheating, or stealing. Traditionally, membership of the honor committee is not based on cadet rank or positions of military responsibility. Universally, it is a symbol of accepting the honor code and the cadet's responsibility to the school. Cadet empowerment is the final step in adoption of the codes and behavioral norms expected of the cadet.

Moral Tradition

The moral tradition of the military school model goes back to West Point's earliest days. In 1818 Sylvanus Thayer hired West Point's first academy chaplain, Rev. Thomas Picton, who was also a professor of ethics. Until 1972, church attendance was an important aspect of the expectation of moral conduct at military academies and schools, but in that year the Court of Appeals of the District of Columbia ruled against mandatory church attendance at the federal military academies and state-supported military colleges and universities. Federal courts went even further when in 2003 they ruled against Virginia Military Institute's nondenominational and voluntary prayers at evening meals. The resurgence of military schools within public education further separated military schools from their traditionally close relationship with religion. On the other hand, most of the private military schools are either church-sponsored or nondenominational Christian. Despite this, the dedication of military schools to character building has enabled the maintenance of a moral tradition.[31]

Texas Military Institute Corps of Cadets marches in to church services, 2015.
(Courtesy of Allan Rupe/TMI—The Episcopal School of Texas)

Sociologist Kim Hays examined three Quaker religious secondary boarding schools and three private military secondary boarding schools and concluded there were similarities. All three military schools were in operation in 2011. Both sets of schools held their moral traditions and provided a set of virtues to emulate. Both types of schools have a strong ethic of service and strong moral nature. The military schools studied by Hayes each had lifestyles that grew from "carefully constructed and purposeful moral world views."[32] These schools gave order to cadet life through a moral tradition based on military virtues of loyalty, competence, selflessness, integrity, and pride. Military schools "by their efforts to communicate a strong moral code, a desire to serve others, and a sense of responsibility, stand out in contrast to the average public school, where few teachers have the time, opportunity, permission, or perhaps even desire to teach such lessons."[33]

Military School Culture
Inclusion Time Line

The military schools initially were closed cultures limited to white males studying technical/engineering subjects who were being prepared at the collegiate level for leadership in the active army or militia. The collegiate military state colleges in the South would extend the educational opportunities with an expanded curriculum to enhance their graduates' futures in

Fig. 1.3. Military School Culture Inclusion Time Line

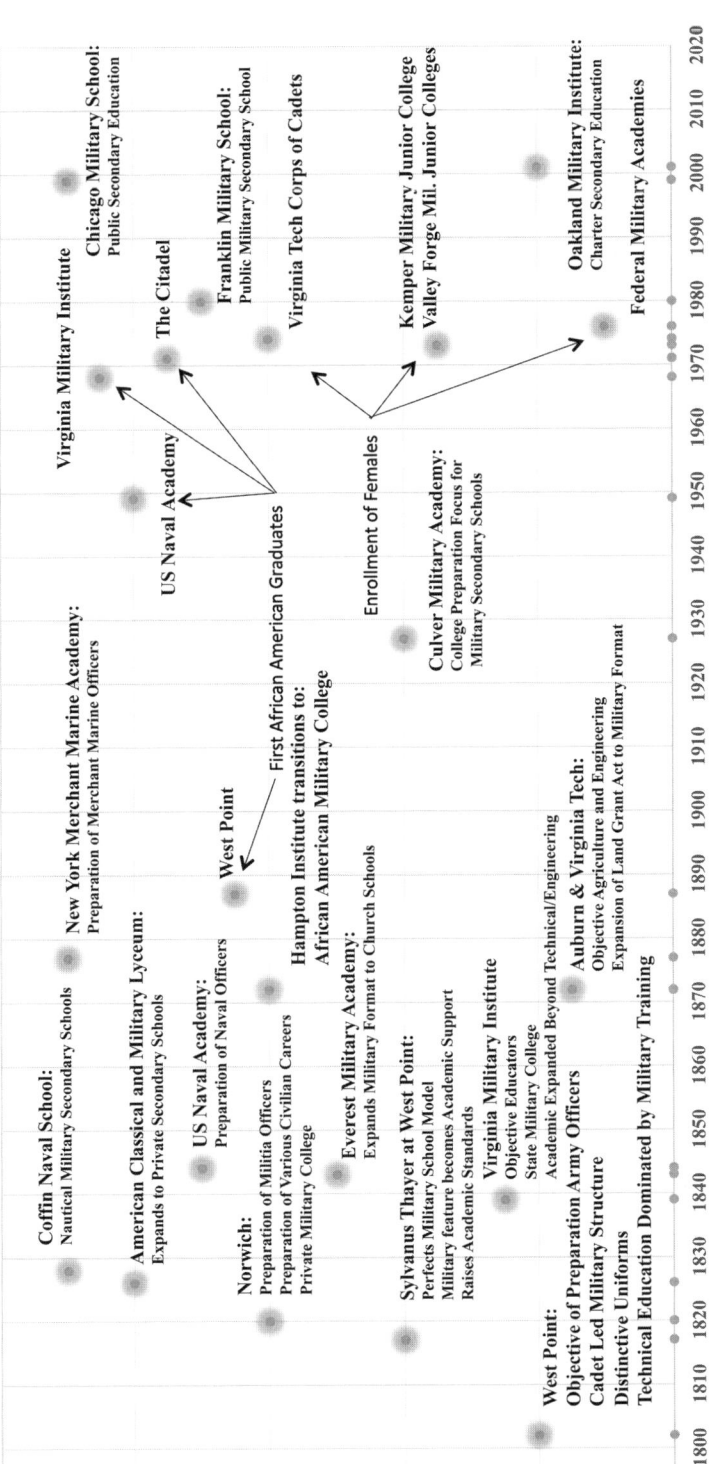

education and other fields. The orientation to naval and maritime leadership further expanded career preparation at the college level and a nautical theme to secondary schools. This was furthered in the expansion of collegiate military experience in the agriculture colleges of the land grant act.

Early in the time-line, the military format was expanded into secondary education, first into private schools, then into church-affiliated institutions, and much more recently into public education at the secondary level, including charter schools. Where once cadets or midshipmen were white males, the culture expanded first to include African Americans and other minorities. Most recently the military school experience was expanded to include females. The graphical depiction in Fig. 1.3 is only a glimpse into a very complicated story of an educational format that is now largely American; includes all races, religions, and genders; and is a fixture throughout the United States.

★ 2 ★

The Long Road to West Point

Initial interest in developing military schools in the United States grew from requirements of the Revolutionary War and the influence of European military schools. Before the Revolution at least two schools taught subjects related to military engineering, artillery, fortifications, and gunnery. These technical military subjects were taught in traditional academic style and along with the curriculum of the time. The school in Westmoreland Potomac, Virginia, was headed by the Reverend Thomas Smith in 1771; the other was the English Grammar School in New York City, founded in 1774.[1] There is no evidence that the schools incorporated a uniformed militarized student body of the type that defines a military school.

European military schools were established not long before the American Revolution, and several of their graduates served in the Continental forces of Gen. George Washington. The most influential was Gilbert du Motier, Marquis de Lafayette, who attended the military academy at Vincennes, near Paris, and left there as a lieutenant of musketeers. The school was established in 1751 and had five hundred young noblemen as cadets. Its location was moved to the Champ de Mars in 1765. Napoleon Bonaparte was a 1785 graduate of Champ de Mars. The current French Military Academy, Saint-Cyr, is often cited as influencing West Point's establishment, but actually it was founded by Napoleon in 1802, the same year as West Point. Another French military school that influenced the establishment of American military schools was the Royal Engineering School of Mézières, established in 1748.[2]

A Polish officer serving with the Continental Army who influenced the formation of the first American military school was Tadeusz Kościuszko, who rose to the rank of general in the Continental Army. He was one of the original members of the corps of cadets of the Polish academy established by

King Stanislaw in 1765. The school, located in Warsaw, addressed military and liberal arts courses.[3]

The greatest influence that encouraged Americans to professionalize the officer corps was the enemy. The British officer corps was highly trained at the Royal Military Academy, Woolwich, which was established in 1741 with the goal of producing "good officers of Artillery and perfect Engineers."[4] Its graduates constituted a highly professional leadership group of British engineers and artillerymen in America that was the envy of General Washington's chief of artillery, Gen. Henry Knox. The memorandum to Congress from Gen. Knox titled *Hints for Improvement of the Artillery of the United States* reflected the respect he had for the British military educational

General Henry Knox (1750–1808), by Alonzo Chappel (*National Portrait Gallery of Eminent Americans: . . . from Original Full Length Portraits by Alonzo Chappel,* vol. 1 [New York: Johnson, Fry & Co., 1862])

system: "An officer can never act with confidence until they are masters of their profession, an academy established on a liberal plan would be the utmost service to the continent, where the whole theory and practice of fortification and gunnery should be taught; to be nearly on the same plan as that at Woolwich, making allowance for difference of circumstances; a place to which our enemies are indebted for the superiority of their artillery to all who have opposed them."[5]

In May 1776, while serving on the Continental Congress's Board of War, Knox wrote to John Adams "suggesting the establishment of academies for educating young gentlemen in every branch of the military art." Adams replied, "I am fully aware of your sentiments that we ought to lay foundation and begin building institutions . . . for promoting every art, manufacture, and science necessary for the support of an independent state." This comment from the future second president of the United States foretold the service military schools would provide, well beyond the instruction of military art to that of science and liberal education.[6]

An early indication of the direction military schools would take was from the report of a Continental Congress committee to study the state of the army. The committeemen arrived at the Continental Army headquarters in September 1776 and spent several days studying the state of leadership and the army. Their report reflected an army "badly officered" and even encouraging "soldiers by their examples to plunder and commit other offences." The committee's recommendation stressed the importance "of having officers of known honor, ability and education." Their resolution recommended "that the Board of War be directed to prepare a Continental Laboratory and a Military Academy and provide the same with proper Officers."[7] The recommendation foreshadowed the role that morality, honor, and gentlemanly conduct would take in the formation of the future military school culture. Expectations well beyond the teaching of basic theory of war and technical skills of the warrior's craft emerged very early.

The memorandum from General Knox and the report and recommendation of the committee that visited the Continental Army arrived at the Continental Congress within days of each other. Two days after the committee's report, and three days after the general's memorandum, the Continental Congress passed a resolution dated October 1, 1776: "Resolved, that a Committee of five be appointed to prepare and bring in a plan of a military academy at the Army."[8]

The war and the establishment of the new republic would delay the formation of the US Military Academy for twenty-six years until 1802. Several measures were implemented short of a military school, including allocation of funds for an officers' course at Carlisle, Pennsylvania, in 1778. This was the Corps of Invalids' school, charged with "propagating military knowledge and discipline" for officers having "good character, both as citizens and soldiers." Also at Carlisle an ordnance school functioned, instructing the fabrication and repair of weapons from 1777 to 1779. Another course for engineers and artillerists operated at West Point for a few years, then terminated in 1780. None of these examples were military schools or suited to fill the needs identified by General Knox.[9]

In 1783, after the war had ended, Congress appointed a committee, including Alexander Hamilton, charged with developing a plan to support General Washington's desire to establish America's first military school. Washington responded to Hamilton's request for his views on the project by soliciting the views of his key generals in New York as well as foreign officers and the governor of New York.[10]

It was no surprise that General Knox was the strongest advocate of military schools: "A perfect knowledge of the principles of war by sea and land is absolutely incumbent upon people circumstanced as we were, and determined to be free and independent. From these considerations arose an indispensable necessity of forming and adopting a complete system of military education." Knox actually advocated the creation of three military schools for the southern, eastern, and middle regions. Gen. Friedrich Wilhelm Von Steuben also supported regional military schools, but his views also called for 90 percent of graduates to return "to civil life to diffuse military education among the citizens and become leaders in a well-regulated militia." Von Steuben's views would foretell a future direction of the citizen soldier concept and military education outside the jurisdiction of the federal government.[11]

Gov. George Clinton of New York went a step further and advocated a military school for each state "where degrees in the arts and science are conferred." Col. Jean-Baptiste Gouvion, a graduate of the French Royal Engineering School of Mézières, submitted a series of detailed recommendations that included practical application of techniques, not just theory. General Washington forwarded his personal recommendation to Congress, calling for an academy of engineers, but without success. Washington again provided his views of military requirements for the young nation in his "Sentiments on a Peace Establishment" in 1783. He felt four actions would be required to "maintain the lasting Peace, Happiness and Independence of the United States." First was a standing force to garrison West Point and

President George Washington (1732–99), by Charles Willson
Peale. (Library of Congress)

selected locations along the frontier. Second was all states' maintenance of
a well-organized militia. Third was the establishments of various arsenals to
supply the nation's forces in time of war. Finally, Washington clearly stated
the need for academies to teach military art, with special emphasis on the
skills associated with military engineering and artillery.[12]

The lack of support by Congress was attributed to two factors: the fear and
disdain of a standing army based the colonies' experiences of the British army
being used as a method of repression, and the question of constitutionality,
which led Thomas Jefferson to oppose the establishment of the federal military

academy because "none of the specific powers given by the Constitution would authorize it." His opposition was vigorous when in 1793 the national military academy was proposed. At the time, Jefferson and others viewed the constitutional powers shared with the states much less narrowly than those assigned to the federal government. The term *necessary* and *proper* when exercising powers shared between the federal and state governments were his point of contention in opposition to the Federalist Party.[13]

Finally, in 1794 an act of Congress authorized the Corps of Engineers and Artillerists and "the necessary books, instruments and apparatus for the use and benefit of the said Corps." Its cadets would number thirty-two. West Point, New York, was the station of the army's Corps of Artilleries and Engineers (also called the First Regiment of Artilleries and Engineers) and a logical location for the training. The school taught military courses, but it was not really a military school. Officers, noncommissioned officers, and cadets took courses, and cannon crews were drilled in duties associated with their assignment with the Corps of Artilleries and Engineers until 1796, when the building they used burned to the ground, a theme that became familiar in the future of military schools.[14]

President Washington made a final appeal in 1796 for the support of the Senate. Vice President John Adams, who had supported the military school cause in 1776, signed the positive response. But actual establishment of the school was still years away. Under the administration of President John Adams, in 1798, Congress expanded the number of cadets to fifty-six and authorized four teachers. There was no military school established, and cadets were assigned to various units of the army as officers in training. President Adams submitted a report prepared by Secretary of War James McHenry that estimated the cost of establishing a military academy. Harrison Otis presented a bill calling "for establishment of a Military Academy and for organizing the Corps of artillerists and engineers," but again, the bill was not acted upon.[15]

Twice in January 1800 Secretary of War McHenry submitted reports to Congress through President Adams addressing the need for a military academy. The secretary's report addressed the concerns of maintaining a large military force, in that "it is important that as much perfection as possible be given to that [force] which may at any time exist," and noted that these troops should be "perfect in organization and discipline and dignity of character." The report called for a military academy, which expanded the military school well beyond that of simply teaching a few courses for artillerists and engineer officers. McHenry's report outlined a plan to pre-

President Thomas
Jefferson (1746–1826),
by Gilbert Stuart.
(Library of Congress)

pare cadets to become "engineers (including geographical engineers), miners and officers for the artillery, cavalry, infantry and navy and the school
would teach all sciences necessary to perfect knowledge of the different
branches of the military art . . . and be provided with the proper apparatus
and instruments for philosophical and chemical experiment . . . for astro-
nomical and nautical observation and surveying."[16] It took the concept of
the military school well beyond that which was considered a preparatory
course for military service. It also pointed out the importance of character
development, which would become a central part of the military school
ethos. This report has been largely ignored in its implications for the development and expansion of the educational concept of military schools
throughout the United States.

By the time Thomas Jefferson assumed the presidency in 1801, his
views on a national military academy had evolved. One explanation is
that he had championed unsuccessful efforts in Virginia to deemphasize
divinity and classical languages in order to add courses addressing law,
chemistry, natural history, and modern languages. In turn, he had become
an advocate for a national university to provide practical education to
civil servants, to which there was significant sectionalist opposition. But

as president, Jefferson saw that a means to create a national university lay in supporting the establishment of a military academy with an orientation towards science.[17]

On March 16, 1802, the House of Representatives passed a bill addressing the recommendations of McHenry to put into law the US Military Academy. The acts that finally led to the establishment of the first military school placed the academy at West Point, New York, designated the senior engineer officer as superintendent, and authorized the secretary of war "to procure, at public expense, the necessary books, implements, and apparatus for the use and benefit of the institution." The establishment of a military academy would fulfill Jefferson's desires for a national institution of science. Additionally, West Point offered a means by which Jefferson could win over the sectionalists, a group of which he was formerly a member, with his support for a national military academy. An indication of this potential support came from Charles Pinckney, who admitted that state militia were ill prepared to handle the challenges of engineering and artillery duties.[18]

Another of Jefferson's motives may have been the fear of Federalist control of the army's officer corps. The bill establishing the military academy afforded Jefferson's Republicans the opportunity to appoint cadets from congressionally sponsored "sons of the country's sturdy Republican stock." The Republicans envisioned West Point's future graduates to "break the upper class monopoly in the officer corps . . . the aristocracy of wealth and birth . . . to be replaced with the aristocracy of virtue and talent."[19]

The establishment of the school was never seen by President Washington, its ranking proponent, as he died in 1799. The academy's supporters included some of the greats of early US history, including Secretary of War Henry Knox, President John Adams, Alexander Hamilton, General Von Steuben, the Marquis de Lafayette, and Secretary of War McHenry.

The process to open the US Military Academy at West Point was a lengthy confrontation between the need for what Washington called a peace establishment and the American fear of a standing army, concerns of overreach by the central government, and fear of the federalist domination of the officer corps. Yet out of those complicated maneuvers came the opening of the nation's first military school. This school at West Point would set the standard for all American military schools and stand on the banks of the Hudson River as the birthplace of an American educational format.

Early West Point and Captain Partridge's Military School Movement

The first superintendent at West Point was Lt. Col. Jonathan Williams, the nephew of Benjamin Franklin and an honor graduate from Harvard in 1787. During the Revolution he accompanied his uncle and served as an arms agent for the Revolutionary cause. A member of Philadelphia's American Philosophical Society, he was a well-respected, published scientist. After coming to the attention of President Adams, he was commissioned a major in the 2nd Regiment of Artillery and Engineers in 1800. But it was President Jefferson, aware of Williams's reputation as a scientist, who appointed him superintendent at West Point.[1]

Williams was described as a "scholar, and an amiable and polished gentleman." At West Point, newly promoted to lieutenant colonel, he was assisted by three officers as instructors and was initially assigned nine cadets. He felt the academy's academic subjects of astronomy, geography, physics, mathematics, and the military skills of fortification and artillery fire and direction were "inseparable and interwoven." The initial faculty at the academy included Capt. Jared Mansfield, a Yale graduate who taught math and physics and had published *Essays, Mathematical and Physical*, considered a significant American work. Another professor was Capt. Amhurst Barron, a Yale graduate with teaching experience at Cambridge.[2]

As a result of the dual nature of the head of the academy as both chief of engineers and superintendent, academic focus suffered. Williams frequently had to leave the academy to inspect harbor defenses, a distraction outside the academy required as the young nation was challenged along the coast. In two incidents in 1807, a French warship entered the Delaware

River, and the British frigate *Leopard* fired on the USN *Chesapeake*, killing three, as a squadron of British vessels lay off the Virginia coast. President Jefferson envisioned a series of fortifications from Louisiana to Maine, but the country's poor state of defense required Williams's extensive attention. Adding to Williams's frustration was the fact that even as chief of engineers he had no command authority over units assigned at West Point in order to solicit support.[3]

Those early years were further hampered by insufficient academic preparation of the cadets, who ranged in age from ten to thirty-four, and significant lack of textbooks and scientific equipment. The curriculum was still developing, as well as the identity of the school as a military school rather than a technical course that cadet members of the Engineer and Artillery Corps attended. Williams himself describe the state of the academy as "a punky, rickety child." That was reflected in the tenures as cadets of the next two superintendents at the academy. Joseph Swift was a cadet for just a year, and Alden Partridge for four months. When the War of 1812 began and Williams sought a wartime command, it was denied, and he resigned.[4]

Col. Joseph Swift took Williams's place and was immediately faced with the same dual challenges. Swift had an interesting start as a cadet at West Point. Arriving at the age of sixteen before the academy was actually functioning, he was taught four hours daily and made a member of the officers' mess. One day he received a directive from his professor through a servant. Swift refused to obey, and when approached by Professor Baron and asked why, he stated, "I refuse to receive any verbal order from a servant." The professor replied that Swift was "a mutinous young radical," and an altercation ensued.[5] This story is told not to reflect badly on Swift, but to point out the momentous task the first two superintendents had in transforming the culture of the pre-academy cadet into the military school cadet that would be the center of the academic and disciplinary efforts of the military school culture.

Faced with war, Swift was often superintendent by correspondence. The War of 1812 lasted two and one-half years, and its demands caused him to be absent, particularly in 1812 and 1813. He was required to travel to North Carolina; Washington, DC; and New York City, and his focus was taken from the academy to fortification and the war's needs. However, he did expand the academy's curriculum with the hiring of a chaplain to teach ethics, history, and geography. Swift also completed plans for much-needed physical improvements. His absences prompted him to select Capt. Alden

Partridge to act in his place while duties took him from West Point. This was a function that the captain had done at least in some respects as early as 1807 for Col. Jonathan Williams as well.[6]

Captain Alden Partridge at West Point

Alden Partridge was born in Norwich, Vermont, and attended Dartmouth College in Hanover, New Hampshire. He left as a senior in 1805 to become a teacher. His departure from Dartmouth was in good standing, with recommendations from the professor of mathematics that described him as having "good moral character and [being] well qualified for instructing a school."[7] But a teaching career was not in his immediate future, as he became a cadet in the Corps of Engineers at the US Military Academy at West Point on July 9, 1806.

After less than four months as a cadet, Partridge was commissioned a first lieutenant. He remained at West Point and was promoted to instructor in mathematics in 1808 and to captain in 1810. It was in 1810 that Partridge proposed correction to the significant flaw in his experience as a cadet. In a letter to Colonel Swift he proposed that cadets not be granted duty outside their academic requirements and that a diploma be adopted and awarded only after professors certified completion of academic requirements.[8]

In 1813 Partridge was made acting superintendent and was promoted to professor of engineering. The mixing of military and academic duties seemed in keeping with the character of the institution, but by the nature of the appointment and Swift's expectations, to oversee the academy in Swift's absence would hinder Partridge's future success. Partridge by that time had been a member of the faculty for seven years, and many of his peers did not think well of him. The special confidence he had secured with the last two superintendents, his authority, and use of cadets' time heightened bad feelings with many. A balance is needed between academic and military demands. Partridge became the legitimate authority over the military demands, but his control over academic ones was not officially confirmed until 1815.

Stephen Ambrose described Partridge at that time at West Point as a man with "the body of a penguin and head of a hawk with sharp, pointed noise, square, jutting chin and tightly set mouth." He walked the academy grounds in a strut with coat—always buttoned—sword, sash, and cocked hat with black silk cockade adorned with a yellow eagle. Most thought he owned no civilian clothing, and he never appeared out of uniform or official demeanor. "Old Pewter," as he was called behind his

Alden B. Partridge
(1785–1854).
(Courtesy Sullivan
Museum and History
Center at Norwich
University, North-
field, VT)

back, loved close-order and cannon-crew drill, and his cadets got their full
share. Intelligent and energetic, he was totally focused on the academy,
and he associated with few outside those halls.[9]

Partridge held significant positions of leadership for eight of his first
fifteen years at the academy. He can be credited with two changes that were
critical to the growth of West Point as a military school: First, no cadet was
to leave the academy for detached duty until his academic studies had been
completed, and second, upon a cadet's completion of studies the institute
would issue a certificate signed by the superintendent and professors.[10]

Partridge was even responsible for the trademark gray uniforms. During
the War of 1812 Gen. Winfield Scott led regular army troops dressed in
gray uniforms in the major victory at the Battle of Chippawa. The uniforms were gray because the issue blue dye was unavailable. To demon-
strate West Point's pride in the regular army's victory, Partridge adopted
the gray uniform.[11]

Even his most ardent critics credit Partridge with "establishing regulations for the operation of the academy that forced a more balanced curriculum, along with emphasis on drill. He also initiated the separation of the
position of superintendent from the position of chief engineer. "The evolution of a military school in the United States came about with no other

school in the country to emulate. It evolved as an institution of higher
education, providing basic engineering training . . . with mathematics
and drawing . . . natural philosophy, French and engineering arts."[12] The
school grew from 40 authorized cadets in 1802 to 250 cadets in 1812.

Partridge was appointed superintendent in 1814 and was in that position until 1817. However, his relations with the faculty, both civilian
and fellow military officers, were not good. Initially, problems at West
Point were twofold: the dual nature of the cadets' responsibilities during
the earlier years coupled with the leadership's lack of experience with the
military school concept. Cadets were assigned duties to support their regiments, which took them away from their academic duties. It was not that
Partridge and Swift did not seek the best for their academy; it was that they
were not sure what a military school should be, and neither Partridge nor
Swift had the background or experience of working in a well-established
military school. The majority of Partridge's educational experience was
at Dartmouth, and Colonel Swift's was in West Point's first year and at
Taunton college prep school.

The experience needed to run a military school could have come from
George Washington's desire for the academy in 1783. Washington clearly
desired a school where French engineers, with their experience of the
Royal Engineering School of Mézières, would teach and initially run West
Point. But the army's officer corps fought this idea, even after the War of
1812. Captain Partridge supported opposition to giving leadership positions to French officers. Interestingly, General Swift recommended that
Gen. Simon Bernard, a French officer, replace Captain Partridge as superintendent at West Point in 1816. General Bernard was a graduate of
another French military school, École Polytechnique, which was founded
in 1794 and was a military school from 1804 to 1970.[13]

Captain Partridge's lack of success with the academy faculty was further compounded when members of the faculty complained not to General
Swift but directly to Secretary of War William Crawford about Partridge's
leadership. Members of the faculty supported the separation of scholastic
and military duties. Partridge's vision was that a cadet's academic and military chain of command should function as one. Much of the faculty who
had worked with Partridge never felt that his control of the academy in
Swift's absence was legitimate or that he should have become superintendent. The ongoing conflict with the faculty led to a court of inquiry and
finally Partridge's court-martial and permanent removal from West Point
and active military service.[14]

The downfall of Partridge at West Point came quickly after a visit to the academy by President James Monroe. The secretary of war involved the President, and, unfortunately for Partridge, he had not made an ally of either. Monroe suggested, in no uncertain terms, that Partridge should be replaced by Maj. Sylvanus Thayer: "I do not think much of your Captain Partridge and would prefer to see Major Thayer in his place."[15] The President directed that a court of inquiry be conducted.

General Swift, who was chief of engineers and still had oversight over the academy, gave Partridge the option of assuming duties away from West Point or taking leave until the court of inquiry occurred. In accordance with the President's wishes, the academy changed command on July 28, 1817, and Partridge departed for Vermont. After thirty-three days, on August 30, Partridge's better judgment failed him, and he returned against orders and personally logged his return to West Point as having "taken upon himself [Partridge] for the present, the command and superintendence of the institution, as senior officer of engineers present."[16]

Until that point Partridge still had a promising career in the Corps of Engineers and was facing only a court of inquiry. He could have been in the army long after Monroe's administration was out of office. However, as a result of his actions of August 30, General Swift placed him under arrest, and a court-martial was convened. Partridge was found guilty of disobeying orders, in that he assumed command of West Point following the appointment of Thayer as headmaster. He also was found guilty of issuing illegal orders. As a small consolation for Partridge, the court president, Gen. Winfield Scott, requested, "in consideration of the zeal and perseverance which the prisoner [Partridge] seems uniformly to have displayed in discharge of his professional duties, . . . leave to recommend him to clemency of the President of the United States." In accordance with the judgment, Partridge resigned from the army in disgrace on April 15, 1818.[17]

There are two basic points of view regarding Capt. Alden Partridge, one being that he "possessed few leadership qualities and as a result alienated the academic staff," and the other, that he was the victim of a political struggle between the secretary of war and the chief of engineers, which influenced the President's ruling against him. The latter argument is supported by the manner in which the faculty circumvented General Swift with their complaints against Captain Partridge. But Partridge's failure was a function of his inability to unite the efforts of academic preparation and military training. In his struggle with the faculty he controlled what he could, military training, at the expense of academic preparation. This further alienated the faculty, who grew too loath and mistrustful.[18]

The degree of imbalance between military and academic standards that resulted is reflected in his successor's initial impression. Sylvanus Thayer instituted a general academic examination of cadets shortly after his assumption of duties. The result was dismissal of twenty-one cadets based on absence of basic aptitude to progress or their failure to progress academically despite having been cadets for several years.[19] This was close to 10 percent of the cadet corps and a clear indication that academics had fallen a far second behind military training under Partridge.

There was no clemency for Partridge, as the army did not want to revisit an issue involving the civilian authorities and further negative public attention.[20] These events, despite their negative connotations, resulted in the emergence of parallel efforts to refine and expand the military school culture and concept in the United States. The first was the beginning of a second career by Alden Partridge, who established and led a successful university. Perhaps of even more historical significance, he started a nationwide crusade for the establishment of military schools outside the federal government's administration.

Alden Partridge and the Military School Movement

As superintendent at West Point, Sylvanus Thayer would lead that institution from 1817 to 1833. In that role he firmly established an ethos and standard of education that led to the expansion of the military school concept well into the current century. Meanwhile, Alden Partridge would lead and champion the military school movement that would expand the concept outside the federal government's purview and enlarge the academic focus to preparation for civil careers and the military objective to the citizen soldier concept and leadership of the militia.

According to Dean Baker, the court-martial left Partridge with "severe emotional scars," and he became suspicious of the federal bureaucracy. He immediately moved forward to find redemption for his reputation through the founding of an alternate military school. The American Literary, Scientific, and Military Academy, later known as Norwich University, was founded in 1819 in Norwich, Vermont, with its first cadet enrolled on September 4, 1820. Partridge opened the doors of the new school with a faculty of six, including himself, a chaplain who taught ethics, a professor of chemistry, two professors teaching languages (English, Greek, and

Latin), and a professor of practical sciences (geometry and topography) who also acted as master of the sword (physical fitness). Partridge taught mathematics, philosophy, and military science.[21]

The American Literary, Scientific, and Military Academy gave Partridge an opportunity to develop cadet regulations and standards that reflected his own philosophy without political interference. The standards in the 1826 prospectus of the school included rules against "prevarication or falsehood," which would lead to immediate dismissal.[22] These standards were an early foreshadowing of what is now a common part of the military school culture: codes against lying, cheating, and stealing.

Partridge's reputation was reestablished through his competence as a scientist and mathematician. Samuel Latham Mitchill, who was considered the father of American chemistry, stated that Partridge's "scientific zeal and proficiency I have often given a very favorable opinion." Mitchill called Partridge's contributions to learned journals valuable and encouraged him to take on a survey of the eastern United States.[23]

Enrollment at the new school indicated that the public, at least in Vermont, either overlooked the West Point scandal or sided with Partridge on it. In the first year of operation there were 100 cadets; in the second year, 140; and by the sixth year, no fewer than 197. That enrollment compared favorably with West Point's, where authority was given to increase the cadet corps to 250 in 1819. Cadets at Norwich were uniformed in dark blue with three rows of bullet buttons, in contrast to West Point's gray attire.[24]

Although in the beginning most Norwich cadets were from New York or New England, by 1826, of the 293 cadets 100 were from the South. Partridge actively sought out cadets from throughout the country and particularly from the South. One reason for the large Southern enrollment may have been the antimilitary attitude in New England after the War of 1812. As late as 1834 there was a strong antimilitary opinion in the North and a tide of contemporary social influence against the military setting of Norwich University. Critics of Norwich regarded it "not only as out of tune with the age, but also as a promoter of that wild war spirit."[25]

Captain Partridge is best known as the father of Norwich University, but he is also known as the father of the ROTC. Norwich University is the second oldest military school in the United States, having operated for more than 194 years. It is also the only privately owned military college in the country, and in 2013 it had a student body of 2,112 undergraduates and 1,200 graduate students, of whom 1,386 were members of the corps of cadets. In forming his first school and those that followed, Partridge was

guided by his analysis of the shortcomings he saw in education and defense
in the United States. This analysis was both a product of his educational
experiences at Dartmouth and at West Point, with its focus on engineering, and a concern for the leadership of the state militias. Outside West
Point there was no military school in the United States before Norwich's
establishment. West Point was strictly focused on providing officers for
the regular army, and until Norwich there was no school providing professional military education for officers of the militia.

Partridge saw five significant shortfalls in educational institutions of his
time. The first was that the classical education was "not sufficiently practical nor was it properly adapted to the various duties an American citizen
may be called upon to discharge." He criticized the liberal colleges and
universities for their emphasis on Latin and Greek languages at the expense
of English and practical technical subjects. Second was the total neglect
of physical education for "preservation of health . . . capable of enduring
exposure, hunger and fatigue." Third was the amount of idle time allowed
students, which he considered "injurious to their constitutions and de-
structive to their morals," when they should have been using that time for
improvement of the mind and body.[26]

The fourth defect in education, according to Partridge, was the freedom
of students to use their wealth to engage in activities that encouraged moral
corruption. His fifth objection was that courses of study were too limited
and did not fully use students' capacity or inclination for the acquisition
of knowledge. His final objection was that the courses of study were based
on a prescribed length of time rather than the student's rate and capability
of progress.[27] At the same time, Partridge saw a critical shortcoming in the
nation's approach to leader preparation in the state militia.

These conclusions were made by Partridge even before his departure
from West Point. In an 1815 letter to Secretary of War Alexander J. Dallas,
Partridge concluded that West Point alone was insufficient to meet the
demands of the nation's military forces, which were largely state militias.
His plan called for diffusion of military science throughout the country in
order to provide sound officers to the militia and educate the midshipmen
of the navy in a more professional manner. The plan also included three
additional military academies patterned after West Point to support the
eastern, southern, and western regions, with an enrollment of 150 federally funded cadets and 250 privately funded cadets, for a total of 400 per
academy. These schools would allow those cadets who desired a civilian
career and association with the militia to enter the militia with the same

preparation as that received by cadets going into the regular army. His plan was not well liked, and the concept of privately paying young men to attend one of the military academies was considered controversial.[28]

Between 1815 and 1816 Partridge sought political supporters for his plan. Congressman Richard M. Johnson of Kentucky, chairman of the Military Affairs Committee, became a supporter. Johnson concurred with Partridge's plans, with the exception of the privately funded cadets not bound for the regular army. This part of the plan was central to Partridge's desire to expand the professional military training to the militias. It was a fundamental issue for Partridge and has since become part of the citizen soldier concept in the United States. Partridge would eventually be associated with the citizen solider concept and years later be known as the father of ROTC. Partridge's work following his release from West Point brought military education out of the confines of the US Military Academy and helped produce military officers for the nation for nearly one hundred years before the federal government in 1916 established a formal program for military officer education outside West Point.[29]

Between 1820 and 1842, forty-three new military schools opened beyond the confines of West Point. This was the beginning of military school expansion in the US education system. Starting in the late 1820s, Captain Partridge traveled across the United States lecturing about his educational philosophy and led efforts to start military schools throughout the country. Between 1819 and 1853 he had a direct hand in the establishment of eighteen military schools by personally starting the institutions himself, sending educators to start them, or visiting to establish military schools or oversee the transition of schools to the military format. These schools included the American Literary, Scientific, and Military Academy (Norwich University) and the seventeen schools shown in table 3.1.[30]

Partridge's negotiations were not always successful in the establishment of military schools. Much of his success was based on his personal reputation and his ability to provide high-quality professors. In 1827 Judge Henry Shippen suggested that military training be part of the Allegheny College program. Shippen was a veteran of the War of 1812 and had been a major with the Pennsylvania Division. Shortly thereafter, Shippen became president of the college board and suggested that a "military academy, similar to that at West Point or one run by Captain Alden Partridge, . . . would be useful to this community and more likely to receive legislative aid." In 1828 Partridge visited the school, and the next year the issues of converting the school and of Captain Partridge's financial commitment were discussed by

Table 3.1. Military Schools Established by Captain Partridge after Norwich University

Opened or *Transitioned* *to Military*	School	Location
1827	Literary, Scientific, and Military Academy	Perth Amboy, NJ
1827	Western Literary, Scientific, and Military Academy	Buffalo, NY
1828	American Institute	Washington, DC
1828	Dinwiddie (Hatch) Military Academy	Danville, VA
1828	New Jersey Institute	Orange, NJ
1828	Virginia Collegiate Institute	Portsmouth, VA
1829	Jefferson Military College	Washington, MS
1830	North Carolina Classical Literary, Scientific, and Military Institute	Fayetteville, NC
1830	North Carolina Classical Literary, Scientific, and Military Institute	Oxford, NC
1839	Virginia Literary, Scientific, and Military Academy	Portsmouth, VA
1842	Pennsylvania Literary, Scientific, and Military Academy	Bristow, PA
1844	Missouri Literary, Scientific, and Military College	Saint Louis, MO
1844	North Carolina Literary, Scientific, and Military Academy	Raleigh, NC
1844	Raleigh Military Academy	Raleigh, NC
1846	Wilmington Literary, Scientific, and Military Academy	Wilmington, DE
1849	Gymnasium and Military Institute	Pembroke, NH
1853	National Scientific and Military Collegiate Institute	Brandywine, DE

the board. James McKay represented Partridge and committed to sharing any initial loss equally between Partridge and the board as a group. As a guarantee, McKay requested a $500 commitment, but the Allegheny College board offered only $350, and "thus by a kind Providence, in due time, Alleghany [*sic*] was delivered from the military experience."[31]

Between 1819 and 1911, seventy alumni of Norwich University were working as faculty in military schools. In addition to the seventeen military schools listed in table 3.1, the alumni of Norwich University founded or helped transform five other military schools: Pennock Military and Classical School, Rutland, Vermont, in 1824; Arrow Military Academy, Arrow Rock, Missouri, in 1839; Mount Sterling Literary and Military Academy in Old Fort Mason, Kentucky, in 1847; Russell Military Academy, New Haven, Connecticut, in 1840; and Highland Academy, which transitioned to Highland Military Academy, Worcester, Massachusetts, in 1857.[32] There were a total of twenty-three military schools associated with Partridge and his students prior to the Civil War.

Before the establishment of the Virginia Military Institute in 1839 and the Military College of South Carolina in 1841, Captain Partridge and the alumni of Norwich University were critical to the expansion of the military school concept beyond West Point. Norwich University educated cadets not only from the northern states and New England, but also the South. After the establishment of Norwich, the reputation of his schools, cadets, and alumni helped expand the military concept throughout the United States.

Alden Partridge has been much maligned because of the focus on his last days and subsequent court-martial at West Point. All too often forgotten are his years of contribution at the academy supporting absentee superintendents and functioning as the military commander of cadets. The latter function, with responsibility for 250 cadets, was assigned while he was teaching engineering. His final assignment at West Point was made without regard to the ongoing internal strife with the faculty or their ability to undermine his efforts through communications beyond that directed to the chief of engineers.

However, Norwich University remains as a monument to him. The establishment and success of both ROTC and the military colleges outside the federal jurisdiction endorse his philosophy. Captain Partridge did much to set the circumstances that Thayer would manipulate and improve into a model military school. Finally, Capt. Alden Partridge's personal efforts to spread the military concept set the stage for a format of military education that would eventually include 842 schools.

★ 4 ★

Sylvanus Thayer and
the Military School Culture

Maj. Sylvanus Thayer replaced Capt. Alden Partridge as superintendent of West Point, and their relationship would become a bitter one. As part of Partridge's campaign to create multiple regional military academies after his departure, he was publicly critical of West Point. This dismayed and turned many, if not the majority, of West Point alumni against him. Thayer likewise degraded Partridge until the day he died. But the two educators had many things in common. Both had attended Dartmouth College. Partridge departed in his senior year, and Thayer left just before the graduation ceremony, where he was valedictorian of the Dartmouth class of 1807. Thayer, like Partridge, attended West Point for a brief time as a cadet, Thayer for one year and Partridge for three months. Both Thayer and Partridge were faculty members together in 1810 teaching mathematics. The War of 1812 ended the similar experience of the two. Thayer left West Point for service and in 1814 distinguished himself "for actions at Canary Island and in the Defense of Norfolk," for which he was promoted to major. Service in combat is no small point of honor among military men.[1]

It was not uncommon for military officers who had not served in combat to be looked upon less than favorably by their contemporaries. Combat experience, both from the practical viewpoint and the bankroll of credibility, helped Thayer with his work at West Point. The other benefit of his service was seeing the poor performance of the American officer corps, largely untrained and without the benefits of West Point, which likely gave him his vision of the importance of the West Point mission.

But the biggest advantage Thayer had was his personal experience with a European military education. Between 1815 and 1817, Thayer and Col. William McRee studied at the École Polytechnique and collected one thousand technical texts, which formed the basis of the West Point library. École Polytechnique was established in 1794 as a civilian engineering school, but it was converted by Napoleon to a French military engineering school in 1804. Unlike the United States, which had a shortage of West Point trained officers, France had no shortage of officers trained at the École Polytechnique or one of the other French military schools. So the school Thayer studied was well advanced beyond West Point as both an engineering college and a military school.[2]

Thayer was a fanatical educator and soldier who fervently believed in the course he would set for the academy. Of medium height for the time, he was privately a shy and humble man, yet he dressed and carried himself like the ideal officer. Highly intelligent, cold, and humorless, he never married, and he lived a Spartan life devoted to the academy. Thayer came to know every one of his cadets, and they felt he monitored the academic and military progress of each of them.[3]

Thayer assumed the duties of West Point superintendent in 1817 and progressed the academy's development with two educational philosophic pillars. The first pillar was, according to Lance Betros, a four-year program

Sylvanus Thayer (1785–1872), by Robert W. Weir. (West Point Museum)

centered on science, mathematics, and engineering. Subject matter also included French, mechanics, history, chemistry, mechanical drawing, geography, and ethics. This demanding curriculum was further supported by a strict schedule of study hours. Thayer established classes small enough to allow daily recitation by each cadet, with the intent of providing practical experience; a focus on subjects that would orient the cadet to be a technically trained military engineer; a requirement of daily grading and resectioning based on those grades; and the institution of rank ordering by academic standing and year group.[4]

The second pillar Thayer brought to West Point was the development of character. Thayer further "regularized the entire cadet schedule," which helped translate West Point's culture to the military school culture of the future. He insisted on limiting departures and leaves, "high standards of appearance, weekly attendance at chapel, and a strict enforcement of rules against lying, stealing, and other irregular or immoral practices." Cadet regulations were managed by the commandant of cadets, and he and his staff acted as role models for appropriate behavior and gentlemanly conduct. The cadets were organized into two cadet companies by height. Led by cadet officers, well equipped with muskets, and well drilled, they were marched to and from the mess hall in formation and were dismissed in front of the North and South Barracks.[5]

Thayer is widely credited with the start of West Point's honor code, which actually became a written code in 1932. What Thayer did was ingrain in the corps of cadets "a code of ethical behavior set forth as the ideal of the commissioned officer corps. Central to the ethical system was the idea that an officer's word was his bond. The penalty for failure to meet this standard, was for the officer [cadet] to be immediately cashiered and disgraced."[6] The combination of his approach to academics and enforcement of cadet discipline became known as the Thayer Method.[7]

The superintendent at Culver Military Academy summed up Thayer's contribution by stating that Thayer "is perhaps more than any other man responsible for that blending of military, intellectual and moral training that has enabled the graduates of West Point to achieve distinction in civil no less than in military pursuits." Thayer actually did more in that he refined the organizational culture of the military school concept and perfected a model for the country's military schools. According to Larry Manning, where Partridge had failed, Thayer had succeeded for three reasons. First, he was building on the work of predecessors who established the initial facilities, curriculum, and policies. Second, he had complete support of three presidents: Madison, Monroe, and Adams. Finally, he possessed the organizational skills to enlist the support of a competent faculty who could have taught elsewhere if they desired.[8]

Thayer's support of the faculty was not without its controversy. The commandant of cadets 1818–19 was Capt. John Bliss, considered the best drill instructor in the army. Unfortunately, he had a temper that resulted in his assault on a cadet. Captain Bliss's actions were inspired by a cadet who purposely marched out of step and repeatedly failed to correct himself. According to Theodore Crackel, Bliss's "heavily [sic] handed methods were not well suited for controlling the high-spirited, bright, and articulate cadets and predictably, he became a focal point of cadet discontent." Five cadet leaders presented their objections to Thayer in the form of a petition of more than 100 cadets and were reprimanded by Thayer. When the five returned with a new petition representing 189 cadets, despite the superintendent's warning, they were dismissed. The ringleaders were court-martialed, and their dismissal was upheld by President Monroe. Captain Bliss departed the academy in January 1819, with the cadets remembering him as "harsh and tyrannical"; however, they also reflected on the high degree of drill they had perfected under his instruction.[9]

More important than the conduct of Captain Bliss was Thayer's drive to instill a higher standard of discipline among the cadet corps. This was a battle of young wills against Thayer's devotion to the military school ideal. In 1821 the cadets' rebellious nature took advantage of a mess hall fire. As the cadets and administration were focused on battling the blaze, unidentified cadets aimed a loaded cannon at Thayer's home and lit the fuse. Fortunately, the cannon failed to fire, but the incident only strengthened Thayer's resolve.[10]

Thayer took three steps to establish the degree of discipline required in building a model military school. The first was his assignment of a commandant of cadets, who, unlike previous military commanders of cadets, had no academic responsibilities and could focus exclusively on the military and character development of the cadets. This is a key element in the military school culture. By assigning one man responsibility for the military operation of the academy, Thayer could expand his personal reach into academic affairs while balancing the two competing interests.

The second principal disciplinary improvement was the establishment of a code of conduct that "fully defined and circumscribed cadet life." This was done with a new set of 134 rules that regulated conduct and the limits of cadet authority. These rules addressed a wide variety of offenses: con-

West Point about 1821–25, by W. G. Wall. (Library of Congress)

duct at church services, unauthorized absences, wearing of civilian clothes, and conduct of unauthorized meetings or communications.[11]

The third measure Thayer took in 1825 was a demerit system with routine and clear penalties. He did this with the help of additional officers (tactical officers), who reported offenses to the commandant. The commandant adjudicated offenses and delivered appropriate punishments such as marching tours and "a specific number of demerits [to] be given . . . for a particular deficiency." The demerits further affected class standing— the class ranking, adjusted by the demerit factor, would help eliminate favoritism and provide a truer picture of the whole cadet.[12]

These measures created the standards of conduct that Thayer desired as successive cadet classes were enculturated to adopt those values. The demerit system is so central to the conduct of military schools in the twenty-first century that anyone familiar with the topic could not imagine a military school without it. Whether the demerit system originated with Thayer is uncertain, but at West Point before its institution offenses were adjudicated by courts-martial, and punishments were awarded by personal preference rather than by written guidance.

Thayer and West Point did not escape criticism during his time as superintendent. Their chief critic was Partridge, as part of his quest for regional academies. Among others was Congressman Newton Cannon of Tennessee, whose resolution in 1820 called for the abolishment of West

Point. Congress rejected that proposal, but again in 1830 the famous congressman Davy Crockett accused West Point of being exclusively for the privileged, educating the sons of the rich and neglecting the poor. The state legislature of Kentucky in 1833 also called for West Point's abolishment.[13]

Despite the criticism, the academy grew in reputation. Its graduates proved to be competent soldiers with outstanding educations provided by the school. In the midst of a poor showing by the army at large in the War of 1812, the West Pointers were a bright spot. No West Point–engineered fortress fell to the enemy. Ten of the forty-seven West Point graduates serving among 3,495 officers in the army were killed. But the West Pointers accounted for 10 percent of the promotions for bravery or exceptional service. By 1823 the reputation of West Point had grown so that there were one thousand applications for the limited openings. Outstanding professors such as Dennis Hart Mahan and Edward H. Courtenay brought the school to new academic heights.[14]

The student body's small size of 250 is deceptive and must be placed into the historical perspective of the early nineteenth century. The nine most influential colleges in the 1800s included Harvard, with an enrollment of 200; Princeton, 150; Dartmouth, 140; and the University of Pennsylvania, 150. West Point in 1819 was the only engineering school in the country until Norwich University opened its doors in 1820, and in 1824 Rensselaer School, a nonmilitary institution, became the third. After 1833 more civilian institutions of higher learning began offering civil engineering courses of study.[15]

But Thayer, like Partridge, had growing conflicts with the President. Andrew Jackson referred to Thayer as a tyrant and overturned a series of the superintendent's disciplinary decisions. The last of Jackson's actions against Thayer's wishes involved the dismissal of Cadet H. Ariel Norris. Norris had a long record of reprimands and courts-martial that included a previously overturned dismissal. When Jackson reversed his second dismissal, Thayer resigned after twenty-six years as superintendent.[16]

Thayer, unlike Partridge, was focused inward on West Point. His demands on the academic institution of West Point established it not only as the country's first engineering school but as the best of his time. His leadership and application of the school standards he saw in France helped create the model culture for military schools to emulate for the next two centuries. Most importantly, he perfected the unity of effort between academic preparation and military training. This became the centerpiece for the American military school model: a complementary relationship with academics placed in a superior role and military discipline as an overarching concept heavily influencing not just military training but also academic performance.

Thayer did not accomplish this to influence an effort to expand the military school concept outside of West Point. However, the reputation of the school and its alumni did help the military school concept to emerge. West Point alumni were involved in the establishment of other military schools, including the first military secondary school in the nation, the American Classical and Military Lyceum, which opened in 1826. Its first superintendent and founder was Augustus L. Roumford, West Point class of 1817. The American Classical and Military Lyceum used the building of a former Catholic seminary in Germantown, Pennsylvania, where Roumford had been a teacher. Roumford uniformed his cadets in gray uniforms with white trousers in the summer, and they wore caps with a seven-inch crown adorned with a yellow cockade and eagle while they conducted daily drill and morning and evening parades and marched to St. Luke's Episcopal Church for religious services.[17]

The American Classical and Military Lyceum's cadet corps contained 150 young men and boys, many of whom were from the South. Among the cadets were at least two from Mexico, and others were from Cuba. Gen. George Meade, Union commander at Gettysburg, was prepared for West Point at the school, as was Gen. Pierre Gustave Toutant Beauregard (CSA) and Adm. Henry DuPont (Civil War commander of the South Atlantic Blockade Squadron). Roumford taught mathematics and languages and employed at least six other professors, including future Mexican War general John Anthony Quitman (1798–1858) and Colonel Kober, a Prussian who taught German. The school operated for ten years, closing in 1836.[18]

After the founding of Roumford's school, at least seventeen other military schools were documented to have been founded by former West Point students or under West Point alumni leadership. Among them was the Huntsville Military, Scientific, and Classical School, which Bradley S. A. Lowe, class of 1814, founded in 1831 in Huntsville, Alabama. The Aiken Classical and Military Academy opened in South Carolina in 1842 with Alfred Herbert, class of 1835, as the principal.[19] Georgia Military Academy, in Greenville, South Carolina, opened in 1847 with William F. Disbrow, class of 1843 (not a graduate), as principal.

That same year the Maryland Military Institute (Military Academy) in Oxford was founded by Tench Tilghman, West Point class of 1832. The school's first superintendent was John H. Allen, also of the class of 1832. Allen would depart after ten years and found his own school in Chillicothe: the Ohio Military Academy, which opened in 1857. St. Thomas Hall in Holly Springs, Mississippi, made the transition to St. Thomas Hall Military Academy in 1849 under the leadership of Claudius W. Sears, West Point class of 1841, as president. The Mississippi Military Institute was founded in 1848 by a Colonel Goldsborough, who claimed to be a West Point graduate, although his claim has not been confirmed. The Alabama Scientific and Military Institute, Tuskegee, founded in 1846, was under Arnoldus Brumby, class of 1846, who in 1851 would become the first superintendent of Georgia Military Institute, Marion.[20]

Kentucky Military Institute, in Farmdale, was founded in 1846 by Colonel Robert T. P. Allen, class of 1834, who later transitioned the Bastrop Academy in Bastrop, Texas, to Bastrop Military Institute as its first superintendent in 1857. Other military schools founded under West Point alumni included the Western Military Institute, Georgetown, Kentucky, founded in 1847 by Col. Thornton Johnson, West Point class of 1822 (not a graduate). Thornton was followed in 1851 by Bushrod Rust Johnson, class of 1840, who led that institution until the Civil War. Caleb Goldsmith Forshey, a nongraduate of West Point's class of 1838, established Texas Military Institute in Galveston, Texas, in 1854.[21] D. H. Hill, class of 1842, was the first superintendent of Hillsborough Military Academy, Hillsborough (sometimes spelled Hillsborro), North Carolina, which opened in 1860. Louisiana Seminary of Learning and Military Academy, Alexandria, opened in 1860, and its first president would become one of West Point's most famous alumni, Gen. William Tecumseh Sherman, class of 1840.

The most influential alumni in the expansion of the military school format in the United States were the first superintendent of the Virginia Military Institute (VMI) and, to a lesser degree, the initial leadership of the South Carolina Military Academy, which is referred to as the Citadel, its official designation since 1910. VMI's first president was a West Pointer, Col. Francis H. Smith, class of 1835. The Citadel's first two presidents directed the school for nine years and were both West Pointers. Capt. William F. Graham, class of 1838, died of tuberculosis in office as president and was followed by Maj. Richard W. Colcock, class of 1826.[22]

The alumni of the academy, including Alden Partridge at Norwich, founded, transitioned, or initially directed twenty of the early military academies. Additionally, two other military schools, VMI and the Citadel, would influence the expansion of the military school concept extensively in the southern United States in the years preceding the Civil War. However, it was the US Military Academy under the leadership of Sylvanus Thayer that perfected the model and much which became the common organizational culture of military schools for the next two hundred years.

☆ 5 ☆

Francis Smith, Virginia Military Institute, and Southern Military Education

The creation of a state-supported military school in Virginia had been discussed in that state before 1835. That year, an article published by Lexington, Virginia, lawyer John Preston stated that the idea of an institute of a military character was not new and had become a popular idea since the establishment of the arsenal in Lexington. The arsenal was guarded by undisciplined state soldiers, and a murder involving one of the soldiers had occurred in the quiet Virginia village.

The reputation of West Point had been a subject of the *Lexington Gazette* only three weeks prior to John Preston's comments. In January 1836 a petition of ninety-four Lexington citizens called for the arsenal to be the responsibility of a new military school. This action progressed through the state legislature until the governor appointed a board of visitors with directions to establish a military school in the town. The board included the state engineer, Col. Claudius Crozet, a Frenchman, a graduate of the École Polytechnique who had served under Napoleon before coming to the United States in 1816 and who had worked at West Point as a professor of mathematics under Alden Partridge. At Crozet's suggestion "the regulations of the West Point Academy were made substantially the code of laws for the government of the institution."[1]

But Virginia Military Institute (VMI) was not to be a clone of West Point. West Point was to serve as a model, but the education was "not to fit its graduates for a single profession . . . but to prepare young men for the varied work of civil life." This required the education at VMI to have a "diminished intensity" but a "comprehensiveness enlarged." VMI was the nation's first state military college, and, like Partridge's Norwich, it looked at an education beyond the engineering focus of West Point to the needs of the state and its militia.[2]

The first superintendent of the Virginia Military Institute was Francis Smith, West Point class of 1833. He had been at West Point the last four years of Col. Sylvanus Thayer's reign as superintendent. Smith's feelings about Thayer were revealed in his address to the West Point alumni in 1879, when he spoke at the annual reunion. Smith described Thayer as "a noble specimen of West Point character" and alluded to his encounters with the superintendent before and after graduation. Smith served in the 1st Artillery Regiment and then returned to West Point as a professor of moral and political philosophy before resigning from the army in 1836 and becoming an educator at Hampden-Sydney College in Virginia, where he taught mathematics.[3]

To Smith, VMI's major contribution was not military leadership. Instead, Smith emphasized the role his graduates would play in the education of Virginians. Service in the classroom was a function he viewed to be more critical than service in uniform. Part of his philosophy was his belief that good teachers must be good Christians, so he ensured that the institute incorporate religious studies and mandatory church attendance reflective of his devotion to his Episcopal faith. His goal, beyond creation

Francis H. Smith
(1812–90). (Virginia
Military Institute
Archives)

Virginia Military Institute about 1858. (Virginia Military Institute Archives)

of educators for the state, was to develop moral men with traits of "self-control, productivity, patriotism, republicanism, and philanthropy."[4]

Smith took charge of the institute's initial twenty-eight cadets and led the institution for fifty years, 1839–89. He built a college based on many of Thayer's West Point academic fundamentals, including much of the science and mathematics core curriculum. But Smith added liberal arts, as he desired that VMI provide for a "thoroughness of education" that would give the school a distinct character and provide for the production of good teachers with command of English and "sound mathematical minds."[5]

These similarities and differences between the West Point and VMI curricula started with Smith's enthusiastic adoption of the cornerstones of West Point curriculum: mathematics and French. Smith incorporated the mathematic courses and textbooks of West Point's Prof. Charles Davis, a man he greatly admired. French at the institute was likewise to enable cadets to translate French math, military, and engineering texts. At that point, though, Smith departed from the West Point model. He added Latin, which was shunned by the anticlassical philosophies of Thayer as well as Partridge. In fact, the cadets were translating the same classical texts as used at the best classical universities. This, coupled with Smith's emphasis on English grammar and literature, made the VMI graduate highly competitive in the field of education.[6]

Educators March Forth from VMI

Through his connections, the superintendent of the Virginia Military Institute recruited a pool of military educators to expand his philosophy of education. Initially those educators were, like him, West Point graduates, but as VMI produced graduates, they became his source for teacher and administrator recommendations. In the first fifteen years of VMI's existence, 54 percent of its graduates became teachers, giving Smith many opportunities to expand his philosophy as well as VMI's reputation. This reflected not only the purpose envisioned by Smith but also the state legislature's desires. Virginia provided tuition for one cadet from each state senatorial district per year. These "state" cadets would repay their obligation not with military service but in performing their guard functions as cadets located with Lexington's arsenal and two years of work as teachers in Virginia.[7]

Col. Francis Smith's correspondence while superintendent reflects an interest in helping both West Point alumni and VMI graduates with advice and employment referrals. West Pointers who requested his advice and assistance included Col. Robert T. P. Allen at Kentucky Military Institute, who requested VMI's regulations. Maj. Richard W. Colcock, who took over the Citadel after the untimely death of its first superintendent, visited VMI and spoke with Smith for counsel on a variety of subjects. Arnoldus Brumby sought Smith's advice as he took over Georgia Military Institute. Col. Tench Tilghman requested Smith's guidance on a variety of issues and requested the VMI regulations for his Maryland Military Academy. And Maj. William Tecumseh Sherman, at the Louisiana Seminary of Learning and Military Academy, also turned to Smith for guidance. Smith used these relationships and those with nonmilitary educators to help the alumni of both West Point and VMI obtain employment.[8]

Like those of West Point, many VMI alumni actually started military schools of their own based on their educational experiences. But the list of pre–Civil War military schools with founders or initial leadership from VMI was even larger than West Point's, as the Virginia school's alumni were more prepared to go into educational careers.

A total of twenty-six military schools were established or created by VMI alumni prior to the firing on Fort Sumter on April 12, 1861.[9] Once the Civil War began, however, many private, male, nonmilitary schools and colleges saw enrollment drop dramatically. In 1860 one of those colleges in Virginia was Randolph-Macon College, which was at the "climax of its antebellum prosperity."[10] The departure of so many students for the Confederate Army by August 1861 prompted the school to look at adoption of the military format. The Board of Trustees of the Methodist college initially appointed Rev. Maj. William H. Wheelwright, VMI class of 1845, as professor of military tactics in August 1861. The following

February the board completed the military reorganization of the college with the president, Rev. William A. Smith, named colonel and commandant of cadets and another VMI man, Maj. J. E. Blankenship, class of 1852, named professor of mathematics and military science. "A regular uniform was prescribed, drills were observed daily, and other things of a similar character were enjoined, all looking towards preparation of the student for the duties that awaited him in defense of his country."[11]

After the war, Randolph-Macon would reopen as a nonmilitary college and thrive. But years later, reaching back to those few years of military operation, the school's two affiliated academies, Randolph-Macon Academy in Front Royal and Randolph-Macon Academy in Bedford, would convert to military prep schools in 1910 and 1920, respectively.

The Citadel Establishment and Influences

The contributions of Francis Smith and the VMI alumni to the expansion of the military school concept explain the popularity of the military school in the southern United States, but they do not tell the whole story. In South Carolina, Governor George McDuffie proposed the creation of a state military school combining "military instruction with the usual subjects taught in state schools in 1836." Military schools by that time were becoming familiar, as about one hundred South Carolina men had attended Norwich, and seven had served in the state legislature. In 1842, Governor John P. Richardson established two schools, one at Columbia and the other in Charleston. The schools were under one board of governors, and the cadets would, like those at VMI, take the duty of guarding the arsenals at each location. In March 1843, twenty cadets reported to the Citadel Academy in Charleston and fourteen to the Arsenal Academy in Columbia. The two schools were known collectively as the South Carolina Military Academy (the Citadel). After 1845, only cadets in their initial year at the academy were housed at the Arsenal in Columbia.[12]

Between opening in 1842 and the start of the Civil War in 1861, the Citadel had four superintendents. Capt. William F. Graham served only one year prior to his death; he was replaced by Maj. Richard W. Colcock, who served as superintendent until 1852. Colcock was replaced by a Citadel faculty member, Maj. Francis W. Capers, who served until 1859 and was replaced by Citadel graduate Maj. Peter F. Stevens, class of 1849. Unlike Francis Smith at VMI, whatever part the four Citadel superintendents played in the establishment of additional military schools prior to the Civil

War was largely lost to history when the Citadel and its records were destroyed by Union forces in 1865. But a number of military schools created by the school's alumni have been identified. The Citadel's alumni, like VMI's, were much more likely to go into education than were the alumni of West Point, with 27.7 percent following that profession.[13]

Nine military schools are known to have been established by Citadel alumni, but others are likely to have had Citadel graduates as founders or key faculty. In Alabama, Gibson Hill was assisted by N. W. Armstrong, Citadel class of 1851, in establishing the Southern Military Academy in Fredonia in 1851. In North Carolina, Wilmington Military Academy was opened in 1856 by James D. Radcliffe, and Hillsborough Military Academy was founded by Charles Tew, class of 1846. Bowden Collegiate Institute was started in 1856 by John Richardson in Georgia. Florence Wesleyan University in Alabama transitioned to LaGrange Military Academy in 1858 under James W. Robertson, class of 1850, as the superintendent, and the Brandon State Military School in Mississippi was founded by Pierre S. Layton in 1860.[14]

In South Carolina, Citadel graduates established Aiken Classical and Military Academy in 1856, with C. W. McCrary and T. H. Mangum, both of the class of 1856, as principals. About that same time, Anderson Military Academy was established with Joseph Manning Adams, class of 1856, as principal. But by far the most successful of the Citadel-inspired schools was Kings Mountain Military School, founded by Michael Jenkins, class of 1854, in January 1855. Its military and academic programs were so well thought of that the school's graduates were permitted entrance into the Citadel as sophomores. The school opened with just 12 cadets, ages eleven to sixteen, but by the end of the first year it had expanded to 60. The curriculum spanned five years and included mathematics, history, chemistry, languages, literature, astronomy, geology, physiology, and philosophy. By 1861 its cadets numbered 140. At the onset of the Civil War the school closed, but it reopened in 1866 and operated until 1886.[15]

Southern Attraction to the
Military School Concept

Between 1831 and 1843, 39 military schools operated in the United States. Regionally, the division between North and South was relatively equal, with the South having 17 schools and the North 19 schools. Between 1843 and 1854, 91 military schools operated in the United States, 52 (57 percent) in states that later entered the Confederacy and 28 in the Union states, with 11 associated with what would be considered border states during the Civil War. However, in the six years before the war the number of military schools in operation had climbed to 138, and regional orientation shifted to the South. Ninety-eight (71 percent) of those schools were in what would become Confederate states.

The explanation for both the growth and regional shift has been largely attributed to a Southern cultural orientation toward militarism and preparation for war. The sectionalism that led to Southern concerns for military preparedness grew out of divergent economic interests of the Northern industrial base and the Southern cotton-based wealth and slave-supported economy. The two largest issues dividing the North and the South grew from questions of slavery and economic policies that favored the North. Both issues were questions of states' rights under the Constitution, with slavery issues going back at least to 1784, when Thomas Jefferson proposed to Congress the prohibition of slavery in all western expansions. The subject of the expansion of slavery would divide the states; the North was opposed to the western expansion of slavery, and the South supported expansion. This division of opinion was for a time addressed by the Missouri Compromise of 1820, but by 1848, with the expansion of US territory, the crises reemerged as additional states were admitted into the Union as either slave or free states. In addition to the political issues associated with slavery, the moral issue grew as the abolitionist movement in the North gained more political influence.[16]

The national discord around slavery was further deteriorated by economic interests and associated tariffs. As early as 1828, with the enactment of tariffs designed to favor the manufacturing Northern states, South Carolina led a crusade that became known as the Nullification Crisis. The resulting compromise, which lowered tariffs, enforced a Southern attitude towards the importance of states' rights and their place in trade and slavery issues. These issues grew to support the need for military preparedness in the South, which claimed that it "lived in a crisis atmosphere . . . menaced through encirclement by a power containing elements unfriendly to its interests, elements that were growing strong enough to capture the [national] government."[17]

Despite the heightened political and military situation, growth of the military school format in the South even as late as a year before the outbreak of fighting was not preparation for war. The vast majority of the increase in military schools was in secondary rather than collegiate institutions. Neither the state governments nor parents saw their sons in secondary school being prepared to be officers in a state militia defending states' rights or slavery.

At the collegiate level the transition of institutions to a military format likewise was not considered preparation for war. In 1860 the University of Alabama converted to a military format. Even at that late date student conduct and formation were the focus of the change. Student misconduct had a long history in Tuscaloosa, dating back at least to 1833. Alcohol consumption was such a problem that in 1847 the Alabama General Assembly passed a law against the sale of alcohol to university students. In 1858 misconduct prompted a grand jury to issue a county warning that 90 percent of misdemeanors and crimes in the town were the result of alcohol consumption by students of the university and public schools. The university president, Dr. Garland, worked for six years to convert the university to correct those evils. The transition of La Grange Military Academy, also in Alabama, was motivated not by military preparedness but by the desire to correct a declining enrollment. La Grange College in its twenty-seventh year in 1857 was faced with threats to move the instruction from La Grange to Florence. In order to save their college, the administrators made the transition from a literary and sciences institute and adopted a popular military format.[18]

In both collegiate examples, explanation for the transition from an educator's point of view was one of student moral formation, enrollment, finances, or institutional reputation. From the state government's point of view, preparation for war or leadership of the state militia may have been a factor for 47 state cadets of the 170 cadets at La Grange who received tuition provided by the state. However, the obligation for Alabama state cadets likely was the same as that for VMI cadets: teaching in their school's state. The resulting state support of the University of Alabama was based on the recovery not only of institutional reputation but also of potential state service by its graduates.

The conclusion that militarism was not the motivation for the popularity of military schools in the South was supported by Webb's study of the origin of nineteenth-century military schools: "The charge that the Southern systems of state military schools were for the purpose of developing leaders for an inevitable civil war is a false, mistaken, and erroneous assumption." His assertion was that these colleges were in effect normal schools for the preparation of teachers for their state.[19]

Further, according to Clarence Mohr the expansion of pre–Civil War military schools in the South can be attributed to Southern parental concern for discipline, character development, and sound citizenship and a means to provide an affordable education to less advantaged white males.[20] The establishment of VMI and the Citadel in 1839 and 1841 helped meet those needs. Those schools were not established out of fear from Northern encirclement.

Jennifer Green found that parents likewise were attracted to sending their sons to military colleges in the South to develop the qualities of self-discipline, obedience, and morality. These attributes appeared to be of greater value to the parents for their sons' development than academics. The second attraction was full state funding offered by twelve Southern state military colleges. Those schools offered select cadets full tuition, which had the advantage over civilian institutions, at which the state scholarships for needy students rarely covered all four years.[21]

Clearly, two factors that strongly influenced the expansion of military schools in the South were Francis Smith's educational objective and the VMI alumni educators he sent out to educate the South. The military format met parents' desires for a moral and structured education and the states' need for educators. Along with the Citadel's contribution at least thirty of the other Southern schools were established by alumni of VMI and the Citadel. Many other schools employed the teachers from the two schools, and others were founded on the concepts demonstrated by the two institutions. The educational objectives of Francis Smith were clearly a success. However, the institute he founded next would be heard from not in its educational role but as a central figure in the preparation of Southern military leadership in the Civil War.

★ 6 ★

Military Education and
the Civil War

As a result of the Civil War there were four major impacts on military schools in the United States. During and immediately following the war, many military schools in the South were forced to close, and most did not reopen. Outside the South, the Civil War influenced Northern states to transition to or establish new military schools. Several military schools also participated as units in active military service and even fought as combat units. Finally, the postwar occupation of the South also limited military school activities for several years.

Between 1855 and 1860 there were at least eighty-five military schools operating in the South. Documentation shows that only 22–26 percent of those schools reopened or continued operation after the war. In 1860, the year prior to the war, the Union states had only thirty-two military schools. By the end of the war in 1865, the number of military schools in the Union states had increased by fifteen, or 40 percent.

Southern Military Faculty and Cadets
Depart to Be Soldiers

Many of the smaller Southern military schools closed when key faculty left to join Confederate military forces. Chuckatuck Military Academy in Virginia is a good example, when President James J. Phillips, VMI class of 1853, left his school to recruit the Chuckatuck Light Artillery for Confederate service. Shelbyville Military Academy in Tennessee, another small military school, closed because of a combination of the loss of its founder, Capt. William H. Keiter, VMI class of 1859, who departed for military service in May 1861, and the enlistment of its cadets. It appears the school's leadership was taken over by Alexander McKinnery Rafter, who "within two months of the beginning of the war enlisted with all the pupils." Rafter and his cadets may have been members of the Tennessee Belmont Artillery Battery who witnessed the death of Captain Keiter. In 1862, Keiter was ordered to fire what he believed to be a defective artillery piece. Against his better judgment he followed orders, and the resulting explosion killed him.[1]

Nathan B. Webster, Norwich class of 1839, left the Virginia Literary, Scientific, and Military Academy for very different reasons. He closed his school and departed for Canada, likely for two reasons: Portsmouth, Virginia, was under Union occupation, making the operation of a military school nearly impossible, and he had conflicting loyalties, having worked in Virginia for twenty-two years but having been born in New Hampshire and educated at Norwich University.[2]

Kentucky Military Institute is an example of one of the larger state military schools that closed. "When the Civil War broke out every cadet and almost every faculty member joined one side or the other, forcing the school to close." The North Carolina Military Institute in Charlotte opened in 1859 and had grown to 150 cadets by April 1861. Three future Confederate generals—D. H. Hill, James Lane, and Charles C. Lee, were among the staff. The institute closed in 1861 as the 1st North Carolina Infantry Regiment enlisted the faculty as the regiment's staff officers, and the cadets were permitted to join with their parents' permission. Another military school in North Carolina was the Statesville Military Academy, established by John Barr Andrews a few years prior to the war. He closed the school while raising C Company, 4th North Carolina Infantry, composed largely of his own cadets. Captain Andrews died of wounds received at the Battle of Gaines Mill in 1862, and his body was returned to Statesville for burial.[3]

The Arkansas Military Institute formed the basis of Company I, 3rd Arkansas Infantry. The school's permanent closure was sealed by Union troops in 1864, when they burned the school buildings. In Arkansas, St. John's College cadets formed Company A (Capital Guard), 6th Arkansas Infantry. Their legacy helped inspire the reopening of the school after the war. In Alabama, Evander Law of the Citadel class of 1856 enlisted the older cadets from a military school in Tuskegee to form Company B of the 4th Alabama and closed his school. Bowden Collegiate and Military Institute/College in Georgia closed as a military school when seven graduates and a large number of cadets formed a 130-man unit, Com-

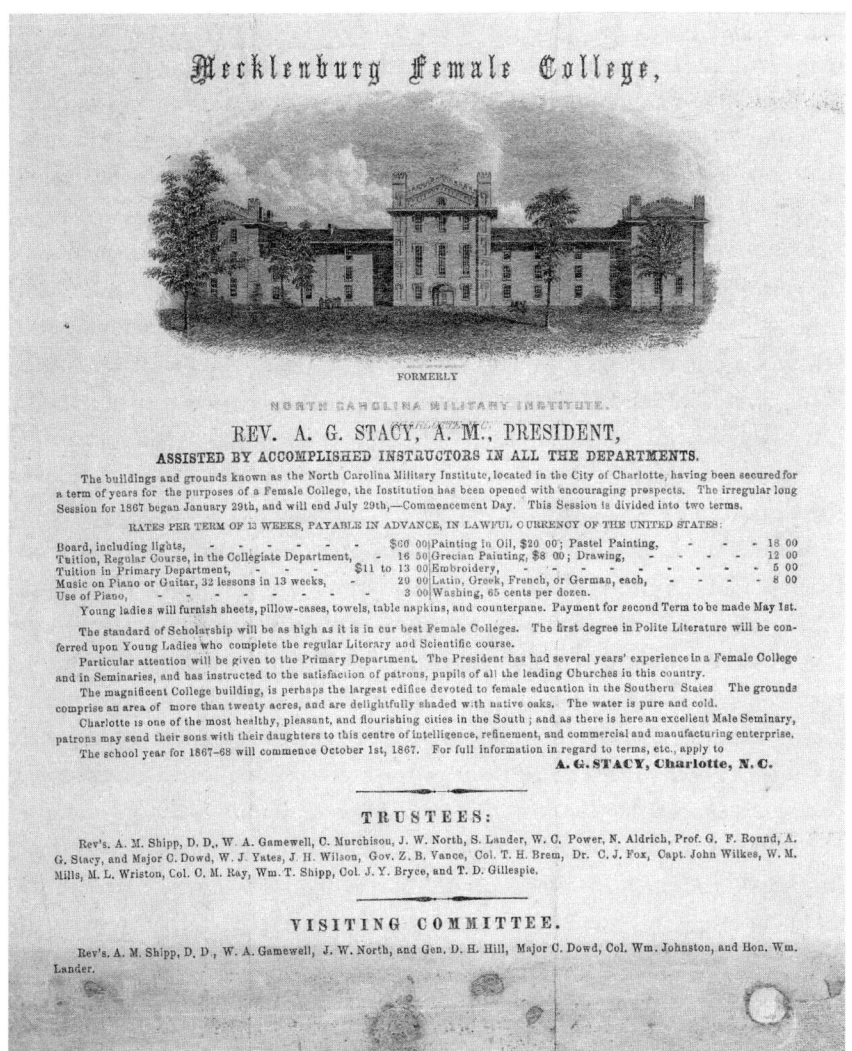

North Carolina Military Institute in Charlotte, engraver Samuel Sartain, from an 1867 advertisement for the Mecklenburg Female College. (North Carolina Collection, University of North Carolina at Chapel Hill)

pany B, Cobb's Legion, under the leadership of the college president as the captain. Other colleges closed as cadets organized Confederate units, including Lynchburg Military College, where faculty and cadets formed Company E, 11th Virginia Infantry. La Grange Military College and Academy in Alabama initially lost 48 of its 170 cadets as they enlisted at the outbreak of the war. In 1862 the school closed, with the remainder joining the 35th Alabama Infantry Regiment, with their superintendent as colonel, a professor serving as lieutenant colonel, their commandant as major, the cadet adjutant as regimental adjutant, and cadets manning the majority of B Company, with representation in other companies as well.[4]

Kings Mountain Military Academy closed when the founders became the commander and adjutant of the 5th South Carolina Infantry Regiment. Company I (Jasper Light Infantry) of the regiment was largely comprised of the academy's graduates and graduating seniors. The school reopened after the close of the Civil War and remained open until 1886. The Franklin Military Institute of Duplin County, North Carolina, formed Company E, 20th North Carolina Infantry, in 1861, with the school's principal, Claudius B. Denson, elected to serve as their captain. Denson was a graduate of Virginia Literary, Scientific, and Military Academy and employed as his coprincipal Richard W. Millard, who was an associate of Nathan B. Webster, founder of the Virginia Literary, Scientific, and Military Academy.[5]

Besides the impact of cadet enlistments on school enrollment, the Confederate Conscription Act of 1862, which ordered compulsory military service for all white males eighteen to thirty-five years of age, had another heavy effect. Although an exception was made for certain school administrators, teachers, and professors, there was no exception made for military school cadets, whether attending a college or a secondary school. Even the state military schools, which were a source for a professional officer corps, were not an exception. This fault in the law led the governors of Alabama, Georgia, South Carolina, and Virginia to protest. The Georgia governor even designated the cadets of the Georgia Military Institute as state employees as a ploy to evade the law. The Alabama governor declared his cadets from the University of Alabama exempt in defiance of the law. The Virginia governor, in coordination with Colonel Smith of VMI, used delaying tactics as well.

But the perception was that being a cadet was not fulfilling one's patriotic military service obligation. As a result, the draft pulled or drew cadets from state military schools in large numbers. For example, approximately 875 cadets left VMI during the war prior to graduation. A group of cadets from the Citadel in June 1862, when ordered to return to classes during the defense of Charleston, departed and formed Company F (Cadet Rangers), 6th South Carolina Cavalry Regiment.[6] Departing cadets at the University of Alabama helped form Company F, 7th Alabama Cavalry, in 1863 and Captain Bascom T. Shockly's Escort Company of Cavalry in 1864.

Symbolic of the fate of most southern military schools, the University High School and Military Academy of Athens, Georgia, surrendered to Union forces in 1865. For Southern military schools the Civil War and the Northern occupation were major setbacks. Northern occupation, with its prohibition against the issue of drill weapons and an environment of suspicion, slowed any military school recovery in the South.

Drill Masters and Cadets in Combat

Of the Southern military schools that remained open during the war, at least ten set aside their academic function for a period of time and conducted military operations in support of the Confederacy. Virginia Military Institute became the most famous of these schools for its charge during the Battle of New Market. The cadets of the Citadel, the University of Alabama, Georgia Military Institute, and West Florida Seminary, along with midshipmen of the Confederate Naval Academy, all engaged in combat.

These institutions would fight in a number of engagements and lose in combat at least twenty killed, seventy-one wounded, and fourteen dead of illness. Although their actions cannot be said to have played a major role in the war, the image of young cadets in combat, especially those of VMI at New Market, became a part of the American military school lore and part of the ethos of patriotic service.

Several military schools, including the North Carolina Military Institute, the Hillsborough Military Academy in North Carolina, Texas' Bastrop Military Institute, the Louisiana Seminary of Learning and Military Academy,

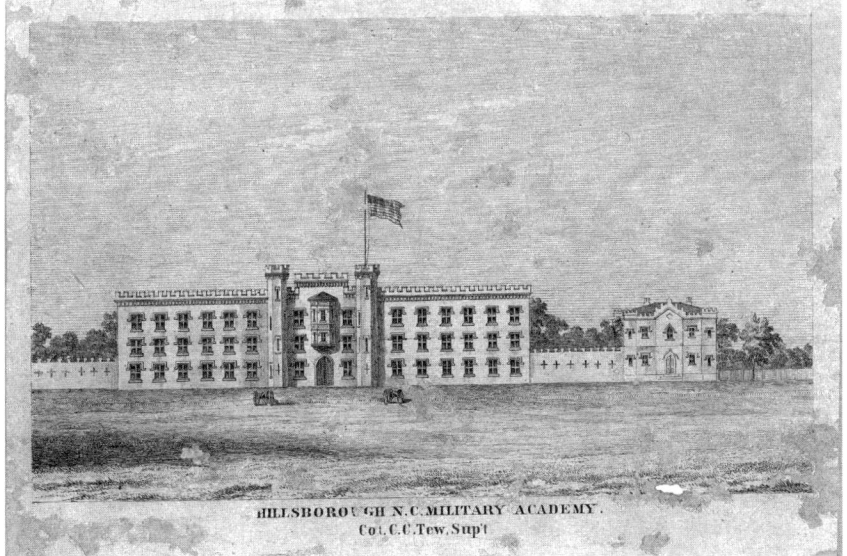

Hillsborough Military Academy. (North Carolina Collection, University of North Carolina at Chapel Hill)

and others helped drill recruits in preparation for military service. Additionally, La Grange Military College and Academy initially sent one cadet company commander, Cadet Capt. John Smith Napier, to help train the 16th and 27th Alabama Infantry regiments. After graduation he joined the latter unit and became its adjutant.[7]

The Hillsborough Military Academy had the influence of both the Citadel in its founder, Charles Tew, and Virginia Military Institute in one of its first two teachers, Maj. William H. Gordon, VMI class of 1852. The academy numbered 130 cadets in 1861, when a large number of newly elected Confederate officers of North Carolinian volunteer regiments came to Hillsborough to ask for help. According to Cadet Capt. William Cain, a company commander of the academy, "it was certainly a novel sight to see the little cadets from ages 13 years and older tramping his squad of grown and sometimes grizzled men over the parade ground." These new officers, having no military experience, took the drill instruction of their young instructors seriously. In hindsight, Cadet Cain saw the Hillsborough contribution as bringing order out of chaos by forming regiments with inexperienced soldiers. In fact, three months of instruction with the aid of the young cadets helped produce "well drilled and disciplined" units. It also resulted in the vast majority of cadets enlisting in the regiments they helped train.[8]

The Citadel

The cadets of the Citadel were called upon numerous times during the Civil War for service in combat and to serve as drillmasters for new Confederate Army recruits, as guards for storage facilities and prisoners of war, and to help man fortifications in the Charleston area. On January 9, 1861, prior to the bombardment of Fort Sumter, a detachment of fifty Citadel cadets manned four twenty-pounder siege guns and fired on the *Star of the West*, an unarmed steamer carrying reinforcements to Fort Sumter. Of the three shells fired from their position on Morris Island, two missed and one struck the ship, turning the vessel around and back to New York.[9] Many consider this action, rather than the April bombardment of the fort, as the beginning of the Civil War.

Apart from Union naval bombardment of Charleston, the cadets' next combat came in December 1864. The Citadel cadet battalion, numbering 343 freshmen organized in Company B from the Arsenal and upperclassmen in Company A from the Citadel, were called to active service again to fight alongside regular Confederate troops at Tulifinny Creek. The cadets assembled early the morning of December 4 and marched from their barracks with faculty and staff and fixed bayonets. Between December 6 and 9 the cadets were in action as part of a Confederate force of one thousand men facing a Union force of fifty-five hundred soldiers, sailors, and marines. On the first day the cadets marched on newly landed Union forces and engaged the enemy's right flank in ten minutes of small arms fire before the Confederate line was forced to withdraw to more defendable terrain along the railway line.[10]

The next morning at dawn the cadet battalion, along with the 47th Georgia Infantry Regiment, formed skirmish lines and started a silent advance on the Union center. The cadets' appearance was described as that of "Dandy-Jim looking kids." Unfortunately, the advance was discovered, and the Confederate force came under heavy fire. The cadet battalion charged with a rebel yell, driving the Union pickets back to their entrenchments. Maj. James White, the superintendent, observing the strength of the Union force entrenched to his front, directed an orderly withdrawal. The engagement was costly, with at least one cadet killed, eight cadets wounded, and one faculty member wounded.[11]

After taking up defensive breastworks alongside the 47th Georgia, the battalion leadership kept the cadets' heads down with calm directives, sounding more like the classroom than the battlefield to the veterans on their flank: "Down Mr. Hagwood, Down Mr. Haynie." Keeping his cadets under cover until the Union counterattack came within rifle range, Major White ordered, "Battalion Attention, Ready Aim Fire." The fire from the more than three hundred cadets had deadly effect and resulted in the enemy's retreat.[12]

On the final day of the Tulifinny action on December 9, the cadet battalion was positioned on the Confederate left and helped repel a battalion of US Marines, who retreated as the 127th New York Infantry Regiment on their left gave way. The cadets were part of the pursuit of the retiring enemy forces. The cadets proved themselves; as one veteran soldier stated, "Them youngsters'll fight like hell." They were also compared to General Hood's Texans, and a Confederate artillerymen stated, "The cadets fought as if they were on a parade ground."[13]

The Citadel cadets ended their participation in the war by defending the fortifications of Charleston, then retreated with Confederate forces before the city fell. In April 1865 the governor of South Carolina furloughed the upperclassmen cadets of the Citadel, and in May, the freshman cadets of the Arsenal. During the period of active service, seven additional cadets

Citadel cadets fighting at the Battle of Tulifinny on December 9, 1864, mural by David Humphreys Miller, in the Daniel Library at the Citadel. (Courtesy of the Citadel)

died from disease. But the last bloodshed was after the official furlough; on May 9, 1865, Cadet McKenzie Parker, in a skirmish with Union troops, was killed near Anderson County. The Civil War cost the Citadel fourteen cadet lives lost in combat and from disease and twice that number wounded. The school was burned by Union troops and would not reopen until 1882.[14]

University of Alabama

Early in the war, the University of Alabama cadets were called out of the classrooms periodically to drill newly formed Alabama regiments and to serve as wagon guards for supply trains. Like many Southern schools, the University of Alabama cadets and faculty left to enlist at such a rate that in 1862 there were no graduates.[15] No cadets were called up under the Confederate Conscription Act, because the school was excepted as a state-supported institution, and it actually grew, with new cadets joining the University of Alabama's ranks.

But on at least two occasions the cadets were called on for combat operations. In July 1864, fifty-four cadets, while traveling from Montgomery as part of a composite force called the Lockhart Battalion, detrained to counter a Union Army cavalry advance on Auburn at Chehaw Station. Initially fighting from behind a rail fence, they advanced to the better protection of a ravine against Rousseau's Union Cavalry, which was armed with the superior Spencer carbine. Two University of Alabama cadets were among the eighty Confederate casualties. Both sides claimed victory—the Confederates because they advanced and secured Auburn, and the Union because they had suffered only eleven to thirteen casualties.[16]

The final military operation for the Alabama cadets started on the night of April 3, 1865, when a force of fifteen hundred Union cavalrymen approached Tuscaloosa. In the late hours the university president ordered, "Beat the long roll! The Yankees are in town!" With that, the roll of cadet drums roused the three hundred cadets from their barracks, and they formed for battle just east of the intersection of University Boulevard and Greensboro Avenue. Heavy fire there in two brief engagements with Union cavalrymen of the 6th Kentucky Cavalry resulted in four wounded cadets, including Cadet Captain John H. Murfee, the brother of the commandant, James Thomas Murfee, VMI class of 1853.[17]

As the cadets waited for dawn and the coming battle against an overwhelming Union force, the decision was made by the president of the university to abandon Montgomery and the university's buildings. The cadets marched out of town and entrenched themselves on high ground. The next day Union troops burned the university to the ground. The cadets then marched to Marion, where a shortage of food led the commandant to place the cadets on leave to reassemble in thirty days. But four days later

University of Alabama cadets assemble for action in Montgomery, April 1865, by John Paul Strain. (Courtesy University of Alabama)

General Lee surrendered, leading to the end of the war. The university would not reopen until 1871, and it did so without a corps of cadets.[18]

West Florida Seminary

West Florida Seminary was another Southern military school whose cadets were called for combat operations late in the war. The school was founded in 1851, and enrollment grew to 79 by 1853. When the war began, the school had 250 cadets, but by 1865 enrollment was down to only 58 students. The cadets were called out for the Battle of Olustee but were relegated to guard Union prisoners, a duty they had performed on several other occasions. In March 1865 three regiments of fifteen hundred Union troops landed by sea at Apalachee Bay and marched north. The small corps of cadets of the West Florida Seminary, numbering only 34–37, was called to active duty, and 25 cadets, led by Capt. Valentine M. Johnson, VMI class of 1860 and wounded war veteran, traveled by train south to Newport. Johnson held several cadets, as young as eleven, back to perform guard duty rather than sending them into combat. In Newport the small company of cadets joined the 5th Florida Cavalry Battalion by running under fire to the trenchline across the Saint Marks River.[19]

The next day, as Union forces marched north, the cadets moved with the 5th Florida to Natural Bridge to counter the threatened river crossing there. At the location where the Saint Marks River went underground, forming a land bridge, the Confederate troops and cadets entrenched and repulsed the eight Union assaults conducted by the 2nd and 99th Colored Infantry Regiments. The Union troops then withdrew, having suffered twenty-one killed, eighty-nine wounded, and thirty-eight captured. The Confederate force lost four men, and twenty-two were wounded. Cadet Tom Frazier was among the Confederate dead, having fallen off the train early in the deployment. During the war the West Florida Seminary was also known as the Florida Military Collegiate Institute. It would reopen after the war as a nonmilitary school, and, through several name changes, ultimately become Florida State University.[20]

Georgia Military Institute

Georgia Military Institute, founded in 1851, grew by the start of the war to enrollments of 150–200 cadets. The campus was 110 acres, including College Hill, with its buildings and parade ground a mile south of Marietta. The cadet battalion as well equipped with short cadet muskets, swords for cadet officers, and four six-pound fieldpieces. The institute was

THE WAR IN GEORGIA—THE MILITARY COLLEGE, MARIETTA.—FROM A SKETCH BY CAPT. D. R. BROWN, 20TH CONN.

Georgia Military Institute about 1865, by Capt. D. R. Brown, 20th Connecticut.

tasked early in the war by the governor of Georgia with training newly arrived soldiers at Camp McDonald. The Confederate Conscription Act initially threatened to close the school, but, through the governor's protests, the policy was changed, and by 1863 the school was similar in size to its prewar enrollment. In 1863 the cadets were called upon to guard a rail bridge during a raid by Col. Abel Streight's Union cavalry. In May of the next year, the cadets moved by train to Resaca, attached to Maj. Gen. William Walker's division, and played a role in repulsing a Union assault with no cadet casualties. They were on active duty for just a few days, then returned to their classrooms in Marietta.[21]

By June the situation for the Confederacy in Georgia had worsened, and the cadets were called upon again and moved by train to West Point, Georgia. After encampment for about six weeks, they joined forces near Atlanta and participated in action at Turner's Ferry by defending that crossing on July 4 and 8. They suffered no casualties during that period, although they came under cannon fire as they marched to Atlanta, where they took up defensive positions. The Battle for Atlanta cost the Georgia Military Institute five battle casualties, including one killed. Life in the trenches would contribute to six cadets' death from disease. After almost two months away from their studies, the corps of cadets was moved to the capital, Milledgeville, on August 20, 1864, and camped on the capitol square, attempting to resume academic work.[22]

The cadets were there when Union troops burned their campus in Marietta on November 13. Soon after, departing Milledgeville, the cadets were back in combat defending the Oconee River railroad bridge for three days against several assaults. Flanked by Union forces, the cadets withdrew under cover of darkness, having suffered three wounded and one killed in action.[23]

Between November 28 and December 20 the cadets defended various positions in defense of Savannah, including Oliver's Station, Ogeechee Station, and the Oconee Bridge. As part of the rear guard of the Confederate evacuation of Savannah, the cadets suffered two wounded and one killed. The corps of cadets was back in Milledgeville when the Confederates surrendered at Appomattox. In their final days as cadets and in the final days of the Georgia Military Institute, cadets functioned as police in Atlanta. On May 20 they were paroled and the institute was disbanded. Nine of their comrades died in Southern service, three of them killed in combat. Nine others carried the wounds of war home; one was missing an arm, and another had lost an eye.[24]

Virginia Military Institute

Cadets at VMI did not fire the first shots of the Civil War, as the Citadel can claim, but they were involved in major events leading up to the war. In 1859, sixty-four cadets escorted the governor of Virginia and provided guards for the public execution of John Brown. In the early days of the war, VMI was called upon to drill assembling Confederate forces. VMI's cadets were the first to be lost in combat. When Union forces started to advance on Manassas in July 1861, the emergency redeployment of Thomas "Stonewall" Jackson's Brigade and the 18th Virginia Infantry Regiment resulted in the attachment of cadets for training purposes, and they accompanied the units they were training. On the fields of the First Battle of Bull Run, three cadets were killed and two were wounded. They were the first Southern cadets to fall in combat, but far from the last from VMI.[25]

In May 1862, a year later, the institute was called to active service with Stonewall Jackson, this time not as individual cadets but the entire corps of cadets as a unit. During the Battle of McDowell, they were not committed to combat but remained in the field for more than two weeks. After the battle they collected the wounded from both sides and buried the dead. The final five days of their active service were spent marching back to Lexington in the rain, which contributed to one cadet's death from illness.[26]

From late April through July 1863, cadets of VMI were working as drillmasters at Camp Lee. In August 1863, fifty cadets mounted on horses were sent southwest of Lexington to search for deserters, while a force of two cadet companies was sent seven miles northwest to defend against a suspected Union cavalry raid. The success of the deserter hunt is unknown, and the Union raid never materialized. Again, three times in November and December, cadets marched out of Lexington in response to cavalry raids, but they encountered no Union troops.[27]

On May 11, 1864, the corps of cadets of VMI marched into events that would make their deeds a centerpiece of military school lore and Southern tradition. Upon their arrival at New Market, the Confederate commander, Maj. Gen. John C. Breckinridge, welcomed them: "Young Gentlemen, I hope there will be no occasion to use you, but if there is, I trust you will do your duty." The cadet battalion became the reserve force for brigade commander and VMI graduate Col. Gabriel C. Wharton. Unfortunately, Colonel Wharton did not formally brief Lt. Col. Scott Ship, the VMI commandant, on the situation, and, as the cadets marched down the hill into an area of relative safety, Union cannon fire caused four casualties.[28]

Initially, the VMI battalion, numbering 226 cadets and faculty, with 3 cadets left behind ill, functioned as an infantry reserve behind the Confederate line. An additional 32 cadets manned two cannons placed along the Confederate artillery line. As the battle was entering a critical stage, General Breckinridge, in an agonizing decision, ordered, "Put the boys in and may God forgive me for that order."[29]

With that order the cadet battalion went into the gap between the 51st Virginia Infantry Regiment and the 62nd Virginia Mounted Infantry Regiment, the latter fighting dismounted (fig. 6.1). Both units had already suffered heavily, and the right four companies of the 51st Virginia had faltered under withering fire. As the cadets moved into the forward line, they passed a Confederate officer trying to regroup his men. "Rally men and go to the front. Here you are running to the rear like a lot of frightened sheep. Look at those children going to the front. Rally and follow those children." Just the initial movement into the line cost the cadets at least six casualties. They endured heavy enemy fire and a faint-hearted attack against the Confederate line by three Union regiments, but as the Union attack faltered, a Union artillery battery behind started preparing the horses to withdraw their cannons from the field.[30]

Shortly afterward, the Confederate counterattack started to the left of the corps of cadets, but with simultaneous commands from Captain Wise,

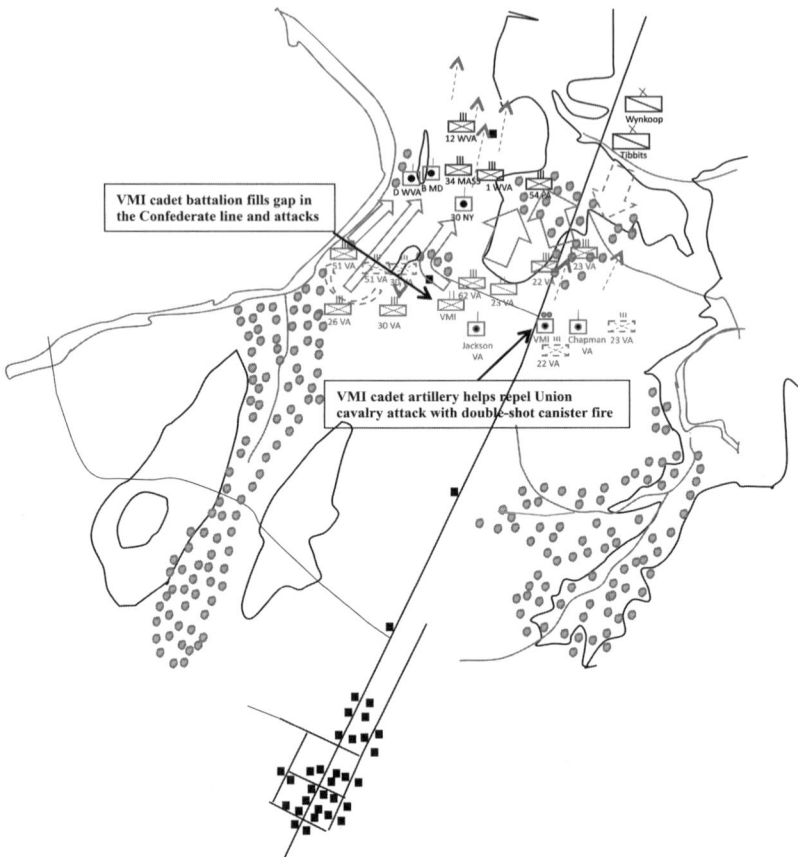

Fig. 6.1. Battle of New Market, 2:45–3:00 pm, May 15, 1864

acting commandant, and Captain Preston, one of the tactical officers, the cadets joined in charging the Union artillery battery. The following was described by a journalist on the scene: "The cadets from VMI moved forward in the charge upon the enemy's battery. Their step was as steady as the tread of veteran soldiers. They never faltered, but went into the harvest of death as though they were accustomed to such bloody work."[31] As the cadets advanced across the open, muddy field, some of them lost their footwear and marched forward barefoot. Across the field, which became known as the field of lost shoes, the cadets' "lines [remained] well formed and continued to advance toward the battery" up to the crest of the hill. On the right of the enemy battery, the Union 1st West Virginia Infantry retreated in disorder. The 34th Massachusetts Infantry and artillerymen of the 30th New York Artillery Battery retreated to the rear under fire from

Charge of the VMI cadets at New Market, by Benjamin West Clinedinst. (Courtesy VMI Museum, Lexington, VA)

the cadets, leaving behind one twelve-pounder cannon as the cadets swept the position. A member of the Union commander's staff, Maj. Theodore Lang, described the charge: "I have never witnessed a more gallant advance and final charge than given by these brave boys on that field. They fought like veterans, nor did the dropping of their comrades by the ruthless bullet deter them from their mission."[32]

In the meantime, the cadets manning the institute's artillery on the right of the Confederate line were ordered to prepare double-shot canister as the Union cavalry was forming to attack. As the Union horsemen advanced, the Confederate artillery, with the cadets firing their cannons, caused confusion and broke up the charge. The Union commander ordered a withdrawal, and the 22nd Virginia Infantry attacked under the command of Col. George S. Patton Sr., a VMI graduate, ensuring that the Union movement to the rear became a rout.[33]

The battle ended in Confederate victory, but the VMI Corps of Cadets suffered ten killed and forty-seven wounded. After the battle, the Confederate secretary of war ordered the corps of cadets to Richmond. In route they stopped in Charlottesville and crossed paths with men of the Stonewall Brigade, the most heralded unit of the Confederate Army. The hardened combat veterans of the brigade had already heard of the cadets of

VMI at New Market, where their actions had become part of the Southern tradition and a testimony to military school devotion to duty. In Richmond, the corps of cadets was reviewed by President Jefferson Davis, West Point class of 1828, thanked by the Confederate Congress, and issued a new battle flag and Enfield rifles.[34]

Sadly, on June 12, 1864, upon the capture of Lexington and VMI by Union troops, Gillespie W. Henry, VMI class of 1862, the adjutant of the 14th West Virginia Infantry, was unsuccessful in his attempt to convince the Union commander not to burn the school.[35]

Confederate Naval Academy

The Union states had two national academies, the US Military Academy at West Point and the US Naval Academy, but in 1861, at the beginning of the war, the Confederacy had no national military schools. This changed in July 1863 with the opening of the Confederate Naval Academy in Richmond. About sixty midshipmen mustered aboard the CSS *Patrick Henry*, which then moved to the location of Confederate fortifications guarding Richmond at Drewry's Bluff on the James River. The midshipmen studied aboard and helped man the cannon emplacements when needed.

But the first combat was seen by the academy midshipmen detached to augment a Confederate naval raid in Bern, North Carolina. On the evening of February 2, 1864, forty Confederate sailors and marines, including eight midshipmen of the Confederate Naval Academy, deployed into the dark water of the Neuse River in two launches and headed towards Union Navy forces believed to be downstream. In the darkness they came upon the USS *Underwriter*, a Union gunboat with eighty-nine crewmen and four cannons. Boarding the vessel under fire, the Confederates fought armed with pistols, rifles, and sabers. During a brief, intense action, five Confederate sailors and marines were killed, including Midshipman Saunders, who was killed with a saber wound to the head. Sixteen others were wounded, including another midshipman. The Union ship was captured with nine Union sailors killed, including the captain of the ship; forty-six Union sailors were wounded and captured. The *Underwriter* was burned, and the Confederates with their captives returned to Confederate lines.[36]

The naval academy aboard the CSS *Patrick Henry*, docked at Drewry's Bluff, became endangered by the advance of General Butler's Union troops in May 1864. With the Union advance, midshipmen were assigned temporary duties manning the cannons at Drewry's Bluff, nineteen midshipmen joined the crews of the ships of the James River Squadron, others

Confederate sailors, marines, and midshipmen assault the USS *Underwriter*, as reflected in *Frank Leslie's Illustrated Newspaper*, February 27, 1864. (From "House Divided: The Civil War Research Engine," courtesy Dickinson Archives and Special Collections)

were assigned to an artillery battery north of Richmond, and several were sent to ships and raiding parties in North Carolina. The fighting did not reach Drewry's Bluff, but the squadron was engaged in limited action on June 20, shelling Union gunboats in Trent's Reach on the James River. During that time Lt. Wilburn B. Hall, commandant of midshipmen, led a group of midshipmen in several skirmishes with Union sharpshooters near Howlett House and Drewry's Bluff. By the end of July most of the midshipmen had returned to duty at the academy.[37]

In the latter part of 1864 the academy functioned as normally as possible, with the "rumblings of war" heard daily. Midshipmen were often called from class to man the fortifications, participating in long-range artillery duels, some lasting for three days. At other times details of midshipmen went downstream to lay mines and anchor spar torpedoes. In October the *Patrick Henry* moved upstream, and the sixty midshipmen and ten officers manned trenches guarding a pontoon bridge across the James River at Wilton Farm. Shortly after their return to Drewry's Bluff, the winter closed in, with three to six inches of snow. Many midshipmen fell ill as conditions aboard ship and in winter huts deteriorated.[38]

In March 1865 the midshipmen and the *Patrick Henry* moved upstream closer to Richmond. They docked at Rockett's Landing, and the midshipmen lived in a tobacco warehouse. On April 2, 1865, the academy received orders to "have the Corps of Midshipmen with proper officers at the Danville depot today at 6 pm." Shortly after arriving and drawing equipment, they learned that Richmond was to be evacuated, and the midshipmen and their officers were to guard the train containing the Confederate archives, the treasury, and President Jefferson Davis and his wife. The trains departed that evening with the midshipmen fully armed and with three days' rations. Between that time and April 13, the Confederate government was escorted by train through Virginia and North Carolina and on to Chester, South Carolina. At Chester the treasury of gold in boxes and silver in kegs was moved by wagon train, eventually arriving in Augusta, Georgia, despite Union cavalry patrols and the chaos in a defeated South.[39]

By May 2, 1865, Robert E. Lee had surrendered, Present Lincoln had been assassinated, and Gen. Joseph E. Johnston had surrendered to Sherman. On that day the Confederate secretary of the treasury relieved the naval academy of responsibility for the gold and silver and ordered the academy disbanded. The midshipmen received written orders detaching them "from the Naval School and leave is granted to visit [their] home. You will report to the Secretary of the Navy as possible."[40]

Never surrendered and never paroled, the unit disbanded after executing its final mission. President Jefferson Davis traveled a different route, arrived in Augusta, and learned that the midshipmen were no longer his guard. He told Captain Parker, "I have no fault with you, but I regret Mr. Mallory [secretary of the navy] gave you that order."[41]

Northern Military Schools

Prior to the first major battle of the Civil War at Manassas, Virginia, the midshipmen of the US Naval Academy found themselves defending their institution from potential attack. Baltimore had rioted at the arrival of Union troops in April 1861, and Annapolis became both the point of embarkation for reinforcements and a target of potential Confederate attack. Cannons were placed at the front gate, and the midshipman battalion was repeatedly called out to defend the installation.[42]

That same month the USS *Constitution* came to the academy and took aboard seventy-six midshipmen. They were placed on strict guard duty both day and night, as secessionists were seen spying on their ship from the opposite shore. The midshipmen were frequently called to action "armed to the teeth," with little rest for two weeks.[43] The reasons for concern were further borne out when the steamer *Maryland*, while towing the *Constitution* further out into the harbor to better safety, towed her aground. It was discovered that the *Maryland*'s captain was a secessionist, and he was arrested.

Because of the unstable situation, a few days later the midshipmen boarded the *Constitution* and moved its operation to Newport, Rhode Island, until 1865. On the day of departure, 10 midshipmen, half of the class of 1861, were order to the fleet. Upon arrival at Newport, the remainder of the class of 1861 was ordered to the fleet, followed just a few days later by the classes of 1862 and 1863. This left just 76 plebes (first-year midshipmen), who in July 1861 were joined by the largest class to that time, the class of 1865, numbering 203.[44]

At West Point, of the 278 cadets present on November 1, 1860, 65 would resign for reasons associated with the Southern cause. West Point, unlike the Naval Academy, limited its cadet active service to an early commissioning of the first-classmen after they were sent to Washington, DC, on May 6, 1861. In 1863, another riot, this time the draft riots in New York City, brought the cadets of West Point to take up arms and increase their guard with rumors that violence would spread to the academy. Cadet guards were issued ammunition, and they manned picket lines with field guns posted at South Dock, North Dock, and Gee Point. The superintendent sent several officers and his fifty-nine regular soldiers to the city for riot control.[45] Although some accounts say the cadets were deployed to the streets of New York, only the enlisted soldiers at the academy, not the cadets, were sent.

Like the military schools of the Confederacy, Norwich University was heavily engaged in the conflict by supporting the Vermont mobilization of drillmasters with various units, including the 3rd Vermont Infantry Regiment. The cadets also drilled students at Dartmouth University and Bowdoin College. The Dartmouth students and Norwich cadets eventually manned B Troop, 7th Rhode Island Cavalry. At Bowdoin College, training was observed by Prof. Joshua Lawrence Chamberlain, a graduate of Whiting's Military and Classical School in Maine. Chamberlain would later rise to the rank of general officer, receive the Medal of Honor for action at Little Round Top at Gettysburg, and become governor of Maine.[46]

The Norwich cadets were also sent to Camp Rendezvous and helped prepare the 4th and 8th through 16th Vermont Infantry Regiments and the 1st Light Artillery Battery. Camp Rendezvous was in Brattleboro, where the Burnside Military School was located. That school's principal, Col. Charles A. Miles, and his cadets drilled officers of the 2nd Vermont Infantry at the beginning of the war. The Highland Military Academy of Worcester, Massachusetts, had cadets function as drillmasters in "neighboring towns" as the war began. The Peekskill Military Academy cadets in

New York drilled newly formed units and marched with them to the war. Likewise, in New Haven, Connecticut, the men of the 2nd Connecticut Infantry Regiment "had to swallow their pride" as the cadets of the Russell Military Academy taught them military drill to the beat of the academy's very young drummers.[47]

Norwich University was far from the battlefields of the South, but on October 19, 1864, Southern raiders entered the town of Saint Albans, Vermont, from Canada. The twenty-one raiders were led by Lt. Bennett Young, CSA. Young had been captured earlier in the war, escaped to Canada, and returned to the South, where he was assigned to organize a company with other escaped prisoners located in Canada and to "create havoc along the border." The raiders entered the United States and robbed the Saint Albans Bank, the Franklin County Bank, and the First National Bank of at least $208,000. They incurred one wounded; one citizen of the town was killed and a visitor wounded.[48]

Vermont's militia had long since mobilized and had left to fight on the Southern battlefields. As "the only source of organized defense was the corps of cadets," Norwich University was alerted by telegram of a potential similar "invasion" of Newport, Vermont. Gen. Alonso Jackson, the Norwich superintendent, gathered the cadets by drumbeat for roll call. The force might have been larger from Norwich, but like their Southern military school counterparts, the Norwich cadets' ranks had been depleted by the departure of cadets from the school as volunteers to fight with various units in the war. At the formation of the small corps, the cadets were instructed to step one

Norwich University campus with corps of cadets in formation, about 1864–65. (Courtesy Norwich University Archives)

pace forward if they wished to volunteer for active service in Newport. The result was a "complete forward movement in response." The school sent a telegram to Vermont's governor offering the services of the cadet corps, and they were ordered to leave for Newport.[49]

The force assembled comprised two faculty members, including General Jackson, a lieutenant of cavalry, an alumnus veteran, two local volunteers, and forty-seven cadets, for a total of fifty-two men. The cadets joined a loosely organized company of citizens in Newport and marched on the wharf to receive a steamer, standing at the ready should the boat contain Southern raiders. No raiders materialized, and after spending the night in Newport, the cadets marched eight miles to the Canadian border.[50]

An unarmed group crossed the Canadian border to find the Canadians ready to help. According to Portus Baxter, Norwich class of 1825, who was present, the cadets had to uphold the honor of Vermont and calmed the anxiety of its citizens. The corps returned to Norwich never having heard a shot in anger. Of the twenty-one raiders, the Canadian militia captured fourteen, and $75,000 was recovered.[51]

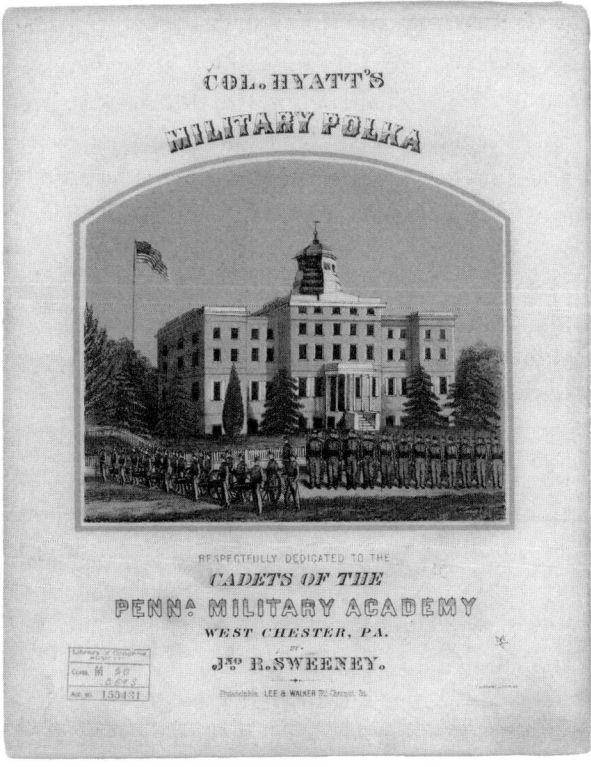

Pennsylvania Military Academy, about 1863, from the sheet music cover for "Col. Hyatt's Military Polka" (Philadelphia: Lee & Walker, 1864). (Library of Congress)

The Pennsylvania Military Academy (Pennsylvania Military College) offered its services to the Union cause when the state governor called for seventy-five thousand volunteers to repel the Confederate advance on Gettysburg in 1863. Col. Theodore Hyatt, the academy president, offered an artillery battery, but the offer was denied due to the age of the cadets. Seven older cadets, led by Cadets George R. Guss, Frank E. Townsend, Johns A. Leslie, and William J. Harvey, requested that Colonel Hyatt appeal for the governor's approval of a battery to serve under their leadership. Governor Curtin agreed, and the cadets recruited the Guss's Independent Artillery Company (Independent Artillery of Pennsylvania). The battery was commanded by Cadet Guss as the captain, with Cadet Townsend as first lieutenant.[52]

After the cadets recruited Chester citizens into the 131-man, six-gun battery, they were mustered into service on June 29 or July 1, 1863. The battery served until August 23 as part of Maj. Gen. Darius N. Couch's Department of the Susquehanna, having moved "at once down the Cumberland Valley Railroad to Carlisle." After their discharge the cadets went back into the ranks of the academy.[53]

Alumni Contributions to the War Efforts

The contributions of military schools to the efforts of Confederate operations have been addressed by Bruce Allardice, but no attempt has been made to look at the military education of Union forces beyond works focused on West Point and Norwich University. This may be because, as Marcus Cunliffe has pointed out, military schools are regarded as a Southern idea and viewed as having no significance to the North.[54]

Military Schools' Contributions to the Southern War Effort

Bruce Allardice's study of Southern military schools in the Civil War provided an estimate that 12,000 alumni of military schools served with the Confederate military forces. This figure was based on his estimate of ninety-six schools operating in the South between 1827 and 1860. These schools included Western Military Institute, which operated first in Kentucky and later in Tennessee and provided more than 1,000 officers, including two generals, thirty-seven colonels and lieutenant colonels, and 18 majors. Virginia Military Institute contributed 1,796 officers, including twenty generals, ninety-two colonels, sixty-four lieutenant colonels, and 107 majors. Georgia Military Institute had 500 alumni in gray, including two generals

Western Military
Institute, about
1852, from *The
South-Western
Monthly 2* (1852).

and thirty-six field-grade officers (colonels, lieutenant colonels, and majors). LaGrange Military Academy in Alabama provided 176 alumni, and Hillsborough Military Academy in North Carolina provided at least 250, including two lieutenant colonels, one major, and two captains.[55]

The Citadel's total contribution has never been documented. The school had 209 graduates in the Confederate forces, including four generals and forty-eight field officers. But a total of 1,699 students had attended the school by war's end, and the majority served the Confederate cause. Kentucky Military Institute had 3,049 alumni serve in the Civil War, of whom at least fifty-three died in Southern service and twelve with Union forces.[56] That school's alumni served in both blue and gray, and their service has never been completely documented.

Jefferson College in Mississippi produced two Confederate generals. The school, established in 1811, made the transition to a military school by 1829 under the leadership of Col. John Holbrook, Norwich University class of 1825. As the Civil War approached, the school numbered about eighty cadets, based on the number of rifles issued by the state in 1859.[57] Other Southern military schools' contributions of Confederate generals include one from Georgia Military Institute, one from Raleigh Military Academy in North Carolina, one from St. Thomas Hall in Mississippi, and one from Western Military Academy in Tennessee.

Additionally, Bruce Allardice found that at least six military schools outside West Point in the Union states contributed small numbers to the Southern cause. These schools were Kenyon College in Ohio, Mount Pleasant Military Academy in New York, Russell Military Academy in Connecticut, Mount Vernon Military Academy in New York, Norwich University in Vermont, and Starr's Military Institute in New York.

Jennifer Green estimated that 11,000 alumni of Southern military schools were prepared to serve in the Confederate military. These estimates were based in part on the identification of ninety-four military schools in the South and North that had played a part in the education of Southerners prior to the Civil War. In addition to Green's estimates, West Point contributed 661 men, of whom 156 became Confederate generals. The US Naval Academy provided 95 officers; the American Classical and Military Lyceum in Pennsylvania likely sent several; and at least 56 alumni from Norwich University served in the Confederate military.[58] Building on the work of Allardice and Green with an additional twenty-six military schools in the South, as well as West Point, the US Naval Academy, and other northern schools, almost 1,700 additional alumni are added for a total of 13,700 men from military schools serving for the Southern cause.

This contribution greatly increased the effectiveness of the Confederate war machine. For example, military school alumni contributed greatly to the population of Confederate staff officers, critical for planning and administration of the army. More than 15 percent of Southern staff officers were military school alumni, with the greatest number from VMI, followed by West Point, the Citadel, and Georgia Military Institute. The representation within the field-grade officer ranks is another indicator of the military schools' impact on Confederate leadership. In the Army of Northern Virginia, 285 of 1,965, or 14.5 percent, of the field-grade officers were products of military schools. These included 156 from VMI, 73 West Pointers, 37 from the Citadel, 14 from Georgia Military Institute, 4 from the Naval Academy, and 1 from La Grange Military Academy. This does not include the "qualified junior officers and noncommissioned officers" that military schools provided.[59]

The highest rank of field-grade officer is colonel, the rank that commands a regiment. The regiment was the basic unit of the army, in which standards of drill and discipline were set, and was the organization with which soldiers proudly identified. Twenty percent of the men who rose to the rank of colonel in the Southern army were prepared at military schools of both the North and the South. State military colleges educated 159;

military schools at the secondary level, 31; West Point, both graduates and nongraduates, 116; and Annapolis, 19 (serving in the army).[60]

At the Battle of Gettysburg, the high point of the Confederacy, Robert E. Lee, a West Pointer, commanded the Confederate Army of Northern Virginia. His four corps commanders were all West Point graduates, and the subordinate nine division commanders were either West Point or VMI graduates. At Pickett's Charge, another West Pointer, Gen. George Pickett, commanded a division of fourteen regiments, of which eleven were commanded by Virginia Military Institute products. The impact of military schools on the Confederate Army was massive, with approximately 42 percent of general officers being former military school students.

Military Schools' Contributions to
the Northern War Effort

Very little has been written about contributions of military schools to the Union efforts apart from those of West Point and Norwich University, despite the fact that there were fifty-five military schools functioning between 1802 and 1860 in Northern states. Additionally, there were fifteen military schools established, or schools transitioned to a military format, during the Civil War in Union states.

Marcus Cunliffe addresses the question of the lack of literature on military schools in the North. First, the popularity of military schools in the South has been assumed to be as strong after the Civil War as before. Second, Cunliffe asserts that "Military academies have come to be regarded as a feature of the South, or rather of the idea of the South. They have hence been regarded as not-Northern, and have no significance attached to them. The men at VMI have been highly visible in the eyes of posterity; those in cadet gray at Mount Pleasant, New York, or Highland Military Academy in Worcester, Massachusetts, or Hamden, Connecticut, have been well-nigh invisible."[61]

The biggest difference between Confederate and Union military schools was that the Union states never adopted state sponsorship. The Confederacy had a series of schools such as VMI, the Citadel, Georgia Military Academy, and others that were supported with state funds, either wholly or partially, and had cadets attending as state cadets with paid tuition. The Union states saw the federal government as fulfilling the needs of military training. Despite this, Norwich University, a private military college providing 705 alumni to the Union war effort, was far from alone. The alumni of the two federal military academies provided the Union about

1,336 men, with 936 from West Point and 400 from the Naval Academy. West Point provided at least 228 generals of the 583 general officers of the Union Army. Norwich contributions included at least 12 generals, twenty-seven colonels, twenty-four lieutenant colonels, and fifteen majors.[62]

Other military schools in the Union states included Russell Military Academy, which numbered 130 to 160 cadets and contributed three hundred officers and likely an additional 150 enlisted soldiers. Highland Military Academy in Massachusetts contributed at least fifty alumni, including one lieutenant colonel, one major, three captains, and a lieutenant. The school had 84 cadets in 1862, and after it converted to a military format in 1856, and prior to 1863, an estimated number of 150 cadets were matriculated.

Mount Pleasant Military Academy in New York operated as a military school starting in 1845, and its cadet enrollment is estimated to have been seventy-five a year. Their contribution to the military manpower of Union forces was close to 425 men. Churchill Military Academy in New York was another well-respected military school. Both Gen. Robert E. Lee and Gen. William T. Sherman's letters reflected a high opinion of that institution. Established in 1841, the school likely contributed 525 Union officers and soldiers, among them Brig. Gen. John B. McIntosh. Another school of significant size was the American Classical and Military Lyceum in Philadelphia, with an enrollment as high as 150. The school operated only from 1828 to 1835 and counted Gen. George Meade, commander at Gettysburg, among its contributions to the Union war effort.

With General Meade at Gettysburg were his eight corps commanders. Every one of those eight generals, in what many believe was the turning point of the Civil War, were graduates of West Point. But other Northern military schools were represented as well. Joshua Chamberlain, colonel of the 20th Maine Infantry at Little Round Top, is one of the most respected examples. Chamberlain's association with Bowdoin College is well known, as is his postwar governorship of the state of Maine. Less known is that he attended Whiting's Military and Classical School, which had operated under a West Point graduate from 1841 to 1847 in Maine. Another alumnus of that school was Gen. James G. Blunt. Robert Collins believes Blunt's attendance could account for his battlefield success: "He could well have been one of the few from Kansas who had any military education when the war broke out. At the very least, Blunt knew the basics of drill, military formations, and disciplined marching." Prior to 1893, Kansas had no military schools. Serving in the Western Theater with him was Gen. Clayton

Powell, who attended the Pennsylvania Literary, Scientific, and Military Academy in Bristol as a youngster.[63]

Other schools with smaller cadet enrollments included Alexandria Institute with thirty; Jarvis Military Academy in Connecticut, thirty-three; Everest Military Academy in Connecticut, sixty-five; Starr's Military Institute in New York, thirty-seven; Delaware Military Academy, eighty-three; Peekskill Military Academy in New York, eighty-five; and, based on postwar enrollment figures, Yonkers Collegiate and Military Institute in New York, forty. In all, there were forty-nine military schools functioning in the Northern states, three in the Midwest, and one in the West. Joining these military schools would be men who supported the Union from the border states' and the South's military schools. Examples of the latter include fourteen VMI alumni who served with the Union Army. Kentucky Military Institute's contribution to Union military manpower has not been determined, but at least two Union generals were alumni.[64]

The resulting 4,800 military school alumni provided the Union with a much smaller cadre for their military forces, but it was significant and too often forgotten. The leadership these schools provided were a critical part of the Union's ultimate victory. In fact, approximately 40 percent of Union generals were products of American military schools.

In-ranks inspection of Yonkers Collegiate and Military Institute, about 1856. (Courtesy American Antiquarian Society)

The United States Naval Academy and Maritime Academies

Prior to 1811 there were no military schools with a naval orientation in the United States. The first evidence of a movement in that direction came with the establishment of courses on nautical subjects. In 1798, Stephen Decatur's family hired Talbot Hamilton to tutor him in navigation prior to his appointment as a midshipman. Decatur went on to become the navy's youngest captain and a hero of the Barbary Wars. Similarly, James Lawrence, a famous naval hero remembered for his dying declaration, "Don't give up the ship," in the War of 1812, was schooled in 1796, by either Samuel Webster or John Griscom (also spelled Griscomb), in navigation and naval tactics prior to his appointment. In 1798, Philadelphia merchant and ship owner John Coulter sent his apprentice, Uriah Levy, to school for navigation and seamanship. Levy attended school for seven to nine months between 1807 and 1808. The school was most likely run by Talbot Hamilton in the Lower Dublin area of Philadelphia. Hamilton was identified as Scottish or English and a former officer of the British navy. Uriah Levy rose to become a commodore in the US Navy—the first Jewish officer—and was credited with ending the practice of flogging. There is no documentation that these schools operated by Hamilton, Webster, or Griscom functioned as military schools with a uniformed student body.[1]

Navigating toward a US Naval Academy

The first steps that led to the establishment of the US Naval Academy took place during the final years of congressional consideration for the establishment of a national military academy. In 1800, two years before the establishment of the US Military Academy at West Point, Alexander

Hamilton recommended that Secretary of War James McHenry include a provision for a school of the navy in future national military academy plans. But as previously noted, the academy at West Point focused on the needs of the Engineer and Artillery Corps of the army, and no naval component was included in the building plan.[2]

Despite this, in 1808 the superintendent of West Point, Gen. Jonathan Williams, recommended expansion of the school to include teaching "nautical astronomy, geography and navigation." His intention was for West Point to serve the needs of both the army and the navy. This did not come to pass, although, according to Edward Marshall's *History of the United States Naval Academy*, during the War of 1812 "many of the cadets of the Military Academy were commissioned as midshipmen in the Navy."[3]

Much of the long, slow progress towards establishing a national naval academy may have come from the traditions of the British naval midshipmen and the lack of a naval academy on the part of the allied French. The American navy faced the British navy in the Revolution, which was officered by men who were often schooled aboard ship as midshipmen. These young men prepared themselves for examination in seamanship and navigation in their sixth year of service in order to advance to the grade of mate. Although the Portsmouth Naval Academy (renamed the Royal Naval College in 1806) was established in 1729 in England, only a portion of the British midshipmen studied its two-year curriculum before going to sea. Many midshipmen learned their skills while serving aboard ship and were mentored and taught by officers and crew. The French navy did not establish their naval schools at Brest and Toulon until 1810.[4]

The US Navy secretary, Robert Smith, ordered the navy's Presbyterian chaplain, Robert Thompson, to organize a course at Washington Navy Yard. The school was voluntary and addressed mathematics and navigation from 1802 to 1804. During that period an event far from Washington spurred the desire to establish officer education. On October 31, 1803, the USS *Philadelphia* and its 307-man crew were captured by Barbary pirates. While in captivity in Tripoli, the US midshipmen studied in preparation for qualifying exams.[5]

The Washington Naval Yard course ended in 1804 with Chaplain Thompson's departure for sea duty. Upon his return in 1806, the course resumed, and in 1808 Thompson traveled to New York and Norfolk to continue teaching. Beginning in 1811, an "informal academy" began operation at the Washington Navy Yard, with midshipmen assigned to the school. This was the first military school in the United States with a documented naval orientation, teaching mathematics, navigation, and astronomy to midshipmen. One of the midshipmen was Josiah Tattnall, who would later rise to captain in the Confederate Navy.[6] Due to the onset of the War of 1812, the Washington Navy Yard was closed, but during that war a course was conducted on Lake Ontario in New York for midshipmen and junior officers.

In 1813 an act of Congress authorized the navy to employ naval school-masters. This act likely influenced Adm. William Bainbridge to establish what has been referred to as the first naval school in December 1815 at the Boston Navy Yard. "His classroom was ship board; which was hands on training, and the students excelled or they were requested to leave." Bainbridge was said to have been an excellent teacher, and the secretary of the navy hoped he would impact the navy's need for promising officers. It is unclear how long instruction continued, but by June 1815 Bainbridge and the crew of the *Independence* were well focused on the problems of Algiers.[7]

The authorization of schoolmasters for the navy helped open additional naval schools in New York as early as 1827 and in Norfolk, Virginia, in 1828. The naval school in New York received assigned midshipmen until 1844. Between 1841 and 1844 they were aboard the USS *North Carolina*. One of those midshipmen, Stephen B. Luce, would play a key role in the expansion of the military school format to a merchant fleet focus.[8] The Norfolk Naval School received midshipmen until as late as 1837.

The First Secondary Military Maritime School

Beyond the US Navy, another maritime military school was operating between 1828 and 1831 on the brig *Clio*, near Nantucket, Massachusetts, with twenty-one boys ages twelve to sixteen. This was the first nautical military secondary school in the country. The school was a product of a financial investment by British admiral Sir Isaac Coffin and was intended to serve his former nation and American members of the Coffin family. Coffin was born in Boston in 1759 and rose to admiral in the British navy. His personal instruction for the school leaves little doubt that it was structured to be conducted as a naval military school. The school was administered first by Lt. Alexander B. Pinkham, who was on loan from the US Navy. The school transitioned to shore and a nonmilitary format in 1831 because of the costs associated with supporting the ship's operation. Before the school went to shore, it did visit Nova Scotia on cruise.[9]

Philadelphia Naval Asylum
and Academy

The establishment of a central and more permanent naval academy was championed by a series of secretaries of the navy: Williams Jones in 1814, Samuel Southard in 1824, and A. P. Upshur in 1841. Southard's efforts were supported by President John Quincy Adams in 1827, but it was not until 1838 that a significant step towards a permanent naval academy was gained with the opening of the Philadelphia Naval Asylum and Academy. Apparently Philadelphia was suggested by Paul Hamilton, secretary of the navy some twenty-eight years before. Initially the school had an enrollment of twenty-two to thirty-four midshipmen and a curriculum of one year, and it was conducted in a wing of what was a home for aged sailors. With the ascension of Prof. William Chauvenet as head of instruction at the school, the course of study was extended to two years, with the content and quality of the course improving through the addition of lectures on maritime law, navigation, ordnance, and gunnery.[10]

The United States Naval Academy

The next major event in the voyage towards the establishment of the US Naval Academy occurred far from a Philadelphia classroom or the US Congress; it was at sea on the USS *Somers*, a brig-of-war sailing from the West Indies to New York. Among the crew was Midshipman Philip Spencer, who had not benefited from the education at the Philadelphia Naval Asylum and Academy. Spencer had a checkered academic career that included failing grades from Geneva College and from Union College, where he had established a chapter of Chi Psi Fraternity. The young Spencer's father was the secretary of war, who convinced his son after the young man had served on a whaler ship that he should become a gentleman by way of becoming a midshipman in the navy. Spencer's career as a midshipman was worse than his academic endeavors. The young man had served aboard several other ships and had a record of drunkenness, and he had a fascination with piracy, He apparently made a list of potential conspirators to take over the *Somers*. The *Somers*'s captain and officers believed Spencer was the ringleader of a potential mutiny to take over the ship and kill the officers, despite the fact that "evidence . . . to this effect, was at best, slim." But he and two seamen were hanged at sea on December 1, 1842, resulting in a reaction by naval leadership and the public that became another catalyst for professionalization of the naval officer corps through the establishment of a naval academy.[11]

In 1844 the navy had twenty-two professors: fourteen at sea, two in Boston, one in New York, two in Norfolk, three in Philadelphia, and three on special service. George Bancroft became secretary of the navy in 1845 and concluded that this arrangement of education was "one of scattering schools, diffusing responsibility, and barren of good results." Secretary Bancroft requested that the board of examiners, who reviewed midshipmen for qualification for promotion, assist "in maturing a more efficient system of instruction for the young naval officers."[12]

As early as 1826 the Maryland House of Delegates had lobbied Congress for a naval school in Annapolis. In October 1845 their long-term desire came to pass as Fort Severn was transferred to the Department of the Navy. Secretary Bancroft received the recommendation from a committee of naval commanders that Annapolis was a suitable location for a naval school, and accordingly an academic staff was organized to include Cmdr. Franklin Buchanan as superintendent; Lt. James H. Ward as executive officer and instructor of gunnery and steam; six other instructors, including Prof. William Chauvenet; and two others from the Philadelphia Naval Asylum and Academy.[13]

Despite the midshipman tradition and naval focus, the school was not without the influence of the military schools that were flourishing in the United States at that time. A. P. Upshur, the secretary of the navy, in his desire for a naval academy, made no attempt to hide his envy of "the advantages which the Army has derived from the Academy at West Point . . . proof that a similar institution for the Navy would produce like results." According to William Leeman, West Point was the naval school's model for scientific and technical curriculum but also for development of "military virtues." Lieutenant Ward served as the executive officer for two years of the newly formed US Naval Academy in a duty that would later be called commandant of midshipmen. Ward was a graduate of Norwich University, class of 1823, and his influence would remain for years through the use of his *Manual of Naval Tactics* as well as his lectures on naval ordnance and gunnery.[14]

Another influential faculty member was Prof. Henry Hays Lockwood, who came from the Philadelphia Naval Asylum and Academy as a professor of natural philosophy and astronomy. He was an 1836 graduate of the US Military Academy at West Point and a former artillery officer. His service at the US Naval Academy spanned from 1847 to 1861 and 1865 to 1871.[15]

The academy opened on October 10, 1845, with fifty midshipmen, of whom the majority were "old hands," ages eighteen to twenty-seven, from the Philadelphia Naval Asylum and Academy and seven were newcomers who were under sixteen year of age. The addition of drill and cannon crew practice to the nautical curriculum and the military trappings influenced by West Point and Norwich, with midshipmen organized into companies, were very different than the relatively "lax control" found in Philadelphia.[16]

During the academy's second year, the war with Mexico began, and with it the academy's first superintendent, Commander Buchanan, volunteered for war service at sea along with fifty-six of his midshipmen. The superintendent was denied, but seven midshipmen were ordered to active service. These included Acting Midshipmen John Adams, W. B. Hayes, and Thomas Houston, who were sent aboard the sloop of war *Dale* and became the academy's first sent to war. With accelerated graduation, ninety alumni saw service, and five of them died—four in combat, with Thomas Shubrick decapitated by enemy fire at Veracruz.[17]

In 1850 the academy graduation standard was extended to four full years. By the Civil War the academy evolved into a much more disciplined military institute than the Philadelphia Naval Asylum and Academy had been. The demerit system was integrated into class standing, and penalties were confinement and dismissal. The midshipmen drilled as gun crews and conducted infantry drill by companies. As in other military schools of that period, the midshipman battalion was commanded by the commandant, who in this case was Henry Hays Lockwood, West Point class of 1836. By 1860 the US Naval Academy had grown to fourteen faculty members and 281 midshipmen.[18]

United States Naval Academy midshipmen in cannon crew drill about 1860. (Courtesy Special Collections and Archives Department, Nimitz Library, US Naval Academy)

The evolution of the US Naval Academy toward the West Pont model cemented nautical education to the military school format. Before that time, nautical education was often training under the senior officers of vessels at sea or limited to navigation and seamanship in small classes with a single subject-matter expert. The growing influence and reputation of the naval academy inspired change; thereafter, nautical schools would be more often in a regimented uniformed format, and secondary schools without a maritime mission would emulate the academy's ethos and navy uniforms and traditions.

The US Naval Academy and its alumni would also influence the expansion of the military school concept. The first expansion was the wartime establishment of the Confederate Naval Academy. In 1862 the Confederate Congress authorized the existing US law addressing the navy to apply to the Southern states. The Confederate secretary of the navy recommended a naval academy in February 1862. The academy was organized by Lt. William Harwar Parker to mirror the US Naval Academy. Lieutenant Parker was an 1848 naval academy graduate and had served as a professor of mathematics, professor of navigation and astronomy, and instructor of seamanship and naval tactics from 1853 to 1857. He was appointed superintendent and served as such until the end of the Civil War. His commandant of midshipmen was Lt. Wilburn B. Hall, who was the US Naval Academy's top graduate in 1859. The school began operation in July 1863 aboard the CSS *Patrick Henry* while it was anchored at the Richmond, Virginia, waterfront, although the official opening of the academy was on October 10, 1863, when it docked at Drewry's Bluff on the James River. During the operation of the school, 124 midshipmen from fourteen states were assigned to the academy. Enrollment usually numbered about 60, and their ages ranged from fourteen to seventeen years.[19] The wartime operations of the school, until the school was disbanded on May 2, 1865, were addressed in chapter 6.

The US Naval Academy's reputation, along with its alumni, influenced the creation and format of thirty colleges, junior colleges, and preparatory military schools, including several that have stood the test of time and continue to operate successfully today. These schools with a naval theme are listed in table 7.1.[20]

Table 7.1. Military Schools with a Naval Theme

Dates of Operation	School	Location
1863–65	Confederate Naval Academy	Virginia
1880–91	Maryland Military and Naval Academy	Maryland
1888–1995	Northwestern Military and Naval Academy	Wisconsin
1905–12	Lakeside Classical Institute	Texas
1910–	Army and Navy Academy	California
1913–20	Florida Military and Naval Academy	Florida
1914–	Riverside Military Academy (Riverside Military and Naval Academy)	Georgia
1915–16	Silver Lake Military and Naval Academy	New York
1916–40s	Talbor Academy	Massachusetts
1917–19?	Pasadena Army and Navy Academy	California
1919–24	Carolina Naval-Military Academy	North Carolina
1920–41	Carson Military and Naval Institute	Michigan
1932–94	Admiral Farragut Academy	New Jersey
1933–77	Elsimore Naval and Military School	California
1935–39	New Jersey Naval Academy	New Jersey
1935–37	Stonehurst Military and Naval Academy	California
1936–53	Admiral Billard Academy, Florida	Connecticut
1937–53	Puget Sound Naval Academy (Hill Naval Academy)	Washington
1939–53	Florida Naval Academy Junior College	Florida
1941–	Leonard Hall Junior Naval Academy	Maryland
1943–53	Hudson River Naval Academy	New York
1945–	Admiral Farragut Academy	Florida
1963–76	Sanford Naval Academy	Florida
1963–	Texas Marine Military Academy	Texas
1981–	Marine Academy of Science and Technology	New Jersey
2003–	Delaware Military Academy	Delaware
2005–	Admiral Hyman George Rickover Naval Academy	Illinois
2006–	Marine Math Science Academy	Illinois
2011–	New Orleans Military Maritime Academy	Louisiana

Stephen B. Luce and the Maritime
Military Academy Movement

Maritime military academies are institutions whose purpose is to prepare officers of the merchant marine rather than the US Navy. The vessels of the merchant marine are civilian owned and crewed, but in times of war these vessels become an auxiliary force to the navy for logistics missions associated with the transport of troops and supplies.

Stephen B. Luce, US Naval Academy class of 1848, would eventually become the father of the maritime military academy movement through his writing, lobbying, and leadership. Luce was born in 1827 in Albany, New York. At the age of six, he and his family moved to Washington, DC, where his father was employed as a clerk for the Department of the Treasury. Based on family tradition, when Stephen was fourteen, he and his father visited Zachary Taylor at the White House and secured an appointment for Stephen as a midshipman in the US Navy.[21]

Luce's first assignment was at the US Naval School, New York, aboard the USS *North Carolina*. The *North Carolina* served as a navy school ship from 1841 to 1844, and Luce spent his first six months as a midshipman aboard. For the next six years he was at sea, first with the frigate USS *Congress* and then with the USS *Columbus*. After nearly six years at sea, he returned to join the second class at the US Naval Academy.[22]

From April 1848 to August 1849, Midshipman Luce attended classes and studied for his qualifying exam to become a "passed midshipman." Unfortunately for him, his pride in his service and the new US Naval Academy cost him dearly. The secretary of the navy authorized the academy to have its midshipmen participate in the inauguration events of President Zachary Taylor, but for reasons unknown, the superintendent of the Naval Academy at the time, Captain Upshur, did not approve of the action. The midshipmen's reaction was "ringing bells, blowing loudly on horns and discharging guns," and in general demonstrating their displeasure in a very unmilitary manner. The resulting punishment pushed Midshipman Luce back seventy-two places on the promotion list and delayed his promotion from passed midshipman to lieutenant by six years. This was an unexpected end to the academy experience by a man who should be considered the father of the maritime academy movement.[23]

After his graduation in 1849, Luce went to sea again aboard the USS *Vandalia*. Serving as a passed midshipman, he spent his off-duty hours in the pursuit of a self-taught liberal education. His journal reveals that among the texts he read were Milton's *Paradise Lost*, Dickens's *Old Curiosity*

Shop, Shakespeare, Grote's *History of Greece*, the Bible and Bible scholars, William Falconer, Lord Byron, Theodor Mommsen, and James Fenimore Cooper. The USS *Vandalia* assignment was followed by another duty that increased his professional knowledge. He served aboard several survey ships on the Atlantic Coast and then on the USS *Jamestown*, which included a voyage to the Central American coast. As a result of the assignments he gained extensive additional exposure to astronomy, oceanography, cartography, and hydrography.[24]

The depth and girth of his academic and nautical experience made him the perfect individual to return to the naval academy, even though he had left under less than stellar circumstances. From 1860 to 1861 he served as assistant commandant, then head of the seamanship department beginning in 1863 and commandant of midshipmen from 1865 until 1869. During that time he revised *Instruction for Naval Light Artillery* by Lt. William H. Parker (alumnus of the US Naval Academy and superintendent of the Confederate Naval Academy), but more importantly, he authored *Seamanship*, a practical guide to the nautical skills.

With the end of the Civil War, Luce shifted his focus from the naval academy to the further professionalization of the navy officer corps. He is best known for his efforts toward the 1873 establishment of the US Naval Institute, with its mission as a forum for scientific and nautical professional debate, and the institute's publication, *Proceedings*. Luce's impact on professional education did not stop with officer preparation at the US Naval Academy or the Naval Institute. In 1885 he was central in the establishment of the Naval War College, the highest professional school for senior naval officers.

These experiences helped develop an educational philosophy reflected in the paper he presented in 1873 at the Naval Institute. He believed the modern naval officer had to be highly cultured and educated. He pointed out the importance of the "study of arts and sciences and the racking of brains and exhaustive innovative facilities of our country." He stated the object of education was the "whole man, the body, the mind, and the heart: its object, and when rightly conducted, its effect is to make him a complete creature. To his frame it gives vigor, activity, and beauty; to his senses, correctness and acuteness; to his intelligence power and thoughtfulness; to his heart virtue."[25] Luce's philosophy, with his ideals of high standards of academic study, physical preparation, and moral preparation reflecting his Episcopal faith, was very much in line with that of the military school movement's leaders. Jennifer Speelman reflects that Luce was deeply influenced beyond the military school experience as a midshipman and on the faculty of the US Naval Academy by Henry Barnard, an education reformer, board member for the US Naval Academy, and first US commissioner of education, 1867–70. Luce drew on Barnard's report from the board of visitors for his 1873 paper at the Naval Institute and surely read Barnard's book *Military Schools and Courses of Instruction in the Science and Art of War in France, Prussia, Austria, Sweden, Switzerland, Sardinia, England, and the United States*, which addressed European naval education.[26]

Luce, like Barnard, saw an additional need for "permanent nautical schools . . . to increase the efficiency and professionalism of merchant seamen. Each man also believed military education was a way to develop character by combining physical and intellectual aspects into the curriculum."[27] With regard to the merchant marine, Luce should also have been

Capt. Stephen B. Luce, about 1909. (Library of Congress)

considered the most influential figure in the maritime military academy movement. His shifting of attention beyond the navy was incumbent on three maritime disasters: the Newfoundland collision of 1854, the Mississippi River sinking of 1865, and the Georgia Hurricane sinking of 1866.

The first nautical disaster occurred on September 27, 1854, when the luxury vessel *Arctic* and the French ship *Vesta* collided off the coast of Newfoundland, resulting in the loss of 322 lives. The collision was a point discussed by the Baltimore Board of Trade in February 1855. That event, like the execution of Midshipman Spencer, inspired the creation of a school to prepare young men to be better able seamen. The school was outside the federal government's purview in the city of Baltimore.[28]

In 1856, the Baltimore board purchased the USS *Ontario*, a retired US Navy vessel, to be their school ship. On September 14, 1857, classes started with just 8 students. "Uniforms were designated with a blue cap, roundabout, and trousers for winter and white for summer." Students were selected to perform officer of the day duties; reading, writing, grammar, geography, algebra, geometry, navigation, and seamanship were all part of instruction. Enrollment grew to 139 students in 1858 and to 243 by 1860. Unfortunately, by 1865 daily attendance dropped to just 5 students, and the school closed. The decline appears to have been the result of the Civil War's effect on Baltimore, changing it from a "commercial to an industrial city."[29]

This school was a foreshadowing of future events that would be prompted by the influence of Stephen B. Luce. The same year as the closure of the Baltimore School Ship, the riverboat *Sultana* on the Mississippi River sank. The cause was a boiler explosion that took fifteen hundred lives, many of whom were recently freed Union prisoners of war on the way to Illinois after the Battle of Vicksburg.[30]

The following year, in 1866, the steamer *Evening Star* sank after encountering a hurricane off the coast of Georgia. The resulting melee between passengers and crew cost 230 lives. In 1874 Luce probably alluded to this very incident when he wrote that "the increasing number of marine disasters, the demoralization, becoming so general on board our merchant vessels . . . all indicate that we can no longer disregard with impunity the examples of other maritime countries in providing technical education for those employed in mercantile service."[31]

The same year as the *Evening Star* disaster in 1866, Luce wrote an article, "Nautical Schools," in the *Army and Navy Journal.* In it he described the advantages of a system of maritime education as established by Great Britain. In 1869 the US House of Representatives appointed a select committee that generated a report titled *Causes of the Reduction of American Tonnage and the Decline of Navigation Interests.* Also called the Lynch Report, the study pointed out that the British shipmaster must "pass a rather rigid examination before a board of examiners . . . , receive a certification from them . . . , [and] for any misconduct or incompetency as master, this certificate may be suspended or taken from him." National concern for the decline of the US maritime capability was expressed in congressional hearings, in the Lynch Report, and in Joseph Nimmo's *Report to the Secretary of the Treasury in Relation to the Foreign Commerce of the United States and the Practical Workings of Our Relations of Maritime Reciprocity.* Both reports identified Britain as the principal competitor to US maritime interests. The Lynch Report said that the American sea captain was a combination of merchant and seaman, whereas his British counterpart was a professional seaman who was certified and held to a higher level of competence by certification, examination, and strict enforcement of standards. The Nimmo Report placed the focus on the need for "upbuilding of our merchant marine . . . [in order] to protect it against British competition."[32]

US Coast Guard Academy

While events continued to illustrate maritime unpreparedness, the Revenue Cutter Service had a minimal formal training program for its officers. In 1808 Capt. Hopley Yeaton added four young men to his cutter *New Hampshire* "so he could train them as seamen, navigators and pilots." Capt. Alexander V. Fraser, between 1848 and 1849, ran an eleven-month school for untrained, politically appointed officers aboard his ship as duties allowed. In 1848 and 1849, Capt. Alexander V. Fraser ran the cutter *Lawrence* like a training ship. He identified "incompetent junior officers" who could not handle the heavy storms and required them to study and take examinations in surveying, law, seamanship, and navigation.[33]

In 1869 the Revenue Cutter Service got its wake-up call for officer education when a commission was appointed to examine abuses in the officer commissioning system and discharged thirty-nine officers as disqualified. At the time there were only two hundred commissioned officers, so almost 20 percent of the service's officers were considered unfit. A report to Congress, dated May 26, 1870, noted that during the Civil War officers of questionable character and professional competence had been commissioned. The report fell short of creating an academy, but regulations were reformed to require physical and professional examinations. The chief of the Revenue Marine Division, Department of the Treasury, Sumner I. Kimball, who had oversight of the Revenue Cutter Service, further reformed the prerequisites for officers. With the support of Chief of Instruction George W. Moore, Superintendent of Construction James H. Merryman, and Capt. John Henrigues, Kimball influenced Congress to approve a school of officer instruction, which received congressional approval in 1876.[34]

The Revenue Cutter School of Instruction opened in 1876 with eight cadets in New Bedford, Massachusetts, serving aboard the cutter *Dobbins*. Captain Henrigues served as the captain, superintendent, and principal instructor. He was assisted by two civilian instructors. Edwin Emery, a graduate of Bowdoin College, had been wounded in the Civil War, where he served as a soldier, color sergeant, and lieutenant in the 17th Maine Infantry. Edwin Emery was the main academic instructor, having taught school and served as a principal. Charles E. Emery (no known relation), who held an honorary PhD from the University of New York, was a consultant to the Revenue Service, designer of many of their cutters' engines, and guest lecturer for engineering instruction. He also was a Civil War veteran, having served first as a soldier and later as an assistant engineer in the Union Navy, where he saw action with the Union blockade.[35]

The first class of eight cadets reported to the training ship, the *Dobbins*, and sailed out of Baltimore in October, arriving at their new station, New Bedford, Massachusetts, in 1876. The academy would close from 1890 to 1894 when the government decided the US Naval Academy could provide services for officers of the Cutter Service. After four years, however, the school reopened at New Bedford, where it operated until 1900, when the academy's second training ship, the cutter *Chase*, docked at Arundel Cove, Maryland. In 1910 the school moved permanently to New London, Connecticut, and in 1915 it was renamed the US Coast Guard Academy.[36]

Stephen Luce, Lobbyist for Maritime Education

Amid increasing concerns of the United States for the quality of its maritime officers, Stephen Luce published his article "The Manning of Our Navy and Mercantile Marine" in 1874. Voicing the need for the education of naval officers and the philosophy of the means to that end, he said the model naval officer should be highly cultured and carefully trained. The training he described, in keeping with his experience at the Naval Academy, was in step with the military school movement of the United States. His article condemned the failure of Congress to include a maritime component to the Morrill Land Grant Act, stating that the legislative halls had "not a single representative voice . . . in behalf of the Navy." Luce praised the legislatures of New York and Massachusetts for their passage of acts in support of maritime education. By that time Luce was fully involved with New York's efforts to establish a maritime educational facility.[37]

In April 1873, New York legislators passed *An Act to Authorize the Board of Education for the City and County of New York to Establish a Nautical School.* By August the committee of the Board of Education charged with that task asked Luce to draft a bill for the US Congress in support of their efforts. Luce's efforts became House Bill 1347, of January 17, 1874, sponsored by Congressmen Henry L. Dawes and Benjamin F. Butler, both of Massachusetts. The draft called for a wide range of actions, including a system of certification of competency for masters and mates, service requirements for maritime school graduates, a national school certification and inspection, and the loan of naval ships and officers for instructional purposes to state schools. Unfortunately, the bill was watered down and passed with only an authorization for the navy to loan ships and officers. Among those Luce had lobbied for further action were Secretary of the Navy George M. Robeson, Charles Sumner (Republican, Massachusetts), Zachariah Chandler (Republican, Michigan), William A. Wheeler (Republican, New York), Benjamin F. Butler (Republican, Massachusetts), boards of commerce and education in eastern seaports, and the president of the General Boards of US Trade.[38]

New York Merchant Marine Academy.

Although the legislation that passed called for schools in New York, Boston, Philadelphia, Baltimore, Norfolk, and San Francisco, a systematic education system for maritime officers was not put into effect. Only New York carried through with actions to establish a maritime state academy. Luce was deeply involved in the establishment of several maritime military colleges, four of which still exist. The New York Merchant Marine Academy was established in 1874. Luce's *The Young Seaman's Manual* was published in 1875, becoming one of the school's principal texts. Luce would assist in the selection, outfitting, and coordination of the academy's delivery of the school's first training vessel, the *St. Mary.* He further influenced the curriculum through his book, which stressed lifesaving skills and drills, and through his association with the school's first superintendent, Cmdr. Robert L. Phythain, a graduate of the US Naval Academy who had worked with Luce in the 1860s at the academy.[39]

As the New York Merchant Marine Academy grew, Luce took command of the USS *Minnesota.* That ship, the USS *Saratoga,* the USS *Pensacola,* the

USS *Monongahela*, the USS *Supply*, and the USS *Juniata* all became training ships for boys as young as fifteen years. From 1875 through 1877 the number of enrollees grew from 260 to 479, of whom 258 boys were passed to the ranks of the regular enlisted navy. This helped Luce and the navy with concerns about the large number of foreign-born sailors; boys entering the Maritime Academy had to be American born. The short duration and scope of the program made it more of a seagoing transition training course than a military school mimicking Annapolis.[40]

With direct help from Luce, the New York Nautical School became the model for those maritime academies that would follow during the next sixty-seven years. The administration of the school recognized its potential and stated in their 1876 annual report that "should this experiment be successful, the cities of Boston, Philadelphia, Baltimore and San Francisco will, no doubt, follow the example of New York, and establish schools of their own."[41]

Maritime School Failures

Unfortunately, San Francisco, which obtained the training ship USS *Jamestown* in 1876, repeated the mistakes made in Massachusetts rather than emulating the New York Nautical School approach. In 1860 the Massachusetts State Reform School opened the rigger *Rockall* (later renamed *Massachusetts*) in Boston, and in 1867, the rigger *George M. Barnard* in New Bedford, to operate as floating reform schools. The goal of the ship schools was to teach the maritime trade to minors who had committed petty crimes. The system failed in 1872 because of the cost and lack of graduates. Likewise, the board of education in San Francisco diverted from their original plan, prepared by Lt. Cmdr. Henry Glass, US Naval Academy class of 1863. The initial class of students included male inmates of the Industrial School, a juvenile reform program. Commander Glass, the ship's superintendent and captain, was immediately hampered by his new recruits, and his ship gained "the appearance of a reformatory which deterred many parents from applying for the benefits . . . because of contaminating influences exerted by these wayward boys." Glass's conduct and management were attacked amid student applications for discharge from the school. In February 1879 the ship was returned to the navy, ending the San Francisco experiment and fittingly returning its students to the San Francisco Industrial School. Henry Glass went back to the fleet and had a distinguished career, retiring in 1903 as an admiral.[42]

Pennsylvania Nautical School

It took another ten years before the Pennsylvania Nautical School was established in 1889. The school rightfully adopted New York's model and experience. This also set the pattern of future maritime academies as postsecondary, two-year, and later four-year college institutions. The Pennsylvania Nautical School's president was Charles Lawrence, who had a long association with the merchant marine industry rather than the navy. He started as a sailmaker before the Civil War and served in the Union Navy using his civilian skills. After the war, he became involved in politics and served on the Board of School Directors for Philadelphia Commons, was a member of the Vessel Owners' and Captains' Association, and was master of the Philadelphia port. In his last role he lobbied for the establishment of the school. He was assisted by Cmdr. Francis M. Green, US Naval Academy class of 1871, who served as a training ship captain from 1889 to 1893.[43]

Both the New York and Pennsylvania schools relied on the traditional sailing ship to provide a platform for the schools, with the New York Nautical School having the USS *St. Mary* and the Pennsylvania Nautical School receiving the USS *Saratoga*. The seagoing classrooms from New York between 1877 and 1888 traveled to Lisbon, Portugal, Cádiz, Gibraltar, Tangiers, Cherbourg, Southampton in Britain, and Queenstown, Ireland. The Pennsylvania school traveled to Europe, South America, and the West Indies as well.[44]

Massachusetts Nautical Training School

In 1891 the momentum of merchant marine education was regenerated because of two events: the passage of the Merchant Marine Act of 1891, which provided subsidies for American-owned and -built vessels officered by Americans, and the establishment of the Massachusetts Nautical Training School. That school elevated the merchant marine academy movement to a new level. Whereas the New York and Pennsylvania schools remained loyal to the sailing ship as the vehicle for educating future officers, the Massachusetts school adopted a more modern, less traditional curriculum, educational philosophy, and training vessel, the bark-rigged steam gunboat *Enterprise*.[45]

The school's first superintendent not only had a naval background, but also had served in the merchant marine. Cmdr. John F. Merry was a merchant marine first officer in 1862, when he was appointed an ensign in the Union Navy. He was a wounded veteran of the Civil War and served on a navy ship with duties in the China Sea, Mediterranean, and Arctic Ocean as well as the Great Lakes prior to taking charge of the school. He remained at the school until December 1895 and later rose to the rank of admiral.[46]

With more modern vessels than the other two nautical schools, the Massachusetts Nautical Training School successfully blended academic and engineering courses with the seamanship curriculum. The other schools would follow suit as they, too, adopted more modern vessels. The Pennsylvania Nautical School acquired the USS *Adams*, a wooden bark-rigged steam gunboat. The New York Nautical School acquired the USS *Newport*, a bark-rigged steam gunboat that it used between 1907 and 1908. It was about that time that both the States of Washington and California contacted the New York Nautical School with interest in establishing their own programs.[47]

As interest grew in the Pacific states, the State of Pennsylvania closed its school in 1913, despite the loud protests of the alumni. A few years later, the dire need for a strong merchant marine program was demonstrated during World War I, and it would result in a third wave of nautical military schools. The Washington State Nautical School opened in 1917, the Pennsylvania State Nautical School was reestablished in 1919, and ten years later, in 1929, the California State Nautical School opened.[48]

Upon reestablishment of the Pennsylvania State Nautical School, its operation was placed directly under the state rather than Philadelphia authorities. The school was firmly reestablished in the military school format, with school regulations stating: "[Student conduct will be] governed, as far as practicable, by the United States Naval Regulations and customs. . . . Formations shall be by classes and will be conducted in a military manner. . . . The rules for military etiquette on board ship are founded on custom and tradition, and their strict observance forms an important factor in the maintenance of discipline."[49]

Washington State Nautical School

The Washington State Nautical School was patterned after the military schools of New York and Massachusetts, with strict military conduct, a two-year curriculum, and the training ship *Vicksburg*. The school was in full operation with uniformed cadets aboard ship by January 1920, and it was run by US Naval Academy graduate Cmdr. Ernest F. Eckhardt (retired). By December 1920 the school had enrolled two classes, and total enrollment was sixty-five, with all but two being from Washington State.[50]

However, what appeared to have been a successful model was closed in 1920. The justification for closure is not well documented, but the cost of ship operation was an issue with state legislatures. According to the congressional testimony of the superintendent of the New York Nautical School in 1921, the Washington Nautical School did not have the record of graduate success needed to continue operating the program. The record of the school's first class in their initial six months backs that conclusion, as it lost fifteen of its sixty enrolled cadets through desertion and withdrawal for inability to adapt to life at sea.[51]

California State Nautical School

The opening of the California State Nautical School in 1929 provided another step in the modernization of the maritime military academy movement. The California school's curriculum was different from that of the other schools. The school conducted classes and housed cadets ashore, with the first class of fifty-six cadets taking residence in barracks on a fifty-acre compound at an old coaling station in California City (Tiburon) in March 1931. But because of lack of state funding, by the time the first class had graduated, the cadets were back on ship for both classes and housing. For the next eleven years the school would be threatened with termination because of funding shortfalls, and cadets remained in the "Iron Mother," as they called their home on the training ship *California*. It was not until 1943 that the school and the cadets resumed academic activities ashore and occupied barracks at a new location, Morrow Cove.[52]

The curriculum at the California State Nautical School was expanded to three years. This allowed the academics to be broadened to address subjects such as meteorology, communications, ship construction, cargo handling, shipboard medicine, physics, and electrical engineering.[53] The extension of the curriculum would become part of the next wave of improvements in the expanding maritime academy movement.

United States Merchant Marine Academy

As far back as 1895 there had been suggestions for the establishment of a national maritime academy on par with West Point and Annapolis. Several publications voiced this opinion in 1913, as did a member of the board of directors of Pennsylvania Nautical School, Joseph C. Gabriel.[54] The twentieth century saw a new champion emerge to extend the maritime academy toward a national maritime school. The maritime academy movement's momentum would increase with the advent of World War II and result in a national academy.

Between 1929 and 1934, Richard R. McNulty, Massachusetts Nautical School class of 1919, wrote articles in the *Marine Journal*, the *Naval Institute Proceedings*, and the *Nautical Gazette* endorsing merchant marine officer training. His concept, unlike that of Stephen B. Luce, depicted a national academy located at New London, Connecticut, replacing not only the state maritime academies but also sending the coast guard cadets to the naval academy. McNulty's proposals made him several enemies in the maritime academies, the US Coast Guard, and the International Seamen's Union. The state maritime academies did not like the idea of their schools being closed in favor of a federal institution, and the coast guard had no desire for their academy to be absorbed into the navy's. McNulty was also a supporter of existing maritime curriculum, with education primarily ashore augmented by training at sea. The union protested this concept, which would devalue their members' experience. Their position was that "the only acceptable training for merchant marine personnel was experience at sea."[55]

At the National Conference on the Merchant Marine in 1930, McNulty found some allies. The chair of the Committee on the Training of Officer Personnel, Adm. Hutchinson Cone, sent out questionnaires to leaders of government, commerce, shipping labor, and education, and sixty-five hundred were returned. These responses endorsed his proposal three to one to replace the state maritime academies with a national academy. In a 1932 article in *Naval Institute Proceedings*, McNulty gained another ally, the superintendent of the New York Maritime Academy, Capt. James H. Tomb.[56]

During publication of McNulty's series of articles, the liner SS *Moro Castle* sank in 1934 due to crew negligence when a fire was not brought under control. Adding to the tragedy, members of the crew filled the lifeboats, and 134 died, including 91 passengers. Only months later, the SS *Mohawk* collided with the Norwegian freighter *Talisman*, with more loss of life and accusations of crew incompetence. But in the latter case, the captain and all but one of his crew went down with the ship.[57]

The next maritime academy may have opened in Virginia if not for congressional opposition. In 1935 the State of Virginia proposed that a maritime academy be established and receive federal funds, but Virginia congressman Schuyler O. Bland helped kill that idea. Bland was a William and Mary graduate and was highly influential in maritime affairs. He saw little connection between maritime education and the military colleges of

Virginia Tech or VMI, which were nestled in the mountains of Virginia hundreds of miles from the ocean.[58]

The Merchant Marine Act, passed by Congress in 1936, included two significant elements that would take the country a step closer to establishing a national merchant marine academy. The first was introduced by the director of the Shipping Board, J. C. Peacock, to "establish a Merchant Marine Academy for the training of citizens of the U.S. as officers for service on vessels . . . of the United States."[59] The second significant element was the establishment of the Maritime Commission.

The commission's personnel division, under the leadership of Rear Adm. Telfair Knight, took on the task of establishing a Merchant Marine Academy. Knight added Richard R. McNulty to his staff. The first concrete step in founding the US Merchant Marine Academy was the establishment of the Merchant Marine Cadet Corps by order of the Maritime Commission in 1938.[60] The sequence of accepting the cadets before having a facility for a military school was rather strange. As a result, the programs for the cadets were not a maritime military school at all, but a reflection of the pre-1838 idea of midshipmen receiving their education at sea at the tutelage of a ship's master. The requirement for merchant vessels to carry cadets based on tonnage was a two-edged sword. It provided training opportunities, but it also brought unwanted employees on board.

McNulty uncovered problems associated with the shipmasters' failure to provide a true training opportunity, and he discovered cases of downright abuse of cadets by shipmasters and crew. In one case union members assaulted an unwelcome cadet on the SS *Laura* without reprimand. On some ships cadets were isolated, and union crew refused to work alongside them. There were sufficient examples of failure on the sea that McNulty pushed ahead with the shore establishment of the academy.[61]

In October 1939 the Maritime Commission defined a four-year course for cadets and directed that shore educational facilities be established in New York, New Orleans, and San Francisco. The three maritime military schools became institutions looking for a home. Harold V. Nerney, Massachusetts Nautical School class of 1925, and the California Nautical School facilitated the San Francisco Maritime Military School's first home in the fall of 1939 aboard the California Nautical School's training ship, the SS *California State*. The California Nautical School was the first of the maritime academies to establish shore facilities. Six months later, the San Francisco Maritime Military School moved to the navy barracks ship, SS *Delta King*, and a few months later, the SS *Delta Queen*. Finally, in January

1942 the school relocated on twenty acres in San Mateo Point north of San Francisco.[62]

Although the New Orleans District had no maritime academy to help them, they set up classes in the spring of 1940 at the Biloxi, Mississippi, Coast Guard Station. In February 1941 the New Orleans District's cadets moved into the Algiers Navy Yard. Sixty years later, in 2011, this became the location of the public charter secondary school, the New Orleans Military Maritime Academy. Unfortunately, by November 1941 the New Orleans District Maritime School was forced to leave again. Quick action by McNulty placed them aboard a 120-foot houseboat, *North Star*, which by chance had sought safety from a storm in Pass Christian.[63]

Pass Christian, Mississippi, was no stranger to military schools. The Mississippi Military Academy operated in that location from 1852 to 1861, and the Mississippi Military Institute moved there from Aberdeen, Mississippi, in 1876 and functioned as a school until 1887. When the cadets aboard the *North Star* moored their boat beside the Hotel Inn by the Sea to avoid a storm, they learned that the hotel was for sale. It was not long until the spacious hotel provided an excellent location for the school.[64]

The New York District Maritime School had the fewest problems, as their first shore facility was initially located at the New York Maritime Academy. They were welcomed by Admiral Tome, the New York superintendent and longtime supporter of the national maritime academy concept. The New York contingent was split due to growth, with one group moving into quarters at the Admiral Billard Academy in New London, Connecticut.[65] The Admiral Billard Academy was a secondary nautical military school that opened in 1936 and would operate as a secondary school until 1953, with a focus on preparing young men for entrance into the US Coast Guard Academy.

Nutley was actively searching for a place in which to consolidate the New York District facilities in order to depart the dual-location operation. In early December 1941 the Walter P. Chrysler estate on Kings Point—the model for East Egg in F. Scott Fitzgerald's *The Great Gatsby*—went up for sale. In the last week of January 1942 the cadets of the US Merchant Marine Academy, previously housed at Fort Schuyler alongside the cadets of the New York Nautical School, rowed in Monomoys (long wooden surf boats) to their new home at Kings Point. When the facility was dedicated in September 1943, Admiral Tomb, US Naval Academy class of 1899, was the first superintendent, and in formation were twenty-five hundred cadets organized into three battalions and eighteen companies.[66]

Admiral Tomb was a beloved figure at both the Merchant Marine Academy and the New York Nautical School, where he had been superintendent since 1927. He served the US Merchant Marine Academy long enough to see the academies establish strong foundations; in retirement, he continued to worked on behalf of the schools. Tomb was followed by Capt. Giles Stedman, known as the Clark Gable of the Seas for his multiple daring sea rescues of stricken vessels. Although he was not the product of a nautical military school, Stedman had served in the navy twice and had captained some of the nation's most prestigious ships, including the SS *America*, flagship of the US Lines.[67]

Among these men the four initial pillars of the academy were established: the Kings Point Campus, the Naval Reserve status of midshipmen, the seagoing requirement for graduation, and the regimental system. The last was in large part Stedman's accomplishment while he served under Admiral Tomb as commandant of cadets. Stedman was a tough discipli-

United States Merchant Marine Academy, 1944. (Courtesy US Merchant Marine Academy)

narian, and by August 1942, under his guidance, cadets took the responsibilities of cadet officers and petty officers; the military school established a demerit system; and twenty-four-hour watch stations were established.[68]

Weeks prior to the initial occupation of Kings Point, the maritime curriculum changed to meet the demands of the war. The four-year curriculum gave way to a program of twelve weeks' basic instruction at any one of the three locations—Kings Point, Pass Christian, or San Mateo Point—followed by six months on a merchant vessel and then thirty-six months at Kings Point. This schedule of three years, ten months remained until 1943, when it was modified to six months of basic instruction at any of the three locations, six months aboard ship, and twenty-four months at Kings Point. The strength of the cadet corps, including those serving aboard ship, was 7,000 in 1944, but it was down to 4,391 by July 1945.[69]

In 1945 the academy superintendent, Giles Stedman, decided to return to his sea career and turn over the superintendent's position to Rear Adm. Richard McNulty, who would become known as the father of the US Merchant Marine Academy. Despite that well-deserved title, McNulty's two-year term had an "abrupt and ugly" end reminiscent of that of Alden Partridge at West Point. According to Jeffrey Cruikshank and Chloe Kline, McNulty was a visionary—an intelligent advocate who avoided the limelight and

Vice Adm. Richard R. McNulty (1899–1980), US Maritime Service. (Courtesy US Merchant Marine Academy)

political demands of the superintendent position. He was a shy man who had even missed the 1943 dedication at Kings Point, likely to avoid the praise that would have been given to him. McNulty's lack of assertiveness resulted in Adm. Telfair Knight's removing him from the academy to a much lesser staff position after just two years. Knight was the man who had encouraged McNulty to take the position and had called him the guiding force behind the academy's establishment.

The US Merchant Marine Academy would end the satellite facilities one by one, with the last, Pass Christian, to close in 1950. McNulty remained a bitter man and did not set foot on Kings Point for twenty-two years, until in 1976 when the alumni association named him "Father of the Academy." Four years later, McNulty died, knowing his contributions were celebrated and appreciated.[70] After World War II the enrollment of the academy ranged from 650 to 1,050. In 2010 there were 964 cadets enrolled.

While the US Merchant Marine Academy was firming up its establishment, a sister institution was on its last legs. In 1940 the Pennsylvania Nautical School's administration came under the auspices of the US Merchant Marine Commission. As a result, the school was renamed the Pennsylvania Maritime Academy, and in 1942 the faculty and cadets transferred to the US Merchant Marine Academy at Kings Point. Pennsylvania resumed administration that September, when the training ship *Seneca* was returned for state use and renamed *Keystone State*. Facilities were open at Morrisville, Pennsylvania, to provide shore support. Like the US Merchant Marine Academy, the Pennsylvania school was oriented to two tracks: deck cadets and engineering cadets.[71]

After the war, plans were made to increase the curriculum from two years to three years, including two cruises. Unfortunately, the state budget review came to the legislature's attention at the same time as allegations of poor management. Further hurting the school were inaccurate reports from the *New York Times* claiming "mutinous behavior" during a cruise. The school had already been closed twice in its history. The fact that the mutinous behavior claim was proven false after an investigation did not prevent the removal of funding and closure of the program in June 1947. The US Merchant Marine Academy "welcomed all cadets who wanted to transfer and received a major infusion of faculty talent."[72]

Maine Maritime Academy

On October 9, 1941, just two months prior to the attack on Pearl Harbor and the entry of the United States into World War II, the State of Maine opened the Maine Maritime Academy with an initial 28 cadets and an eighteen-month curriculum. The academy was initially led by Rear Adm. Douglas Dismukes, US Naval Academy class of 1880. By 1945, at the war's end, 384 men had graduated and served at sea. After the war the curriculum was expanded to three years and then four years.[73] In 2010 enrollment was 932 cadets.

Texas Maritime Academy

The most recent maritime academy was established in 1963. Efforts to start a maritime academy in Texas began in 1920, when the assistant secretary of state of Texas made three different attempts to get a bill passed in the state legislature to establish a Texas nautical school. All the attempts failed. In the 1950s, Rear Adm. Sherman Wetmore was researching the possibility of establishing a maritime academy in Galveston when he discovered a bill that had been passed in 1931 establishing a charter for a nautical school but on which no action had taken place. In 1959 the charter came back to life, and the state legislature authorized some minimal funding to support it.[74]

The Texas Maritime Academy was initially part of an expansion of Texas A&M College, and the Texas Maritime Academy cadets came directly from the Corps of Cadets of Texas A&M. In 1962 the Texas A&M Corps of Cadets assigned twenty-seven cadets (later reduced to twenty), destined to be the cadre of the Maritime Academy, to Company I, 3rd Brigade, a cadet company with leadership oriented toward the US Marine Corps. Assignment to the Marine Company was an idea of the maritime academy's future superintendent, Capt. Bennett M. Dodson, who wanted the best-trained and -motivated cadets for his school. Dodson had a long association with the maritime academy movement, having been the commanding officer of the US Merchant Maritime Academy's Pass Christian cadet training facility and later the superintendent of the Pennsylvania Maritime Academy. Following the closure of PMA, he became the executive officer at the California Maritime Academy.[75]

The twenty sophomore cadets in Texas A&M Corps of Cadets uniform and a small faculty moved to the Galveston campus in 1963. There they were joined by thirty-five freshmen, and the small organization became the corps of midshipmen of the Texas Maritime Academy.[76] The academy maintained close ties to Texas A&M and its corps at College Station for many years. The maritime institution grew to become Texas A&M University at Galveston, and the Texas Merchant Marine Academy functioned as the branch's corps of cadets. There were three hundred midshipmen in 2010.

In Texas the maritime academy movement met and blended well with the military origins of Texas A&M. The formation of the Agricultural and Mechanical College of Texas had been heavily influenced by Virginia Military Institute graduates, who had been schooled by Francis Smith, who in turn had studied with Sylvanus Thayer at West Point. The history of America's maritime academies had created a historical pattern of close association and cooperation. Examples in the 1940s include the New York Maritime Academy's housing the US Maritime Academy and the integration of cadets and faculty of Pennsylvania Maritime Academy into the US Merchant Marine Academy when the former school closed. That pattern continued with the Texas Maritime Academy's first two cruises, which were conducted aboard the *Empire State IV*, the New York Merchant Marine Academy's training vessel, and the *State of Maine II*, the Maine Maritime Academy's vessel.[77]

The story of the US Naval Academy and the merchant marine academy's military school format trace directly to West Point's model. Clearly, this was the case with the US Naval Academy's early development and formation. The US Coast Guard Academy's choice of educational format likewise gravitated naturally to a military school structure. Further, because of Stephen B. Luce, US Naval Academy class of 1848, and other naval academy graduates, the education of merchant marine officers was conducted predominantly in the military school format, as illustrated by the fact that of the seven merchant marine academies, only one is not a military school. This also directly impacted the manner in which numerous secondary schools conducted themselves and the traditions they adopted.

⋆ 8 ⋆

The Lost Cause and the Grand
Army of the Republic

During the aftermath of the Civil War, four significant trends affected America's military schools. The first was the postwar Southern culture's embracement of the Lost Cause and its impact on the recovery and expansion of military schools of the South. Second was the manner in which the South enacted the Morrill Land Grant Acts, which resulted in corps of cadets at agricultural and mechanical colleges. The third was the influence of Union Army veterans on the expansion of military schools in the northern states. And fourth and last was the establishment of African American military schools in Virginia by Union veterans and their influence on the integration of American military schools.

Between the end of the Civil War and 1898 there was a shift in the origins of the establishment of military schools. Up until the beginning of the Civil War in 1861, a total of 311 military schools had been established, with Partridge and Norwich alumni associated with the establishment of at least 23; West Point alumni, 22; Francis Smith and the alumni of VMI, 27; and alumni of the Citadel, 6. Counting an alumnus of Virginia Literary, Scientific, and Military Academy who established the Franklin Military Institute in Kentucky, a total of 78 military schools established through 1861 are documented to have been founded by alumni of other military schools. This does not account in any way for the many teachers from military schools who staffed the newly established military schools and helped pattern them after the school traditions they knew. By the Civil War, Partridge, Thayer, and Smith had created institutions that served as models for educational institutions and were a source of teachers, particularly at the secondary level, in both the North and South. In VMI's initial fifteen years, 54 percent of its graduates became teachers, and Norwich University listed seventy-four of its alumni as military school teachers prior to the class of 1910.[1] Indications are that many of the alumni of West Point and the Citadel also went into education. These were likely military school movement multipliers as well.

Up until the Civil War, the military format in education was significantly influenced by men who had experience with the military in an educational format. Starting after the Civil War and up to the twentieth century, that pattern shifted. Men whose military experience was not in an educational setting but in the ranks of the Confederate and Union Armies became founders of many schools. And men with no military background embraced the military school format as both a successful business and an educational vehicle.

Former Confederate soldiers and educators in the South were influenced toward the military school format through the positive image of Civil War military service and the heroic image of Southern Confederate leadership. Also associated with the Lost Cause was the manner in which the South enacted the Morrill Land Grant Acts of 1862 and 1890, creating multiple college-level military schools exclusively in the South.

The Lost Cause

The Lost Cause encompassed the beliefs held by the white population of the South that helped them deal with defeat and change. It presented the "wartime sacrifice and shattering defeat in the best light."[2] Its objectives were to justify the actions of the Southern states and to find positive historical interpretation of the war. It focused on the military rather than the less positive political and social aspects of the war. Gen. Jubal A. Early is cited as the former Confederate who was most influential in the shaping of the Lost Cause with a reverential view particularly of Gen. Robert E. Lee.

For several decades the social conscience of the post–Civil War South was influenced by the memories of Confederate veterans, magazines such as *Confederate Veteran*, the Southern Historical Society, and memorial associations. These individuals, groups, and publications had a common desire for the Southern cause to be viewed by the world in a positive light. Whereas vexing issues such as slavery were minimized, the nobility, virtue, and sacrifice of the Southern people were used to justify their cause to others and to themselves.

Gen. Robert E. Lee (1807–70), West Point class of 1829, superintendent of West Point 1852–55, about 1865. (Library of Congress)

Central to the Lost Cause was the "firm connection in the minds of
southerners [*sic*] between the martial virtues (courage, patriotism, selflessness and loyalty) and moral rectitude."[3] A symbol of those virtues is en-
graved on Stone Mountain outside Atlanta, Georgia. The rock face of the
mountain contains the largest bas-relief sculpture in the world, depicting,
on a surface of twelve thousand square meters, three mounted heroes of
the Confederate lost cause: Gen. Robert E. Lee, Gen. Thomas "Stonewall"
Jackson, and Confederate president Jeff Davis. These men particularly, and
many other Confederate military figures to a lesser extent, were connected
to Southern military schools. General Lee was educated at West Point and
had served as that school's superintendent. General Jackson likewise had
been educated at West Point and had served on the faculty at VMI. Finally,
the Confederate president was also a West Point graduate. These men,
particularly Lee and Jackson, were sainted in the South as the perfection of
Southern manhood.

Gen. Thomas Jonathan "Stonewall" Jackson (1824–63), West Point class of 1846, professor at Virginia Military Institute 1851–61. (Library of Congress)

James Morrison described how growing up in Virginia amid the influences of the Lost Cause contributed to his decision to attend Virginia Military Institute:

> Still subtle influences may have been at work, even though I was unconscious of them. My grandmothers, both of whom I adored, were daughters of Company I, 3rd Virginia Cavalry. To them, the courage and virtue of the soldiers who served in the Army of Northern Virginia were without parallel in human history, a conviction fully shared by my Great-Grandmother Morrison, the widow of a member of Company A, 12th Virginia Infantry. These ladies, along with parents, teachers, and other older people, instilled in me the belief that the aged Confederate veterans who occasionally could be seen tottering along the streets of Petersburg deserved special attention and respect.[4]

Likewise, Harry Temple and his brother, from Petersburg, Virginia, were bound for the corps of cadets at Virginia Tech. Harry spoke of his childhood memories of walking the Petersburg Battlefield with a former Confederate battery commander in his gray overcoat who recounted the memory of the bloody battle. Temple was well versed on the part his ancestors had played in Virginia and North Carolina regiments in the Civil War, and one of their muskets hung from the wall as a constant reminder.

In Virginia Tech's second year, 1873, the literary society adopted the name the Lee Literary Society. In doing so, they stated, "We can only endeavor to emulate him [Robert E. Lee]. He was far too above us for us to hope to be like him, but if our aspirations are high and noble, we can accomplish much. Although we may not leave behind us the record which others have done, if we have endeavored to maintain the spotless reputation borne by our beloved leader, we shall have been worthy of the name Lee."[5] The Richmond dedications of the Jackson Monument in 1875, the Lee Monument in 1887, and the Soldier and Sailors Monument in 1894 all reflected the central place that the Lost Cause, especially of Robert E. Lee and Stonewall Jackson, had for the South and its military schools. The 1875 event drew forty thousand people to Capital Square, and the parade and ceremony included both VMI's and Virginia Tech's corps of cadets along with Confederate veterans. The 1887 and 1894 events were attended by the corps of cadets of both VMI and Virginia Tech, who had traveled by train from their Virginia campuses in the mountains. The 1894 event had an estimated one hundred thousand people in attendance and a representation of ten cadets and several officers from the Citadel.[6]

Southern educators were said to have embraced the "traits of manly bearing, courage, loyalty, patriotism and morally correct behavior" encouraged by the military school format.[7] Between the end of the Civil War and 1897, 108 military schools were established in the South. These joined the 39 Southern schools that had reopened in a military school format after the war. Their cadets dressed in gray often served in annual parades and monument dedications as echoes of a valiant past.

The Lost Cause resulted in a union of religious and military virtues. Preachers commonly used Southern military leaders such as Robert E. Lee and Stonewall Jackson as examples of Christian virtue.[8] The military school movement and school ethos blended perfectly with the moral goals of the Lost Cause. These schools demanded moral behavior, lived by codes of obedience, and maintained their cadets as much as possible in an environment free from alcohol, the immoral distraction of the period. This is not to mention the strong religious character of many nonsectarian military schools that required church attendance.

Civilian educators who saw schools' transition to military format in the South included Edwin P. Cater, an Oglethorpe University graduate, who in 1881 oversaw the transformation of East Florida Seminary to a military school. The college later evolved into the University of Florida after twenty-two years as a military college. C. H. McKeen, a Washington and Lee College graduate, founded Elmington Military and Classical School in Virginia in 1872. Archibald Campbell, a Hampden-Sidney graduate, founded Wytheville Military Academy in Virginia in 1881.

Also graduating from Washington and Lee College was James Abbott Fishburne, who attended there while Gen. Robert E. Lee was president. General Lee encouraged Fishburne's desire for a career in education and inspired him with the "General's love of duty, honor, and his fellow man." Fishburne first taught at Horner Military Academy in North Carolina after graduation. These experiences inspired him to return to Virginia, where in 1879 he started a day school that in a period of four years evolved into Fishburne Military School. Fishburne Military School gained Col. Morgan Hodgins, VMI class of 1901, who was commandant for fifty-one years (until 1952). The two educators grew the school to 85 cadets by 1914, and by 1925 the school had achieved a historical steady enrollment of between 170 and 200 cadets that continues today.[9]

But perhaps the best example of an educator of the Lost Cause who influenced the establishment of Southern military schools was someone who did not actually found one himself. James DeRuyter Blackwell, graduate of Randolph-Macon College and Dickinson College, joined the faculty of Bethel Academy in Virginia. Blackwell was a Confederate veteran and poet who had among his poetic works "The Dead Drummer Boy," "Our

Fishburne Military School about 2014. (Courtesy Fishburne Military School)

Native Land," and "Forget Not the Dead." He convinced the founder of
the academy, Maj. Albert G. Smith, a hero of Pickett's Charge and postwar
graduate of the University of Virginia, to transition to a military format in
1869, and the school operated as such until it closed in 1911.[10]

The Morrill Land Grant Act

Justin Smith Morrill, a congressional representative from Vermont from
1855 to 1867 and senator from 1867 to 1898, authored the Morrill Land
Grant Act, which became law in 1862. Despite the fact that Morrill lived
in Stafford, just nineteen miles from Norwich, a relationship between him
and Alden Partridge has never really been documented. In fact, Morrill
was a Whig, and Partridge a Democrat, so the two would not have been
political allies. But between 1862 and 1863 Morrill was a member of the
Board of Trustees of Norwich University.[11] Having been a short distance
from Norwich University and becoming a member of the board of that
military institution so quickly after passage of the act suggests there must
have been influence there, with Partridge's citizen-soldier concept and the
idea of military academies supporting the state militias.

The Morrill Act's goal was to help fund the establishment of colleges
under the oversight of the state legislatures, which would address the need
for practical education of industrial classes in agriculture and engineering.
Importantly, the act specified the teaching of military tactics.[12] The act did
not dictate the manner in which military tactics would be incorporated
into the colleges, so different states and regions approached this subject in
very different ways.

By 1890 thirty-nine states had used federal funding under the Morrill
Land Grant Act of 1862.[13] The act influenced the establishment or transition of eleven military colleges, of which four were still in operation as
military schools in 2017. Between 1872 and 1914, two military colleges
became associated with the land grant act, six military colleges were established, and eight others were converted to military format while associated with the act. With the exception of two schools transitioned in the
twentieth century in the western United States, all these schools were in
the South. The influence of the Lost Cause and the Southern view of value
in military education is clearly demonstrated by the manner in which the
South enacted the Morrill Land Grant Act. On the other hand, no college
in the northern United States became a military school under the Morrill
Land Grant Act, although military training was an element of most land
grant colleges to varying degrees.

Virginia Tech and Texas A&M

In 1872, the first three military schools under the Morrill Land Grant Act were the Virginia Agriculture and Mechanical College, Agricultural and Mechanical College of Alabama, and Hampton Normal and Agricultural Institute in Virginia. Hampton Institute was established in 1868, brought under the Morrill Act in 1870, and transitioned to military format about 1872. Texas Agricultural and Mechanical College followed four years later in 1876; it would become first Texas A&M College and then Texas A&M University.

The Virginia Agricultural and Mechanical College, which became better known as Virginia Tech, and Texas Agricultural and Mechanical College, now Texas A&M, have maintained their military school status to this day. The stories of Virginia Tech and Texas A&M describe the changes and influences required of Morrill Land Grant Act military schools that enabled them to survive the many social changes over time. Established in 1872 and 1876, respectively, both schools were located in rural areas, and

Gen. James H. Lane (1833–1907), Virginia Military Institute class of 1854, "Father" of the Virginia Tech Corps of Cadets. (Harry Temple Collection, Digital Library and Archives, University Libraries, Virginia Tech)

their leadership involved former Confederate officers and VMI graduates.

Virginia Tech's military program almost ended a few years after the establishment of the college. The president was Charles Landon Carter Minor, a University of Virginia graduate, who had served as a Confederate staff officer in the Civil War. The professor of engineering and commandant of cadets was Brig. Gen. James H. Lane, VMI class of 1854, who had served on the faculties of three military schools prior to the war: Armstrong Military and Classical Academy in Virginia, West Florida Seminary, and North Carolina Military Institute.[14]

Very different educational philosophies collided on March 23, 1877. President Minor led a faction of the faculty that desired a classical civilian college, and General Lane naturally desired a military college. A fistfight in a faculty meeting resulted in both men being charged with disturbing the peace. The incident was widely reported in the state newspapers, and as a result, Minor left Virginia Tech. The school's board of visitors issued a statement strongly supporting the military character of the school.[15] Lane left the following year and took a teaching position at Auburn University. His establishment of the corps's high standards endeared him as the Father of the Virginia Tech Corps of Cadets. However, General Lane's altercation with Charles Minor would serve as a foreshadowing of future conflicts between the military and a future college president.

The military system during Texas A&M's initial years was secured by another VMI graduate. John Garland James, Virginia Military Institute class of 1866, was the college's president from 1879 to 1883. James had been a cadet corporal in the VMI battalion at the Battle of New Market. After graduating, he was a professor for a short time at the Kentucky Military Institute. He took the presidency of Texas Military Institute in Bastrop in 1867 and moved that school to Austin in 1870. When he moved to Texas A&M as its first president, he took the majority of the faculty of Texas Military Institute with him, and that school closed.[16]

Both Virginia Tech and Texas A&M would, at the end of their first twenty-five years, enter periods under very popular presidents who fully supported both the military nature of the schools and evolution of the institutions toward expanded educational opportunities. The first of the popular figures at Virginia Tech was John M. McBryde, who was president from 1891 to 1907 and considered the father of Virginia Tech. He was a graduate and former president of the University of South Carolina, a lawyer, and a longtime professor of agriculture. Although he did not have experience with a military school, he had left the University of Virginia

Virginia Agriculture and Mechanical College (Virginia Tech) in 1887 had 110 cadets. (Harry Temple Collection, Digital Library and Archives, University Libraries, Virginia Tech)

to fight in the Civil War as a soldier with the 1st South Carolina Infantry until he was struck by typhus.[17]

The second president in the building of the military college was Julian Burruss, Virginia Tech class of 1894, who as a cadet had been the commander of the college's elite artillery battery. After graduation he taught at Speers-Lanford Military Institute and Searcy Female Institute and was principal of a public school. His postgraduate work included a master's degree from Columbia and a PhD, magna cum laude, from the University of Chicago. Prior to starting a twenty-six-year term (1919–45) as president of Virginia Tech, he had been president at the college that became James Madison University.[18]

Burruss successfully integrated women as civilian students on campus. He reduced the mandatory participation in the corps of cadets to two years, while at the same time endorsing greater discipline, which helped build higher morale and esprit de corps among the cadets. He is also credited with increasing the quality of the faculty and expanding the academic opportunities and reputation of the college.

Texas A&M had similar figures, among them Lawrence Sullivan "Sul" Ross, who served as president from 1891 until his death in 1897. Ross's résumé was impressive as the leader of a military school despite not being a military school graduate. He was a graduate of Florence Wesleyan in Alabama, he led Indian scouts, he served as a captain in the 2nd Cavalry Regiment, and he had served in the famous Texas Rangers in the pre–Civil

Texas A&M College Corps of Cadets in 1910 numbered 833 cadets organized into a regiment of two battalions. (Cushing Memorial Library and Archives, Texas A&M University)

War period. During the Civil War he went from private to colonel with the 6th Texas Infantry Regiment and experienced 135 engagements. He is credited with the capture of thirty Union regimental colors. After the war he was elected to the Texas legislature and served two terms as governor.[19]

The second Texas A&M leader of note was William B. Bizzell, Baylor University graduate and former president of the college that later became Texas Woman's University. Bizzell did not have a military background, but his contributions to Texas A&M as president from 1914 to 1925 were significant: improved scholastics, expanded religious and athletic activities, improved facilities, and a moral and cooperative tone.[20]

These men at Virginia Tech and Texas A&M set the stage for cadet corps, which by World War II numbered over 9,000. A significant difference between the two colleges was the manner in which change came to their campuses as they evolved. Texas A&M transitioned to a volunteer corps of cadets in 1963 with a popular and promilitary president, James Earl Rudder, Texas A&M class of 1932. Rudder, a former major general and World War II hero of D-Day, was beloved by the alumni and cadets. In its volunteer status the corps of cadets has thrived through the years, and in 2014 it numbered 2,450 cadets in its ranks. This is as much a reflection on the history of the campus as it is on the very conservative nature of the student body.

On the other hand, President T. Marshall Hahn at Virginia Tech was distrusted and openly disliked by many of the alumni and the corps of cadets in his drive for the greater Virginia Tech. His goal was for the military to be reduced to the ROTC program seen at most universities. The reaction of the Virginia Tech alumni stopped him from achieving that goal but not from changing the cadet corps to a voluntary option starting in 1964. The tone and style of leadership of these two universities as they evolved militarily were very different.

Friendlier administrations would follow at Virginia Tech. First, in 1975 the alumni of the regimental band formed the Highty-Tighties Incorporated, under Charles Cornelison, class of 1967, and in 1992 the Virginia Tech Corps of Cadets Alumni Association was established under Henry Dekker, class of 1944, to ensure the preservation of the corps. After many years of low cadet enrollment, these two alumni organizations joined with two successive commandants, Maj. Gen. Stan Musser and Maj. Gen. Jerry Allen, to provide a constant, outstanding military leadership through focused leadership training programs. These efforts were backed by a growing endowment devoted to the corps, which helped grow the cadet corps to 1,036 cadets in 2014, the largest number since 1967.

Auburn University.

The Agricultural and Mechanical College of Alabama was established in 1856 as the East Alabama Male College in Auburn. Upon its inclusion in the Morrill Act, it changed its name and transitioned to a military school in 1872. It is now better known as Auburn University, a title not officially adopted until 1960. The school maintained its military character for thirty-four years until 1906. According to the 1901 catalog, the military system provided "good order, promptness and regularity in performance of academic duties." In 1890, 203 of the 234 students were members of the corps of cadets. The cadets were uniformed in cadet gray and equipped with 150 Springfield rifles and two three-inch cannons. Although the corps was reported to have good discipline, the college did not have barracks, and cadets lived off-campus. The lack of barracks space weakened cadet discipline and adoption of military requirements and may have been a factor in the end of military school requirements in 1906.[21]

Louisiana State University

In 1874 the Louisiana State University Agricultural & Mechanical College, formerly known as Louisiana Seminary of Learning and Military Academy, was brought under the Morrill Act. The school was established as a military college in 1860 with Col. William Tecumseh Sherman, West Point class of 1840 and future Union general, as its president and Francis Smith, VMI class of 1856, as commandant and professor of chemistry and mineralogy. The enrollment initially numbered fifty-nine cadets and quickly grew to eighty. Sherman departed the campus prior to the outbreak of the Civil War when the Louisiana State Militia captured the federal arsenal at Baton Rouge and sent captured federal property to the academy for its use.

The war helped close the school in 1863, but it reopened after the war in 1865. By 1869 enrollment had climbed to 133 cadets when a fire caused the college's relocation from Alexandria to Baton Rouge. Louisiana, facing a budget crisis, cut funding to support tuition for "State Cadets" in 1873. As a result, the administration had to furlough those who could not pay full tuition, and enrollment dropped to 35 cadets.[22] The demise of the military system was a combination of low enrollment and a merger. Four years later the school merged with the nonmilitary Louisiana State A&M College, and the military school was no more.

Florida Agricultural College

Another Morrill Land Grant Act college that ended its military status because of a merger with a nonmilitary school was Florida Agricultural College, which was established in 1884 and transitioned to a military school in 1887. The college was coeducational, with eighty-one male cadets in 1892. It was said that the cadets and faculty had an exceptional spirit of support for the military program, but in 1905 that came to an end when the Florida legislature organized the University of Florida through the merging of the college with the nonmilitary Petersburg Normal and Industrial School; the East Florida Seminary, which had ended its military school aspect in 1903; and the South Florida Military College.

North Georgia College

Georgia was the home of three military schools, all established under the Morrill Land Grant Act and considered branches of the University of Georgia. The first was North Georgia Agricultural College (North Georgia College), established as a civilian coeducational institution and opened in January 1873. From the very early years the college conducted military training, but it remained organized as a civilian institution. The initial faculty included B. Palmer Gaillard as professor of mathematics, military tactics, and civil engineering. Enrollment in the initial years was about ninety-five males and thirty-five females. The military training was so enthusiastically received that the male students petitioned the college president to allow them to wear uniforms. In 1876 the War Department issued the school rifles and equipment to support that training and in 1877 detailed an officer.[23]

The first three years of the school's catalogs reflect a desire for a uniformed male student body, and a VMI graduate became president in 1877. The military school concept was abandoned in that year, but military training was retained for male students. In 1904, David C. Barrow, a University of Georgia graduate, became the new college president. He had nothing in his background that would lead one to believe he would convert the school to a full military format for its male students. The new commandant in 1905 was Capt. E. J. Williams, who was described as "one of the most valuable officers that the institution has ever had." It appears that he designed a plan to provide officers for the Georgia National Guard and was the guiding influence in the school's conversion. The school now required a system of discipline "military in nature," in which "cadets [wore] uniforms at all times during the school term"[24]

The work of Captain Williams received high praise from the board of trustees in 1908 upon his reassignment, and the board stated that the "life of a student at this institution [North Georgia Agricultural College] very closely resembles the life of a cadet at the U.S. Military Academy." Unfortunately, the challenges of cadet discipline became an issue from at least 1913 to 1921, and as a result, in 1923 Prof. Marion D. DuBois rose to the office of president of the college. The Princeton University graduate acted quickly to end "rigid military discipline." By 1925 the alumni and the student body had mobilized against the president over their displeasure with the new change of direction for the college. First, a resolution of alumni condemned the president, and second, a student petition was made to the board of trustees to change the president. DuBois resigned, despite a vote of confidence from the board, expressing his sentiments: "The citizens of Dahlonega desire a strict military college here. I do not approve of strict military training in an institution of college rank. Under conditions as they now exist, I do not feel that I care to remain at the head of this college."[25] The next year, North Georgia College resumed its military college character, and it has retained it since, with a corps of cadets of over seven hundred in 2014.

South Georgia Military and Agricultural College

The South Georgia Military and Agricultural College in Thomasville was established in 1878, a few years after North Georgia College. Officially, this school, like North Georgia, was a branch of the University of Georgia. Enrollment in 1885 was eighty-five students. The school had a military character from the outset, but like North Georgia it was coeducational. Unlike North Georgia, South Georgia was considered a preparatory school for the University of Georgia, enabling entrance as a college freshman or sophomore there. The small military school operated until at least 1911, but it was likely overshadowed by the success of North Georgia.[26]

Georgia Military College

The final of the three Georgia land grant military schools, Middle Georgia Military & Agriculture College, opened in Milledgeville in 1880. Until 1893 the school was another feeder school for the University of Georgia. Cadets wore gray wool uniforms, and female civilian students were dressed in blue or white dresses as prescribed by the college. Among the influences on the early college were the second commandant, Richard Tyler Crawford, a Citadel graduate, and Confederate general D. H. Hill, West Point class of 1841, who was president 1886–89. Barracks were not provided for cadets until 1884. Enrollment in 1885 was an impressive 407 cadets. In 1900 the school was renamed Georgia Military College, and in 1931 a military junior college was added. In 1934 the school became all male, and it remained as such until 1969. By 2014 the college had evolved, with a coeducational military junior college of approximately 275 cadets and a military program of seventh to twelfth graders numbering 500.[27]

Mississippi Agricultural and Mechanical College

The Mississippi Agricultural and Mechanical College opened in 1880 with former Confederate general Thomas D. Lee, West Point class of 1854, as president. The school's initial regulations were taken from the Alabama Military Institute. General Lee was associated with the college, which became known as Mississippi State University, for nineteen years as president and another nine years as a member of the board. The disciplinary concept at the outset of the college was to "furnish the machinery of discipline."[28]

The school from the outset appeared to have addressed the military program not as a character or leadership development tool, but as a means to discipline the challenging student body. Reflecting on the 1902–3 school year, the second president of the college, John C. Hardy, voiced concern about the school's gaining a reputation of a place to send "bad boys" and went so far as to say, "If the state wants a reformatory . . . it must build one." In the first ten years of the college, the average enrollment was 314 students, but of the 1,832 students who had attended, only 104 graduated. Up until 1908 the rate of returning students was never more than 45 percent. The military system was not fully supported by barracks space, with only 660 of 847 cadets housed in 1899. The dismal record of academic success there suggests a low level of cadet esprit de corps and bonding with the institution and their military organization. This is further reflected in the general state of discipline over the college's first twenty years.[29]

John Kraus emphasized that the military system at the college was inconsistent, with the military department frequently being undermined by the faculty. The military department set policies, such as uniform regulations and attendance to formations, while members of the faculty were empowered and active in giving exemptions to individual cadets. This resulted in a situation described in 1912 as "a lack of co-operation between the Heads of the Departments and the Commandant's office [resulting in] great abuse of privilege during the year."[30] The type of privilege abuses included authorizing civilian clothing, class absences, and shirking of military duties.

To complete the picture of the military setting of the college in its first twenty years, a single set of uniforms was required of cadets and expected by policy to be worn at all times on campus. With labor and lab activities required of agriculture cadets, a single uniform was not enough to allow students to maintain military discipline and uniform standards. The new president, J. C. Hardy, refused to embrace the strict uniform regulations, and the faculty and president had a series of confrontations. With a questionable quality of cadets and a discipline system weakened by faculty interference, grave indications of a lack of traditional military school culture were reflected by acts contrary to the military school ethos. "The average cadet does not as a general rule, have that high sense of duty and justice to make a military system of control successful." Reflective of this were the common practices of lying, theft, and even vandalism. Related to the last was a telling comment in the 1899 annual report that described the damage from four fires on campus: "In the previous years the college had not a single fire involving any loss. This alone would indicate that the burning was incendiary."[31]

The beginning of the end of Mississippi Agricultural and Mechanical College as a military school occurred in 1912, when the senior class presented their demands associated with the college vice president's order restricting conversations with the few female students in the school. The seniors presented their petition at chapel and then walked out on the new president, George R. Hightower.[32] Hightower expelled sixty-one seniors for the walkout and thirty-five others for misconduct on an athletic trip to Birmingham. Some expelled cadets refused to leave campus, and a college-wide strike resulted, with only 325 of 808 students attending class.

Although the dismissals broke the back of the senior revolt and the student strike, the lax military system, for the most part, continued to decay. In 1920 the War Department ended the school's classification as a military college. Ten years later the student newspaper summed up the feeling of the end of the military environment at the college, in part saying that the students were finally "granted the privileges of gentlemen . . . and . . . the right to govern themselves."[33] Considering all the school's problems with cadet quality, barracks space, discipline, and uniforms, it is surprising that the military school aspect of the college lasted for thirty years. But the principal problem was that the military school ethos was never embraced. Honor, duty, and leadership were not aims of the system; the military was simply a method chosen to control young male farmers. The college went on to evolve into Mississippi State University.

Clemson University

Clemson Agriculture College, which officially became Clemson University in 1964, was established as a military college as part of the Morrill Land Grant Act, opening in 1893 in South Carolina. The school was patterned after the Mississippi Agricultural and Mechanical College founded thirteen years before, but the aim of its military system was far greater than just disciplining students. The all-male college cadets lived in barracks, except for a small number of "day cadets" from local families who were required to reside in the barracks at least six months of either their freshman or sophomore year. By 1915 the military spirit at Clemson had "developed and nurtured to the best possible extent."[34] In his 1978 dissertation, John Kraus questioned the discipline of the corps at Clemson based on the antics of their senior privates. However, historically this is common with other military schools, including VMI and Virginia Tech. At the college level there are many examples of those cadets not promoted beyond cadet private banding together in a fraternal group and trying their best to enjoy their senior year to the maximum extent possible.

The college catalog of 1924–25 stated the purpose of the military system: "Clemson is operated as a military school not for the purpose of making soldiers, but in order that students may learn important life lessons."[35] In the 1930s the cadets were uniformed in West Point style gray, with senior officers in riding boots and Sam Browne belts. The corps numbered 2,359 in 1940 and peaked in 1942 at 2,364 in the first year of US involvement in World War II. Clemson military men became well known for the quality of officer they made in World War II—especially cavalry and armor officers.

But World War II was actually the catalyst of the end of Clemson as a military school. After World War II, returning veterans and new-student veterans of the conflict changed the makeup of the student body. By 1949 the corps was back to 1,740, but it was only 55 percent of the student body. According to Kraus (1978), the school dropped its standards of corps discipline to reflect more of the life of a country gentleman or military fraternity and abandoned the pursuit of life lessons. Despite an enrollment of 1,583 cadets in 1954, the college felt that the changes underway, such as the enrollment of women that year, meant departure from the past, and in September 1954 the corps of cadets became voluntary. In 1955 a college-sponsored management study called for "the college to undergo normal expansion without being hampered by the military program."[36] The college opened in September 1955 without the military school requirements. Unlike the reaction of the alumni and cadets of North Georgia College in the 1920s and Virginia Tech in the 1970s–90s, the change did not result in a major backlash.

North Carolina College of Agriculture and Mechanic Arts

Two land grant colleges ended their adoption of military school requirements after a short period—one because of the efforts of the students, and the other, the actions of a new president. North Carolina College of Agriculture and Mechanic Arts was established in 1889 and transitioned to a military format in 1894. Between that year and 1906, cadets wore gray uniforms, marched to chapel and meals, drilled three hours per week, and conducted periodic field training. But the military system appears never to have been embraced completely by the students. In the fall of 1905 the senior class conducted a strike, and the president cut off senior privileges. "Petitions followed to adjust all matters pertaining to discipline and libertics." The results of these actions united the trustees with the students to endorse military training but "not allow the college to become a military school like Clemson Agricultural College. Nor would they permit the administration to enforce outmoded military codes."[37]

University of Tennessee

The history of the University of Tennessee goes back as far as 1794, when it was founded as Blount College. In 1875, after being designated a land grant college in 1869, the school adopted a military format. The cadets were organized into a regiment, uniformed, required to hold drills and parades, and assigned a commandant. According to Michael Dennis, "The school offered dashing uniforms and taste of Confederate military glory to young southerners bred on the lore of Beauregard and Lee. . . . Although the paramilitarization of the campus suggests much about the grip of the Lost Cause on southern culture in practical terms, military discipline was an effective method for controlling adolescents ill prepared for university studies."[38]

Charles Dabney, who became president of the University of Tennessee in 1887, was a Hampden-Sydney graduate as well as alumnus of the University of Virginia and a college in Göttingen, Germany. Dabney was no stranger to the question of military school adoption to secure funding under the Morrill Land Grant Act. He had been a strong faculty supporter in Walter Barnard Hill's fight as president of the University of North Carolina to keep it from becoming another land grant military college. Hill's view was that the Morrill Land Grant Act was intended to create a balanced approach to liberal arts and industrial training and not meant to establish military colleges. Dabney backed that point of view with the example of Robert E. Lee's postwar leadership at Washington College. He took on the military school structure as being "inconsistent with true university life and work." His elimination of the military aspect was like the pattern at other land grant military colleges but unlike the reaction to change at North Georgia, Texas A&M, and Virginia Tech. An alumnus observer noted, "We have come to a turning point . . . where it [University of Tennessee] has outgrown the . . . barracks system of a military academy and we must either give room for expansion or congestion will result."[39]

Eight of the twelve land grant military colleges' military requirements ended in six ways. Two were by means of changes in college leadership, one through the change from a traditional military male enrolled student body, two from a lack of instructional support in order to maintain the military ethos, one through the desires of the cadets, one from consolidation or reorganization with nonmilitary colleges, and one may simply have been overshadowed by a more successful military college. Of the remaining four, three continued as military colleges, and one evolved into a military junior college.

Grand Army of the Republic

The historical image of the South as the principal home of the military school calls for a closer examination when one is faced with the fact that New York state has had seventy-eight military schools in its history and California, ninety-three. The period between the close of the Civil War and 1898 saw the military school format continue not only to expand in the northern United States, but also to spread west. During that period fifty-one schools were established or transitioned to military format in the north, thirty-three in the Midwest, and eight in the western United States.[40]

The Grand Army of the Republic is a generic name for the Union Army as well as the Union veterans' organization of the post–Civil War era. It was symbolized by the May 1865 victory parade of the Union armies of the east and west down Pennsylvania Avenue in Washington, DC. The number of Union Army veterans numbered close to 1.8 million. Like the Lost Cause of the South, former soldiers in blue and their legacy impacted the military school movement in the northern, midwestern, and western states.

Charles Jefferson Wright was one educator who returned from military service and incorporated the military into his educational philosophy. Colonel Wright was a Holbart College graduate who served in the 16th New Hampshire Infantry. He was wounded at both Fort Fisher and at the siege of Petersburg and ended the war as a brevet colonel. In 1869 he was first the principal at New York's Peekskill Military Academy, which had been a military school since 1857. When he took charge of that school there were only seven cadets enrolled. He turned the school around and charted a course that enabled it to operate until 1968. In 1887 Wright left the school after eighteen years and founded New York Military Academy with 78 boarding cadets. Enrollment in 1980 was 516 cadets, and the school was still functioning with more than 141 cadets in 2014.[41]

In 1894 Wright left New York Military Academy and took over as president of Matawan Military Academy in New Jersey. Matawan was another well-established military school, having been founded in 1836 and transitioned to a military format in 1857. An article in *Printer's Ink*, an advertiser's journal, in 1902 described Wright as a man who could take

New York Military Academy Corps of Cadets, 1889. (Courtesy New York Military Academy)

a school near ruin and resurrect it: "Colonel Wright has been identified with several military schools since he left the service at the close of the Civil War, and has more than once used advertising as a stimulant for run-down establishments."[42]

Wright had a financial interest in Cayuga Lake Military Academy in New York. He was listed as the point of contact for that school from 1891 to 1896. Churchill Military Academy, founded in 1843, was another school that would fall under Wright's influence. It changed names to St. John's School in 1869 when Reverend J. Beckinridge Gibson took over the school. Gibson died in 1899, and enrollment waned quickly. Colonel Wright remodeled the school and advertised from Boston to San Francisco. St. John's School continued operation until 1948.[43]

In 1900 Wright founded the New Jersey Military Academy in Freehold. It appears he maintained a close association with the Matawan Military

Col. Charles Jefferson Wright, founder of New York Military Academy and New Jersey Military Academy and head of Peekskill and Matawan Military Academies. (Courtesy New York Military Academy)

Academy, as he died in Matawan in 1910. Prior to his death, Colonel Wright founded two military schools and was associated with the leadership of three others. New York Military Academy remains his legacy.[44]

Capt. Joseph S. Rogers enlisted as a private in the 2nd Maine Infantry and participated in the initial battles of the Civil War, including First Bull Run and the Peninsula Campaign, and he received a head wound at Second Manassas, which took him out of service. He returned home to recover, completed his education at Bucksport Seminary in Bangor, Maine, and reenlisted as an officer with the 31st Maine Infantry. Rogers fought in the final battles of the war, including Petersburg and Appomattox, and received a brevet promotion from captain to major for bravery. In 1877, having stayed in uniform with the regular army after the war, he resigned and founded Michigan Military Academy.[45]

Another example of a returning veteran was Peter De Graff, who was first a principal at several schools before 1868, when he established the DeGraff's Military and Collegiate Institute. DeGraff had been at the Canandaigua Academy of New York before the war as a teacher for four years and principal 1857–61. Enlisting in 1861 in the 33rd New York Infantry Regiment, he fought at Antietam, Fredericksburg, and Chancellorsville before mustering out of the army and returning to education.[46]

The DeGraff's Military and Collegiate Institute started with 25 cadets, and in three years the cadet corps numbered 280. The school's stated goal was "to prepare boys to become healthy, intelligent Christian men." Cadets wore uniforms patterned after West Point's and took academics that were considered "correct preparation for college." Other veterans included Wilbur F. Miles, a Binghamton Commercial College graduate, who returned to Deposit, New York, to take charge of the Deposit Military Academy after recovering from wounds received while serving with the 13th New York Heavy Artillery.[47]

Northern Educators

In part, returning Union veterans explain the expanding popularity of military schools outside the southern United States, but the popularity was much wider. The *School Journal* was a northern-oriented education periodical (New York, Chicago, and Boston) that in 1903 described the educational setting for the post–Civil War military school well: "An increasing number of secondary schools are adopting the military system as a medium for supplying in the education of boys that training which cannot be derived from books alone. The important question is, of course, whether or not the soldier's discipline is an effective means to the desired end. Observation shows the military system, when properly used, fulfills better than any other mode of school discipline, the condition of being to character what study is to mind and exercise to physique."[48] This helps explain the adoption of the military school concept by those outside the ranks of veterans.

A good example of the nonveterans who started schools is Henry Harrison Culver, who in 1894 founded Culver Military Academy in Indiana. He was the son of an Ohio farmer, and his background was largely occupation in the stovepipe business. The school stemmed from his desire to provide high-quality education for young men, and perhaps the school's format was inspired by his feeling for his brother, Litellus Culver, who was killed in the Civil War. Henry Culver, as a young man of fifteen in 1855, started working with his other brother, Wallace, in Saint Louis and remained in that city until 1881. A prominent educator and a military school in Saint Louis could have been factors that influenced Henry Culver's future establishment of a military school.[49]

Saint Louis had a very influential educator during that period. Edward Wyman had conducted Wyman's English and Classical High School, a military school, from 1843 until 1853. At his second school, City University, which opened in 1861, the more than six hundred cadets, organized into three full companies, were known for their uniform appearance and skill at drill under arms. Supported by the school's drum and fife corps, they marched behind the school colors, emblazoned with *Palma non sine Pulverre*, meaning "victory not without toil," instead of the US flag, since many of the young men were attached by birth or sentiment to the Southern cause. The school gained much attention not only on parade, but also when Wyman's loyalty to the Union was questioned. In a gesture of loyalty to their superintendent, all the cadets, including those with Southern leanings, unanimously agreed to parade behind the US colors, which afterwards was explained as a demonstration of their "loyalty to and love of Professor Wyman."[50]

Another example of a noted northern educator was Prof. Orvon Graff Brown, a graduate of Mount Union College, the University of Cincinnati, and the University of Denver. Brown founded Miami Military Institute of Ohio in 1894. His philosophy matched that of the emerging military school culture, as he sought to shape the diverse talents of his cadets in a challenging moral, physical, mental, and social school structure. Brown's background gives few clues to his selection of a military format. His childhood near Gettysburg battlefield and his familiarity with the four military schools operating in Ohio at the time could have been factors that led him to found his school in a military style. There were two military schools in Cincinnati at the time: the Ohio Military Academy, founded in 1832, and the Ohio Military Institute, founded in 1833.[51]

Two other northern educators who adopted the military school format were Stephen G. Wentworth and John Michael Birch. Wentworth was born in Massachusetts and was home-schooled. He was a businessman in Virginia for six years, then moved to Missouri, where he was a public administrator in Lafayette and president of Farmers Bank and later Morrison-Wentworth Bank in Lexington. Upon the death of his son, William, he opened Wentworth Military Academy as a memorial. Neither Stephen Wentworth's education nor his occupation gave clues to the reasons for the military nature of his school. Lexington was the site of two Civil War battles. The graduation address of 1899 at Wentworth stated: "Wentworth is a boy's school and therefore a military school. A boy's school without the military would be as preposterous as a military school without boys in it."[52]

John Michael Birch, graduate of Washington and Jefferson College in Pennsylvania, was headmaster at the Linsly School in West Virginia 1874–81, 1889–90, and 1910–11. In the interims he was superintendent of public schools in Wheeling, West Virginia, and a diplomat in Japan. He converted the Linsly School to military format in 1887 and enrolled about 114 cadets through the 1890s. Although the school dropped its military school format in response to board of trustees' action in 1907 over alumni protests, the military format was returned when Birch took the school's leadership again in 1910. The school continued as a military school until 1979.[53]

One of the most dramatic and least known expansions of the military school format came about through the leadership of Enoch Henry Currier. As a young man Currier suffered an injury that caused loss of sight in one eye and health problems and delayed his education. He became a professor at the New York Institute for the Deaf because of his early interest in deaf education through his interactions with Dr. Harvey P. Peet. Currier taught all grades at the institute over a twenty-year period and was credited with the 1884 invention of the duplex conical hearing tube. Considered an authority on the subject of defective hearing, he was selected as the principal of the institution in 1883.[54] Shortly thereafter, he transitioned the school to become the only deaf military school in history. The deaf cadets were initially organized into three, then four companies and a cadet band. Male faculty wore army uniforms, and Currier functioned militarily as the school's colonel, the military tactics instructor as the school's major, and cadets as company officers. The corps dressed in West Point–style uniforms and marched to class and meals. The cadets' silent drill was heralded from New York to Virginia.[55]

New York School for the Deaf, organized as a battalion of three companies and a band, preparing to form up for parade, 1900. (Gallaudet University Archives)

Currier explained this concept: "[The intentions were not] to make soldiers of the boys. That is a matter of secondary importance. We do not even teach them marksmanship. The military drill was introduced because of its gymnastic value and in order to inculcate into the boys the lesson of discipline. . . . They are proud of their proficiency and compliments they have received from army experts." Currie felt that the first step in education was obedience and self-control. For that purpose he felt nothing was as effective as a military environment. A further benefit was the military's part in "character building, the basic principle of all true education, [which] is most advanced when obedience to authority is most implicit." The school continued its military format with its cadets until 1952.[56]

African American Military Schools

Although military training was conducted at several African American colleges, two schools adopted a military school format. Both were established in the period between the end of the Civil War and the start of the Spanish-American War. Both schools were located in Virginia and were founded by white former Union Army officers. Neither officer was a graduate of one of the nation's military schools, but their service in the Grand Army of the Republic influenced their choice of educational formats.[57]

The first school was established by Gen. Samuel C. Armstrong as Hampton Normal and Agricultural Institute in 1868. General Armstrong was a graduate of Williams College and had served with the 125th New York Infantry at Gettysburg prior to commanding the 9th US Colored Infantry Regiment in the final battles of the Civil War. By 1872 the school had transitioned to a military format for its male students. Male cadets drilled at five in the morning, followed by breakfast and a day devoted to academics. Cadets were uniformed in military blue and marched to the noon meal. In 1901 the male students were organized into a battalion of six companies: two companies of day students, three companies of boarding students, and one company of boarding Native Americans. The barracks were regularly subject to room inspections. The battalion was supported by a twenty-five-cadet band. There were also extensive periods devoted to religion and Bible study, with church attendance required daily and each cadet assigned a specific seat.[58]

Army officers from nearby Fort Monroe supported military training, although their repeated requests for rifles were consistently rejected. In 1892 school disciplinary proceedings included conduct courts-martial, fines, extra drills, and confinement. The cadet routine included two

Hampton Institute cadets, 1899. Hampton Institute was one of only two African American military schools. The corps of cadets consisted of a band and six companies, one of which was made up of American Indians. (Library of Congress)

company drill periods daily, inspections, and morning and evening formations. The next year, General Armstrong died and was buried on campus. The military system appears to have continued until sometime after 1901. The Hampton legacy was that 90 percent of its graduates, including Booker T. Washington, became teachers and played important roles in the historical black Hampton University and the education of African Americans.[59]

The second African American military school was founded as St. Emma's Military School in Powhatan, Virginia. The school opened in January 1895 as the St. Emma Industrial and Agricultural College. In its first eight months there were only nineteen male students. Military requirements were always an integral part of the school. Possible influences include the school's benefactors, Col. and Mrs. Edward Morrell of Philadelphia. Colonel Morrell was a Civil War veteran who enlisted as a private and rose to lieutenant in the 5th New York Cavalry. He was a Union College graduate, and after the war he was a professor of Greek and Latin at Fort Edward Collegiate Institute in New York. By 1898 he had been promoted to brigadier general in the Pennsylvania militia. Another influence was

Saint Emma's Military School, 1925, showing the cadet battalion in formation in front of the Claver Hall barracks and the chapel. (Courtesy Sisters of the Blessed Sacrament Archives)

the success of the Hampton Institute. That school's popularity, along with that of Tuskegee Institute in Alabama, made it a model for educational efforts with African American youth. The third influence was the Catholic Church's successful adoption of the military school format.[60]

By 1907 the cadet corps of St. Emma's was described as being clothed in regulation cadet uniforms and drilled regularly. Also, the corps by that time had a brass band as part of its organization. The enrollment continued to grow, with 150 cadets in 1929. The curriculum, at least up until the late 1940s, was focused on cannery, farming, equipment repair, engineering, accounting, and management. As the only other agricultural school in Virginia, St. Emma's was second only to Virginia Polytechnic Institute and State University (Virginia Tech).[61]

In 1919, after a War Department inspection, St. Emma's gained federal funding through inclusion of an ROTC unit. In 1947, under the Holy Ghost Fathers, the trade school became a more academically oriented military high school in which college preparation and military bearing were central. In keeping with this, the school officially became St. Emma's Military School. With Father Egbert J. Figaro leading the school, enrollment

Saint Emma's Military School observance of the seventy-fifth anniversary. Escorting Bishop John J. Russell of Richmond in uniform is Father Egbert J. Figaro, commandant and principal 1958–72. (Courtesy Sisters of the Blessed Sacrament Archives)

Saint Emma's Military School Corps of Cadets, organized with a band and four companies, during the seventy-fifth anniversary observances, 1963. (Courtesy Sisters of the Blessed Sacrament Archives)

peaked at 370 cadets in 1964. The corps included African Americans from throughout the nation as well as cadets from Africa and the Caribbean. A demanding new cadet training program required uniform excellence in their gray jackets with white caps and trousers. Equipped with M1 rifles, the cadet battalion included a band. But by 1972 declining enrollment forced the school to close, not because of antimilitary sentiment, but as a result of school desegregation and an increasing number of African American families choosing to enroll their sons in integrated schools.[62]

Desegregation is a big part of the military school story. After the Civil War, military schools were not welcoming to African Americans. The first of the military schools outside of Hampton Institute to enroll African Americans was West Point. Between 1870 and 1889, twenty-three African Americans enrolled at West Point, of whom only three graduated: Henry Flipper, class of 1877; John H. Alexander, class of 1887; and Charles Young, class of 1889. It was not until after World War I that more were admitted, and not until Gen. Benjamin O. Davis Jr., class of 1936, that another African American graduated. Between 1932 and 1947 twenty-one African American men were admitted, of whom seventeen would go on to graduate. All the black West Point pioneers came under unfair physical and psychological abuse, which was more often than not the case at other military schools as well. But in 1948 at West Point attitudes in the corps started to change, in part because African Americans were represented in all four classes.[63] This was the same year in which President Harry Truman ordered desegregation in the US Armed Forces. In 1979 Vincent K. Brooks, class of 1980, at the time of this writing a general and commander of US Forces in Korea, became West Point's first black cadet brigade commander.

Norwich in Vermont graduated their first African American in 1916, Maj. Harold "Doc" Martin. During World War II Martin was the director of the ground schools in support of the famous Tuskegee Airmen. He was killed in an aircraft accident in 1945. The first African American cadets at Virginia Tech numbered four and enrolled between 1951 and 1957. All were day cadets not permitted to live in the cadet barracks. Charles Yates, class of 1958, was the first graduate. True integration of the cadet corps came with Cadet James Whitehurst, class of 1963, who demanded to join his company living in the barracks and did so as a junior in 1961. That year he rejected the university president's request that he not attend Ring Dance, the most important event for junior cadets. Upon stepping onto the dance floor, he was cheered by the corps. He later became the first African

American on the board of visitors. In 1985 Calder Derek A. Jeffries, class of 1986, became the school's first black regimental commander.[64]

The first African American at the US Coast Guard Academy was enrolled in 1955, and the first graduate was Cmdr. Merle James Smith, class of 1966, who later received a law degree and served in that field as a coast guard officer during his twenty-three years of service. Between 1936 and 1945 three African Americans were admitted to the US Naval Academy. The third midshipman, Wesley Brown, became the first graduate, doing so with the class of 1949. Brown was a former army enlisted man who had served in World War II; after being commissioned in the navy, he served in the Korean and Vietnam Wars.[65]

At Texas A&M the first African American cadet was enrolled in 1964, and in 2012 Cadet Marquis Alexander became corps commander. The first African American cadet at the Citadel was enrolled in 1966. Cadet Joseph Shine was the second enrolled in 1967 and graduated as a cadet officer on regimental staff with the class of 1971. As a sign of the Citadel's adjustment to integration, an informal boycott was conducted in Shine's support after a Charleston tavern refused to serve him. The first African American cadet enrolled at VMI entered the Rat Line in 1968. In 1982, Cadet Darren McDew became the institute's first African American regimental commander.[66]

The desegregation of public schools starting in 1954 impacted secondary military boarding schools, with some schools benefiting from "white flight" in the late 1950s and early 1960s. However, no new military schools were opened to take advantage of that situation, and a survey of boarding secondary military schools in 2014 reveals that they are diverse: Fork Union Military Academy, Virginia, had a black enrollment of 20 percent; St. John's Northwestern Military Academy, Wisconsin, 7 percent; and Riverside Military Academy, Georgia, 12 percent. The successful integration of the military colleges and federal military academies beginning in 1932 and extending through the late 1960s was followed by that of military private schools. Those early cadets paved the way for a succession of African American cadets who rose to command the cadet corps of most of the military schools in the United States.

Of Sabers and Scripture

The first military school sponsored by a religious denomination appeared in 1834, but the majority of military church schools were established after the Civil War. Of the approximately 842 military schools that have operated in the United States from 1802 to 2014, at least 140, or 17 percent, have been affiliated with various Christian denominations, including the Catholic (45), Episcopal (46), Presbyterian (25), Baptist (9), Methodist (6), and other denominations (9).

The association between religion and military school culture goes back to the first two military schools. Under the leadership of Thayer, West Point required weekly church attendance of cadets and "enforcement [of] rules against lying, stealing and other irregular or immoral practices." American higher education was heavily influenced by religious activities and religious faculty during the eighteenth and a good portion of the nineteenth century. Before 1850, revivalism was a common event at most colleges in the United States. But outside the walls of military schools, the influences of religion decreased significantly prior to the Civil War as a result of two influences initially. The fraternity movement and a collegiate society that valued success over salvation pushed religion, as a central figure of student life, to the periphery.[1]

Secondary free or public education in the late nineteenth century saw a wane in heavily Protestant religious influences in faculty and textbooks and a decline in religious exercises for two reasons. The initial cause was the objections of other denominations, particularly Catholics and Lutherans, and the second was the growth of bureaucratic specialization and standardization and, later, conflicts associated with scientific concepts such as Darwinism. This decline is in contrast to the experience of military schools, which included mandatory church attendance at the federal military academies until 1972, and a majority of secondary military boarding schools, which still require attendance.[2]

Norwich University was established with a chaplain on the faculty, and from its earliest days cadets were required to attend church services on Sundays and encouraged to read the Bible during their leisure hours afterward. The Citadel enforced church attendance on Sunday from its first year. Through the personal involvement of Francis Smith and local clergy at VMI, religious training became an important feature for cadets there.[3]

John Kraus reflects on the similarities in military colleges or schools and church-related colleges, with both having a "strong commitment to purpose, values and to tradition." He viewed the military college culture as expressing the messages of the church college in a modified form. As an example he cited the Virginia Military Institute's 1974 planning document, which explained cadet life in terms of "conviction that the development of honor, integrity, a mature sense of values, responsibility, self-respect and physical wellbeing are essential to the development of man's ability." Norwich University's annual report of 1976 states, "The corps is the instrumentality by

Cadet First Sgt. Douglas MacArthur, West Texas Military Academy (Texas Military Institute) class of 1898, was also captain of the school's football team.

which a Norwich education develops honor, patriotism, self discipline, respect for constituted authority, character and leadership."[4]

Rev. Allan L. Burleson was the first headmaster of West Texas Military Academy (later Texas Military Institute). Among his cadets was Douglas MacArthur. In 1896 Reverend Burleson summarized his views on military schools in a speech in San Antonio on Education Day: "Military training in the American schools is a great moral agency for good which tends to give us better sons, better neighbors, better citizens." More than one hundred years later, in 2002, the president of Randolph-Macon Academy summarized the church military school: "The Church . . . does a masterful job of blending three cultures: military, religion [sic] and academic. In all areas, including school literature and chapel services, no student would ever have to wonder, 'Is this a church-related school?' In ministry of teaching, the environment reflects the United Methodist Church values. This is a military school, but sees the military as a part of the methodology of enforcing support, and discipline, and to give children structure."[5]

Episcopal Church

The Episcopal Church embraced the military format to the greatest extent by far, with a total of forty-six schools being established as or transitioned to military schools by Episcopal clergy. In addition to these, other military schools characterized as nondenominational, such as Valley Forge Military Academy and Culver Military Academy, have found they were best supported by the tradition of Episcopal chaplains. In 1901 the Episcopal Almanac and Parochial List included fifty-nine schools for boys. Of those, eighteen were military schools, reflecting almost 31 percent of the boys' schools.[6]

The first Episcopal-affiliated school was founded by the Reverend Charles William Everest, the minister of both the Grace Church in Hamden and St. John's in North Haven, Connecticut (Episcopal Church Diocese of Connecticut, 1847). In 1843 he established a school in Centerville, New Haven County, and completed its seminary building on Main Street in May 1844. The rectory school's enrollment grew quickly from nine cadets the first year to twenty-one at the end of five years and sixty-five by 1853.[7]

Gray West Point–style uniforms were adopted from the early days of the school, and later Maj. James Quinn and Col. John Arnold acted as military instructors. The cadets were armed with lances rather than rifles and were organized as a military company with a cadet captain, lieutenants, sergeants, and corporals. Drill was conducted two to three times

Cadet officers of the Rectory School about 1855. (Courtesy Hamden Historical Society)

weekly, and Reverend Everest said he was "surprised to see how much can be done in so short a time by systematic work, and this not to interfere with the school duties, but aid them." Because of the military trappings, the school was often called Everest Military Academy. For several years after Reverend Everest's death in 1877, the school operated as the Atlantic Military Institute.[8]

Before the Civil War the Episcopal Church established seven other military schools with two Yale graduates and a West Pointer as founders: Burlington Military (Academy) College, Burlington, New York (1846); St. John's Classical and Military School, South Carolina (1856); De Veaux School, Niagara Falls, New York (1857); Starr's Military Institute, New York (later known as Port Chester Commercial, Collegiate, and Military Institute; 1854); Vermont Episcopal Institute, Burlington (1854); and St. Thomas Hall Military Academy, Mississippi (1844). In addition, two Episcopal schools converted to a military format during this period. St. Timothy Hall Military Academy, Catonsville, Maryland (also known as Catonsville Military Institute) was established in 1845 and transitioned to military in 1854, and Highland Military Academy, Worcester, Massachusetts, was established in 1856 and transitioned about a year later. At a time when there was an explosion of military schools in the South, the Episcopal

Church was founding them primarily in the northern states. This would continue with two additional military schools established or transitioned in Connecticut and New Jersey during the war, including Cheshire Academy, established in 1794 and transitioned to military in 1862.

Even with those additions, the real cement that brought the military format to become a central characteristic of Episcopal schools would emerge after the war through the philosophy of William Augustus Muhlenberg (1796–1877) and the activity of James Lloyd Breck (1818–76). According to Rev. Walter "Chip" Prehn, Muhlenberg established the standard for Episcopal education. This standard was not military, but it blended with the concept of a military school well. For example, Muhlenberg felt that the education of the time was focused on development of "men of letters" and neglected the most important "formation of Christian character." That focus on character was not the only theme blending the military and Muhlenberg's school concept. In terms of school discipline, Muhlenberg was a reformer and emphasized "moral education [with a] steady and firm system" rather than that driven by corporal punishment. Both character and discipline easily translated into the military school, with character central and a disciplinary system supported by demerits and associated punishments.[9]

James Lloyd Breck was one of the Muhlenberg school–style crusaders. Prehn calls him "one of the most inspiring and romantic figures in the entire history of the Episcopal Church." His years of work as a missionary with American Indians supports this image well. Breck's writings take the Muhlenberg philosophy a step closer to the military school concept with its "spiritual warfare and Christian knighthood."[10] Breck worked closely with Henry Benjamin Whipple, the first bishop of Minnesota. They established the Shattuck School in 1858 and St. Mary's School for Girls. The Shattuck School started military drill in 1866, the same year Beck went to California, and transitioned to a full military format in 1904. St. Mary's became a boy's military school from 1884 to 1915. Bishop Whipple also established St. James School as a military school in 1901, and it merged with the Shattuck School in 1964.

The influence of Breck's philosophy encouraged those schools' conversion to a military format years after he had gone. But in California he started a military school in his first years in San Mateo County: St. Augustine Military College in Benicia in 1867. It joined the nearby St. Matthew's Military School, founded by Rev. Alfred Lee Brewer in 1865. Reverend Brewer's cadets, uniformed in West Point gray, were organized into two small cadet companies and a fourteen-cadet band. The school conducted field training for five or six days in May and was equipped with rifles and a Parrott field gun.[11]

The Episcopal Church would continue some orientation toward military schools well into the twentieth century (table 9.1). Prominent among those schools were Racine College in Wisconsin, founded in 1852 and transitioned to military in 1899; Bishop Scott Academy of Oregon, founded in 1852 and transitioned to military in 1887; and Sewanee Military Academy in Tennessee, founded in 1868 and transitioned to military in 1908.

The Bishop Scott Academy in Oregon transitioned under the leadership of J. M. Hill, a Yale graduate. The school received high praise in a survey of military schools of the Pacific Coast: "The academy is more than a preparatory school; for it fits young men for various callings of life . . . , graduates of this school filling with honor the highest positions of trust and responsibility in the power of corporations or communities." Apparently the school prided itself in the degree of responsibility given its cadet officers, as they were "required to assist the faculty in enforcement of rules and regulations, and in so doing [gain] a high sense of personal honor and splendid esprit of corps." This school operated until 1904 and was replaced by the Hill Military Academy, with Dr. Hill again at the helm in 1901. The nonsectarian Hill Military Academy operated successfully for over sixty years, closing in 1962.[12]

The final two military schools were the S Bar H Ranch School of the Wyoming Episcopal diocese, which converted to a military format in 1938,

St. John's Military Academy cadet battalion about 1920. (Courtesy St. John's Northwestern Military Academy)

Table 9.1. Episcopal Military Schools

Opened or *Transitioned to Military*	School	Location
1843	Everest Military Academy	Connecticut
1844	St. Thomas Hall Military Academy	Mississippi
1854	Catonsville Military Institute/St. Timothy Hall Military Academy	Maryland
1854	Vermont Episcopal Institute	Vermont
1856	St. John's Classical and Military School	South Carolina
1857	De Veaux School	New York
1857	Epworth Military Academy	Iowa
1857	Highland Military Academy	Massachusetts
1858	Kenyon Military Academy/College Preparatory School	Ohio
1858	Starrs Military Institute (later known as Port Chester Commercial, Collegiate, and Military Institute)	New York
1862	Cheshire Military Academy	Connecticut
1865	St. John's Military Academy	New Jersey
1866	Cambridge Military Academy	Maryland
1866	St. Matthew's Military School	California
1867	St. Augustine Military College	California
1869	Jarvis Hall Military Academy	Colorado
1870	Worrall Hall Military Academy	New York
1875	St. James Military Academy	Missouri
1879	St. John's Military School	New York
1879	St. Paul's Military School	New York
1880	Croton Military Academy	New York
1881	Bordentown Military Institute	New York
1884	St. John's Military Academy	Wisconsin
1886	Kemper Hall Military Academy	Iowa
1887	Bishop Scott Academy	Oregon
1887	St. John's Military School	Kansas
1888	Northwestern Military Academy (church assumed control, 1911)	Illinois

1889	St. John's College	Maryland
1890	St. Alban's Academy	Kentucky
1890	St. Alban's School	Illinois
1891	Porter Military Academy	South Carolina
1892	St. Austin's Military School	New York
1894	St. Mary's Hall	Minnesota
1895	Howe Military School	Indiana
1897	Burlington Military (Academy) College	New York
1898	Kearney Military Academy	Nebraska
1899	Racine Military School and College	Minnesota
1900	Harvard Military School (purchased by church, 1911)	California
1901	St. James School	Minnesota
1904	Shattuck Hall Military School	Minnesota
1907	Hitchcock Military School/Military Academy	California
1908	Garden Military Academy	Texas
1908	Sewanee Military Academy	Tennessee
1929	Midwest Junior School	Illinois
1938	S Bar H Ranch School for Boys	Wyoming

and the Northwest Military and Naval Academy, a school established in 1888 and procured by the Episcopal Church in 1941 after it had moved from Illinois to Wisconsin. The church in 2014 could boast of no fewer than four prominent military prep schools: Texas Military Institute (TMI), established in 1893; St. John's Military School in Kansas, established in 1887; St. John's Northwestern Military Academy, Wisconsin, established in 1884; and Howe Military School in Indiana, transitioned to a military school in 1884.

The Texas Military Institute in San Antonio is a military school that became coeducational in the 1970s. The school evolved into an Episcopal prep school with two student bodies attending the same classes, one military and one civilian. Between 2000 and 2015 the enrollment for cadets grew from 86 to 155. St. John's Northwestern Military Academy in Wisconsin is the consolidation of St. John's Military Academy, established in 1884, and Northwestern Military and Naval Academy, established in 1888.

The two schools combined in 1995 and took the St. John's Military Academy campus in Delafield, Wisconsin. The school is all male, grades seven through twelve. Enrollment has grown considerably in recent years, from 104 to 120 in the 1990s and to 294 in 2010.[13] Howe Military School in Indiana is a coeducational boarding school with enrollment of around 200. The school has both a high school and grades five through eight. St. John's Military School is a boarding school with grades six through twelve and an enrollment of 193 in 2010.

Catholic Church

The Catholic Church has contributed forty-five military schools to the military school movement since 1847 (table 9.2). The church's association with military schools in the United States came about by a much slower process in the nineteenth century than did that of the Episcopal Church. This was likely a result of the early makeup of the US population as well as the timing of the surge in immigration of Catholics, particularly those from Ireland. St. John's Academy was founded in 1833 in Alexandria, Virginia, when the Catholic education movement in the United States was in its

Table 9.2. Catholic Military Schools

Opened or *Transitioned to Military*	School	Location
1847	St. John's Academy	Virginia
1880	St. Leo Military Academy/College	Florida
1883	Sacred Heart Military Academy	New York
1886	All Hallows College	Utah
1891	La Salle Institute	New York
1894	Christian Brothers Academy	New York
1895	St. Aloysius Academy	Pennsylvania
1895	St. Emma Military School	Virginia
1898	La Salle Military Academy (Clason Point Military Academy)	New York
1898	Xavier Military High School	New York
1901	Marist Military School for Boys	Georgia
1902	Benedictine Military School/High School	Georgia
1903	St. Joseph's Military Academy	California

1905	St. John's Military Academy	California
1905	St. Thomas Military Academy	Minnesota
1906	College of St. Thomas	Minnesota
1908	Stella Niagara Cadet School	New York
1910	De La Salle Military Academy	Missouri
1911	Benedictine High School	Virginia
1913	St. Mary's College (Institute)	Texas
1915	St. Aloysius Academy and Military School	Ohio
1915	St. John's College High School	Washington, DC
1917	Cretin (Military) High School	Minnesota
1918	Hall of the Divine Child	Minnesota
1918	Spring Hill College	Alabama
1922	Linton Hall Military School	Virginia
1922	St. Clement Military School	Massachusetts
1923	Marymount Military Academy	Washington, DC
1923	St. Catherine's Military Academy	California
1924	Mount St. Joseph Semi-military Academy	New York
1932	St. Joseph's College and Military Academy	Kansas
1935	Marmion Military Academy	Illinois
1935	Nazareth (Nasareth) Hall Military School	Ohio
1936	Bishop Quarter Junior Military Academy	Illinois
1936	Christian Brothers College Military High School	Missouri
1937	St. Joseph Junior Military School	Pennsylvania
1938	Moye Military School (Academy)	Texas
1938	St. Aloysius Military Academy	Ohio
1940s	St. Patrick's Military Academy	New York
1941	Barbour Hall Junior Military Academy	Minnesota
1941	Leonard Hall Naval Academy	Maryland
1942	Cardinal Farley Military Academy	New York
1942	St. Edward's Military Academy	Texas
1952	Nazareth Hall Cadet School	New York
1955	Sacred Heart Military Academy	Wisconsin

infancy. Mob violence against Catholics and their associated immigration occurred between 1834 and 1854 in Massachusetts, resulting in the death of a nun and the destruction of a convent and two churches. In 1842 riots also occurred in New York, and in 1844 a Philadelphia riot caused thirteen deaths and the burning of five Catholic churches.[14]

But Alexandria, Virginia, was a cosmopolitan port town with four Methodist churches, three Episcopal churches, two African American churches, a Baptist church, and a Quaker congregation. Rev. John Smith, a Jesuit and an "eloquent, whole souled Irishman," established St. John's Academy, which functioned until 1841. Whether it was a military school from the start is unclear, although at least two of the school's alumni served in the Mexican War. The school was military when it reopened in 1847. That change may have been under the parish's stewardship of Rev. Joseph M. Finotti, an Italian Jesuit who had studied to be an officer in the Austrian army. Among his publications were *Diary of a Soldier* (1861) and *The French Zouave* (1863). Or the change may have been influenced by the school's principal, Richard L. Carney, who had been a student at the school prior to its closure. Carney expanded the school from seven students to seventy-two in four years. The extensive curriculum included Greek, Latin, French, and English. Carney's brother served as mathematics professor. In 1853 the mayor of Alexandria received complaints about the school's use of the streets, likely for drill. He regretted having to write Carney, but this led to the school's looking into the purchase of the lot next door for drill.[15]

In 1861 "the cry of battle resounded through the land . . . [as] a hundred St. John's men armed the Gray." Close to one hundred alumni and students enlisted in the 17th Virginia Infantry, and at least eight died in service. With the advent of the Civil War, Alexandria was occupied by Union troops. Whether the school maintained its military character with "cadet gray coat and pants trimmed in black" and blue cap with the letters SJA is not known, but the school remained open during the war. Enrollment swelled as many schools in Alexandria closed, and new students included the sons of Union officers, including two Union military governors.[16]

The school was the only Catholic military school in the United States for over thirty years. In its initial class in the 1830s were the Civil War commander of the Mount Vernon Guards, Company E, 17th Virginia, and Congressman Bernard G. Caulfield. Caulfield had left Virginia for Chicago in 1852, and there he became part of the Sons of Liberty plot during the Civil War to liberate seven thousand Confederate prisoners of war at Camp Douglas and in turn take Illinois into the Southern cause. Caulfield fled Illinois after the plot was discovered by the Secret Service and remained with Confederate exiles in Mexico until 1867. Upon returning to the United States, he served as a US senator from Illinois twice and was one of the founders of the Chicago Democratic Party.

St. John's Academy also conducted an elementary school while maintaining the military requirements for the older cadets. St. John's Academy claimed to be the oldest boarding school in Virginia and the only military school providing cadets with full field equipment, which they started using on summer trips by the Chesapeake and Ohio Canal to Shepherdstown and Harpers Ferry, West Virginia, 1873–77. The cadets were present at the dedication of the Washington Monument in 1885. Richard L. Carney led St. John's Academy for forty-five years, and in 1892, upon the death of his wife, he left the school and became a priest. Turning administration over to Capt. William H. Sweeny appeared to be a fatal error for the school, as it closed in 1895 with only seven students.[17]

By the time St. John's Academy had closed, four additional Catholic military schools had been established. Among those was St. Leo Military Academy/College of Florida, one of the only two Catholic military colleges ever established, founded in 1880 by the Benedictines. The school functioned as a military college for twenty-three years consecutively, after which it adopted that orientation for only short periods, the first 1908–9 and the second 1918–20. The other Catholic military college was Spring Hill College in Mobile, Alabama, which adopted the military format from 1918 to 1920 as a result of World War I. "The colleges yielded to the army bugle, and dining rooms were transformed into mess halls." This action was done at the same time the Department of the Army authorized the college to have a Student Army Training Corps (SATC) unit and later an ROTC on campus. But the college took the additional step of requiring every student to be in uniform at all times while on college grounds and emphasized military discipline during defined periods. This included the school's high school students, who were organized into a cadet company, as well as college students who were physically able to pass the ROTC requirements. In 1920 the school enrollment was down to 103, and declining enrollment caused a loss of army support for ROTC, so the college reverted to its prior nonmilitary format.[18]

The All Hallows College was a secondary military school established in the Mormon-dominated Salt Lake City in 1886 by the Catholic bishop, Lawrence Scanlan. It was meant to serve not only Catholic boys of the city, but also the growing Catholic population of ranchers and miners in Utah. In the first year, the gray-clad cadets numbered 115, of which 49 were boarders. Within three years the school almost failed when an outbreak of diphtheria struck 30 to 35 cadets and hospitalized 8 to 10. The school quickly recovered and, with the support of the Society of Mary (Marist Order), grew to 225 cadets in 1909, with about half of those being boarders. The college was known not only for the superior military training of its cadets, but also for its athletics, in which it did very well against other prep schools and universities.[19]

The circumstance of the school's closure appears to have been a combination of competition and discord between the diocese and the Marist Order. By 1915 other western states had established a number of boarding schools that drew off many potential boarding students from Montana, Oregon, Washington, and Colorado. The next year the diocese wanted to purchase the school, which was under the corporate ownership of the Marist Order.[20] They wanted the school to weather the declining student population crisis, but it appears the financial burden shared between the two church organizations was never resolved satisfactorily, and the school closed.

The Marist Order would later establish Marist Military School for Boys in Georgia in 1901. The cadets numbered 125 in 1917 and were uniformed in gray, with military drill and discipline an essential aspect of the school. The foundation of the school was "the principle that religion is the primary element of life, and that without religion as a fundamental[,] education in the ordinary sense has little moral efficiency."[21] This school successfully operated as a military school until it became coeducational and demilitarized in the 1970s.

The Society of Mary (Marianists), as opposed to the Marist Order, converted their boys' secondary boarding school in San Antonio, Texas, St. Mary's College, to a military format in 1919. Until 1932 the school would continue as a military school in a city that had at one time four other military schools. That year, the school became Central Catholic High School, which, although no longer a military school, maintains a very strong JROTC program.

The remaining schools founded in the late nineteenth century were all located in New York and founded by the Christian Brothers. That order would embrace the military format more than any other, with the establishment of seven schools. Their first military school was La Salle Institute of Troy, which transitioned to a military format in 1891. La Salle was followed by Christian Brothers Academy in Albany, which transitioned in 1892, and Sacred Heart Military Academy in Brooklyn, which transitioned in 1898.

La Salle Military Academy was established to meet the demand for a Catholic boarding school near New York City in 1883 and was known then as the Westchester Institute. Three years later, the school was renamed the Sacred Heart Academy. In 1898, amid the patriotic emotion of the Spanish-American War, Brother Hilarion, a Civil War veteran, and Capt. T. Moynihan started military training, and the school converted to a military school. In 1903 its name was changed to Clason Point Military Academy. With the cadet corps reaching 225, the school moved to new facilities in Oakland, New York, and was renamed La Salle Military Academy. During the cultural upheaval of the 1960s, enrollment suffered, but Brother Louis DeThomasis, class of 1958, took over the leadership from 1977 to 1984 and revived academic and military standards. The resulting increased enrollment required the organization of a sixth cadet company. Unfortunately, after his departure subsequent administrations did not show the same devotion to the school. The military format became optional in 1999, and the school closed in 2001.[22]

New York's last Christian Brothers military school, Cardinal Farley Military Academy, was established during World War II in Rhinecliff. Outside New York, the order founded the De La Salle Military Academy in Kansas

La Salle Military Academy Corps of Cadets eyes right as a platoon passes the reviewing stand in 1963. (Reprinted with permission of Peter E. Dans, MD, from *LaSalle Military Academy: Pro Deo Pro Patria, the Life and Death of a Catholic Military School*)

City in 1910. They transitioned St. John's College High School in Washington, DC, in 1915; Christian Brothers College Military High School in Missouri in 1936; and Cretin (Military) High School in Minnesota during World War I.

In many ways, the most successful Catholic order in the establishment of lasting military schools was the Benedictines. They established three military schools, Benedictine Military School/High School in Georgia in 1902, Benedictine High School in Virginia in 1911, and Marmion Military Academy in Illinois in 1935. Marmion's military program became optional in 1993, and with it, the requirement for the habitual wearing of military uniforms ended, although a strong JROTC program continues. Both Benedictine Military School/High School and Benedictine High School continue today as military day schools with single-sex education. The Georgia school's military program is mandatory for freshmen and sophomores. It not only continues as a military school but also maintains over 90 percent of its students as cadets. The enrollment of Benedictine High School in Richmond, Virginia, in 2014 was 272 cadets; it occupied a new fifty-acre campus in the fall of 2012.[23]

The Brothers of the Holy Cross established Sacred Heart Military Academy in New York in 1883, and, much later, Sacred Heart Military Academy in Wisconsin in 1955. The latter school was renamed Le Mans Academy and moved to Indiana, where it functioned until 2003. Leonard Hall was established by the Xaverian Order as a Catholic college preparatory school for boys. Destroyed by fire in 1920, the school was rebuilt and expanded to include a space for boarding students, but the Great Depression almost destroyed the school financially, and it moved its focus to elementary education. Perhaps as a means of attracting more students, the school became a unique combination of Catholic and military junior naval school on the eve of the entry of the United States into World War II. In 1968 enrollment had grown to 200, with 160 as boarders. The antimilitary culture of the 1960s quickly took its toll, and in 1972 the Xaverians closed the school.[24]

But the Leonard Hall story did not end, because the parents of the day students banded together and formed a corporation that opened the school the following September, and the Leonard Hall Junior Naval Academy continued in operation as a private school.[25] In 2010 the school had a modest fifty-one cadets in grades six through twelve. Supervised militarily by a retired noncommissioned officer, its students carried on its traditions in their naval-style uniforms.

The story of the Catholic military school would not be complete without a description of the continued contribution of various orders of nuns. Eight orders established military schools. The Benedictine Sisters established Linton Hall Military School in Virginia in 1922. The school discarded the military aspect for a few years in the 1930s but returned to the military format until 1989. The Sisters of Divine Providence established the Moye Military School (Academy) in Texas in 1938 and operated it as a military school until 1959. The Sisters of Mercy opened St. John's Military Academy in Los Angeles in 1905, and it remained open until 1968. The Sisters of the Blessed Sacrament opened St. Emma Military School as the St. Emma Industrial and Agricultural College in Virginia in 1895; it operated as an African American military school until it closed in 1972 (see chapter 8). The Dominican Sisters converted Bishop Quarter Junior Military Academy, Illinois, to military format in 1936, and it operated until closing in 1968. In 1945 the Pallottine Sisters established St. Patrick's Military Academy, a boarding school for boys, in New York, and it operated until 1983 or 1986.

The Sisters of St. Francis of Penance and Christian Charity to this day maintain an association with the military school concept. They operate one of only two exclusively grade school–oriented military schools in the country: St. Catherine's Military Academy in California, established in 1889 as a Catholic girls' school. In 1894 the school transformed to a boys' orphanage. By 1916 the number of orphans had decreased to the point that the sisters hired male teachers as positive role models and changed the focus to a Catholic male day school. In 1923, Capt. D. M. Healy was hired as commandant of cadets, and the transition began to create the St. Catherine's Military School. This transition was undertaken after the sisters had conducted "extensive research." In 2010 the school had 150 male cadets, grades four through eight.[26]

Historically, the order had also founded St. Aloysius Academy and Military School in New Lexington, Ohio, in 1915 and St. Aloysius Military Academy, Fayetteville, Ohio, in 1938. The schools operated until the 1960s and 1980s, respectively. The Stella Niagara Cadet School of New York was founded in 1908 and operated as a military school until 1971. Enrollment ranged from ninety-one in 1937 all the way to three hundred in 1963.[27]

The Sisters of St. Joseph established the final Catholic military school in 1952 when they transitioned the Nazareth Hall Cadet School in New York. The sisters felt that the military format "provides an excellent incentive and practical means for the realization of the ideals which inspired the school's foundation." Sister Marie Paulus explained that Nazareth Hall's program differs from that of a military academy because "a military academy is run thinking of war. Our cadet program is to train the boys in leadership and discipline, give them the opportunity of responsibility, and to maintain high scholastic standards."[28]

Another motivation was linked to the influence of the threat of communism during the Cold War. Mother Rose Miriam Smyth, head of the congregation, wrote to all the sisters in January 1951: "The world crisis alarms all who think. We have an obligation to help to the utmost of our ability. If we remember that whatever is in the mind and heart of a zealous teacher will be in the minds and hearts of her pupils, we will not dare to fail our hourly and momentous opportunities."[29]

Presbyterian Church

The Presbyterians established twenty-five military schools, including two that opened before the Civil War (table 9.3). The oldest among them was the Classical and Mathematical Academy of Bedford, Pennsylvania. This was the first of any denomination-sponsored school to adopt a military format. The school was opened by Rev. Baynard R. Hall in about 1834, and it operated for about five years.[30]

The next Presbyterian military school was the Alexander Military Institute in New York, which was described as a "Classical, Commercial, and Military Boarding School" with a capacity for thirty cadets. It was first opened in 1845 under the direction of William S. Hall and was known as the Alexander Institute until 1857. Gen. Munson I. Lockwood, former commander of the 7th Brigade, New York State Militia, ran the school after that for six years under the name Hampton Military Institute. In 1863, Oliver R. Willis took charge, and the school was again officially renamed the Alexander Institute, although it was often called the Alexander Military Academy. The school was a short distance from the Presbyterian Church and provided, as part of its curriculum, military drill and all courses needed to prepare its cadets for college. The school would continue until at least 1919.[31]

The other pre–Civil War school was the Yonkers Collegiate and Military Institute in New York. This school was established sometime between 1852 and 1854 by Dr. Washington Hasbook. He was followed by a series of superintendents, including M. N. Wiseman and Frederick Norton Freeman, an 1856 Norwich graduate who would later lead Englewood

Table 9.3. Presbyterian Military Schools

Opened or *Transitioned* *to Military*	School	Location
1834	Classical and Mathematical Academy	Pennsylvania
1854	Yonkers Collegiate and Military Institute	New York
1863	Alexander Military Institute	New York
1863	Warring's Military Boarding School	New York
1865	California Military Academy	California
1871	Montrose Classical and Military School	New Jersey
1874	Sweetwater Military College	Tennessee
1875	Suffolk Military Academy	Virginia
1880	Wentworth Military Academy and Junior College (under control of church, 1919)	Missouri
1885	French Camp Military Academy	Mississippi
1890	Clinton Liberal Institute and Military Academy	New York
1890	Danville Military Institute	Virginia
1890	Putnam Military Academy	Ohio
1890–91	Presbyterian School (Greenbrier Military School)	West Virginia
1891	Greenbrier Military School	West Virginia
1892	Mount Tamalpais Military Academy	California
1895	Chamberlain-Hunt Academy	Mississippi
1895	Hoge (Memorial) Military Academy	Virginia
1902	Auburn Military Seminary	Kentucky
1904	Alabama Military Institute	Alabama
1905	Anniston University School	Alabama
1917	McCallie School	Tennessee
1917	Onarga Military School	Illinois
1922	Alabama Presbyterian School for Boys	Alabama
1923	Schreiner Institute	Texas

Military and Collegiate Institute of Perth Amboy, New Jersey. Benjamin Mason was the principal from about 1863 through at least 1879. The school may have closed for a time starting in 1880, but it reopened and had an enrollment of forty cadets in the early 1890s under the direction of Col. H. S. Farley.[32]

During the Civil War, Warring's Military Boarding School was established in Poughkeepsie, New York, by C. B. Warring, PhD. Enrollment in 1879 was thirty-seven boys and two girls. The school operated at least as late 1903. In the closing days of the Civil War, Rev. David McClure opened the California Military Academy in Oakland with the help of his brother, Maj. Stewart McClure. The school was successful, and enrollment reached 125 cadets in 1872, but the following year fire destroyed the school. The McClures rebuilt, and by 1875 the school employed nine professors and had 176 cadets. The cadet corps was even large enough to maintain a small cadet band. Ownership passed to a long-time professor, Col. W. H. Obrien, in 1884, and the school operated until at least 1908.[33]

Eleven more Presbyterian military schools were established in the nineteenth century. New military schools were Montrose Classical and Military School, New York, 1871; Sweetwater Military College, Tennessee, 1874; Suffolk Military Academy, Virginia, 1875; French Camp Military Academy, Mississippi, 1885; Danville Military Institute, Virginia, 1890; and Putnam Military Academy, Ohio, 1890. Transitioning to military format in 1890 or 1891 was the Presbyterian School (Greenbrier Military School), and Clinton Liberal Institute established a military prep school in New York in 1891. Tamalpais Military Academy, California, transitioned in 1892; Chamberlain-Hunt Academy, Mississippi, in 1895; and Hoge (Memorial) Military Academy, Virginia, in 1895.

The Clinton Liberal Institute in Fort Plain, New York, was actually four schools in one: the Fitting School, the School of Fine Arts, the School of Business, and the Fort Plain Military School. The military school was a boys' prep school associated with the college under the direction of a regular army officer, Lt. Harry L. Hawthorne, a hero of the Indian Wars, detailed to support the school. The school opened in 1891 and was fully equipped with 150 rifles and all the associated equipment. The military school functioned until it was destroyed by fire in 1900.[34]

Rev. Arthur Crosby founded the Tamalpais Academy in 1890 and converted it in 1892 to a military format. His objective was "to instill in boys those habits which alone insure success in whatever walk of life they may wander." Enrollment in the late 1890s was approximately sixty cadets, who were uniformed in West Point style and equipped not only for armed drill but also with two cannons for training. The school conducted an annual encampment and training supported by an active-duty army artillery lieutenant. It was identified in 1897 as one of the two best military prep schools on the West Coast.[35]

In the twentieth century, the remaining seven military schools were established, including the final Presbyterian military school, Schreiner Institute in Texas. Schreiner was founded in 1879 and transitioned to a military format in 1923 with 95 cadets. Under the new military format the school had over 300 cadets by 1933. The school expanded to include both high school and junior college military programs. As the United States entered World War II, there were 74 high school cadets and 293 college cadets at Schreiner. The institute would continue as a military school and peak in enrollment at 505 in 1946, but with enrollment challenges it dropped the military school environment in 1972, and later the school evolved into Schreiner University.[36]

Baptist Church

The Baptist Church has had associations with nine military schools (table 9.4). The oldest was the Vermont Academy, founded as a "direct instrumentality" of the Vermont Baptist Convention in 1871. The school converted to a military format in 1885, and an active-duty officer was assigned to the school in September 1891. Enrollment in 1882 was 138 students, of whom 62 were cadets and 76 were nonmilitary females. The cadets were organized into two companies. They wore blue uniforms and were reported to have good moral character. The school continued to function as a military school until 1908.[37]

Howard College of Alabama was established in 1841 in Marion. Col. John H. Murfee, a VMI graduate and former commandant of cadets at the University of Alabama who was wounded while leading his cadets in Civil War combat, took leadership of the college in 1871. Murfee quickly adapted the college to a military format much like the two institutions with which he was most familiar. He was no stranger to the military conversion of schools, having done so at Westwood Military Academy and Lynchburg College. His charm and energy captured the people of Marion, and he earned their loyalty through his delivery of "discipline, methods of instruction, moral culture, and practical education."[38]

The Baptist Convention of Alabama did not see any justification for divinity students having to live the life of a military school cadet, and relations

Table 9.4. Baptist Military Schools

Opened or Transitioned to Military	School	Location
1885	Vermont Military Academy	Vermont
1887	Howard College	Alabama
1891	Bailey Military Academy	South Carolina
1897	Buie's Creek Academy	North Carolina
1903	Fork Union Military Academy	Virginia
1907	San Marcos Baptist Academy	California
1911/1920*	Pillsbury Academy	Minnesota
1918	Bethel College	Kentucky
1919	Locust Grove Institute	Georgia

*Sources conflicted on the date of transition, which may have been a lengthy process.

between the convention and the school soured. In 1884 a group of divinity students moved off campus. When the convention demanded that the college support their academic needs, Murfee refused and was backed by the board of trustees. The Baptist Convention resigned its relationship with the college in protest. The military college continued until 1887, when the Baptist Convention, after meeting in the economically booming Birmingham, found a way to get around Murfee and hold on to the college. They moved the school to Birmingham.

Murfee would not move, and Marion embraced him and his plan to continue the school as the Marion Military Institute. The school kept almost its entire faculty and an unknown number of its cadets and was still open in 2014 as one of the nation's five military junior colleges, with enrollment of 439. By contrast, Howard College's corps of cadets continued and operated as a military school with a voluntary corps of cadets, but numbers waned. In 1913 the new president, James H. Shelburne, stating that the program competed with athletics, ended the corps of cadets.[39]

Fork Union Military Academy in Virginia, established in 1898, converted to military school requirements in 1903. The school still operates as a military school, educating sixth through twelfth graders and offering a fifth year of high school. In 2014 it had 535 cadets, operating in a rural location with a Baptist foundation.

Methodist Church

Methodist Church involvement with military schools was primarily in the South (table 9.5). As the Civil War approached, the church established two Methodist military schools of note. The first, Lynchburg Military College in Lynchburg, Virginia, was a direct result of the coming conflict and sectional divide that even affected the church. Madison College was established in Uniontown, Pennsylvania, in 1830. By 1855 it had evolved into a Methodist college, with a Southern faculty, a Northern board of trustees, and a student body split between the two regional factions. As tensions rose between the faculty and the board, a student fight occurred between a young man from Ohio and the Bailey brothers from the South. The resulting hard feelings spread to the faculty and caused a rift with the trustees that never healed.[40]

At the 1855 commencement, the college president announced plans to open a Methodist College in Virginia the following September. The entire faculty resigned, along with eighty-five of ninety Southern students, or about half the student body. Madison College had military training in which students wore uniforms at drill, but the new school was a military school down to the design of its building. The resulting college in Lynchburg, Virginia, was housed in a castlelike structure much like that of Virginia Military Institute. The president, Samuel Cox, a Yale graduate, used his old natural science teacher, James Thomas Murfee, VMI class of 1853, to function also as commandant of cadets. Also joining the faculty as a teacher in both mathematics and military departments was another VMI alumnus, James E. Blankenship. Captain

Table 9.5. Methodist Military Schools

Opened or Transitioned to Military	School	Location
1832	St. Charles Military College	Missouri
1855	Lynchburg Military College	Virginia
1857	Bastrop Military Institute (Texas Military Institute)	Texas
1861	Randolph-Macon College	Virginia
1910	Randolph-Macon Academy, Front Royal	Virginia
1920	Randolph-Macon Academy, Bedford	Virginia

Blankenship would enlist his cadets into the 11th Virginia Infantry Regiment at the beginning of the war.[41]

Lynchburg College's first commandant, James Thomas Murfee, had a long association as a distinguished educator in military schools. After two or three years in Lynchburg, he became the coprincipal of Westwood Academy in Lynchburg and transitioned that school to a military format. In 1860 he became the commandant of cadets at the University of Alabama and led its cadets in combat during the war. After the war, he continued as commandant there until taking the presidency of Howard College and later Marion Military Institute, as described earlier in the section on Baptist military schools.

The other Methodist military school that opened with the approach of the Civil War was Bastrop Military Institute in Bastrop, Texas. The school opened in 1851 as the Bastrop Academy under Rev. Martin Ruther, a Methodist, and with the support of the Bastrop Educational Society. The academy's coeducational enrollment declined to ninety-four male and female students with the advent of the Common School system, which offered nearly free education.

Col. Robert Thomas Pritchard Allen, a West Point graduate, the founder of the Kentucky Military Academy in 1845, left that institution in 1854 due to poor health and went to Bastrop.[42] In 1857 he converted the male department of the school into a military school. Initially, the academy occupied the first floor of the school building and the Bastrop Female Institute the second floor. The attraction of a military school took the male enrollment to ninety-two the first year. So successful was the military institute that the cadets enrolled from as far away as Kentucky, Tennessee, and Mississippi in 1859. The cadets carried rifles while on dress parade every evening of the weekdays. The uniform was blue with brass buttons and red-striped pants. The academics were demanding, with a four-year course that included calculus, natural history, botany, mineralogy, surveying, mechanics, and Latin.[43]

During the Civil War the school struggled after Colonel Allen departed to command the 17th Texas Infantry and was replaced by his brother, Robert. Although many of the older boys left for military service, the school continued with declining enrollment until 1864, when it closed. The school remained closed until 1867 and then operated for one year as a nonmilitary school. In September 1868, under the leadership of John Garland James, VMI class of 1866, as president and his brother, Fleming Wills James, VMI class of 1868, as commandant—both of whom had fought as VMI cadets at the Battle of New Market—the school was renamed the Texas Military Institute. In 1870 the school moved to Austin atop a hill in a castlelike structure. There it operated as a nonsectarian school with enrollments of one hundred cadets until 1879. That year the school closed, and the faculty, along with John Garland James, departed for employment at the newly established Texas A&M College, where James became Texas A&M's second president.[44]

With the commencement of the Civil War, Randolph Macon College, Boydton, Virginia, had severely declining enrollment. "So great was the depletion in the number of students, and so great the excitement that prevailed throughout the country, that the College authorities deemed it inexpedient to hold the regular commencement exercises [early]." The college hoped to prepare its students for military duties and survive as an institution by reinventing itself as a military school. "A regular uniform was prescribed; drills were daily observed, and other things of a similar character were enjoined, all looking to the preparation of the student for the duties that awaited him in defense of his country."[45]

The college president, Rev. William A. Smith, was appointed as colonel, and James E. Blankenship, previously mentioned with Lynchburg College, was appointed as major and professor of mathematics and military science.[46] At the Battle of First Manassas, Major Blankenship's short-lived service in the field proved that he was a better asset in a school environment than in leading troops in combat. The college functioned on a military footing from 1861 to 1863, with enrollment down to fifty-six cadets when the school closed.

When the college reopened in 1866, it did so without its military trappings. But with the popularity of the military school concept, particularly in Virginia boarding schools, two of the college's associated prep schools, Randolph-Macon Academy in Bedford, established in 1890, and Randolph-Macon Academy in Front Royal, transitioned to military schools in 1920 and 1910, respectively. Although the Bedford branch closed in 1934, Randolph-Macon Academy in Front Royal remains a military school to this day. The school has a nonmilitary lower school and a military upper school with grades nine through twelve, which numbered approximately 312 cadets in 2014.

Prior to the transition to a military format of the Randolph-Macon Academies, Rev. George W. Bruce reopened St. Charles College in Saint Charles, Missouri. In 1901 the school reopened as St. Charles Military College. The school was a secondary boarding school with an average of about seventy cadets annually. Reverend Bruce was followed by Col. Herbert F.

Walter, who served as president for seven years and was a graduate of South-western University in Texas but had also been a cadet at Texas A&M for a year. The school continued operation until sometime after 1916.[47]

The final Methodist military school was established in 1919, when Carlisle School, Bamberg, South Carolina, founded in 1892, became an all-male military school. The Carlisle School merged with Camden Military Academy in 1977 and was no longer associated with the Methodist Church. Camden Military Academy's website states that the school incorporates the traditions of Carlisle Military School, whose students, which had numbered close to three hundred, transferred with the merger.[48]

Other Denominations

Four other Christian denominations established churches, principally in the northern states, in the late nineteenth and early twentieth centuries (table 9.6). The Congregational Church established four military schools, starting with the transition of the Betts Military Academy of Stamford, Connecticut, in 1860. This school operated until 1908, but in the meantime the Congregationalists converted Bunker Hill Military Academy in Illinois, under Rev. Samuel L. Stiver, to a military format in 1883. The school prospered with the change to a military format and in 1913 had sixty-five cadets from grades six through twelve. Also converted to military format by the Congregational Church were the Kamehameha School in Hawaii in 1914 and the Allen-Chamber School of Massachusetts in 1917.

The Reformed Church established two military schools, including Massanutten Military Academy, which was still in operation in Virginia in 2014. Massanutten Military Academy was established by the church in 1899 and converted to the military format in 1917 during World War I. The school's coeducational enrollment of sixth through twelfth graders has varied from 236 to 158 between 2000 and 2014. The other Reform Church military school was Riverview Military Academy in Poughkeepsie, New York. This school also converted during a time of war in 1862 and operated until 1920.

The Lutherans established the Collegiate Institute in Mount Pleasant, North Carolina, prior to the Civil War as a nonmilitary school. After it was closed for two years, Rev. L. E. Busby reopened it, and five years later the school adopted military uniforms and discipline. The school continued until financial problems brought on by the Great Depression caused its closure in 1933.[49]

Table 9.6. Congregational, Reformed Church, Moravian, and Lutheran Military Schools

Opened or *Transitioned to Military*	School	Location
Congregational Church Military Schools		
1860	Betts Military Academy	Connecticut
1883	Bunker Hill Military Academy	Illinois
1914	Kamehameha School	Hawaii
1917	Allen-Chamber School	Massachusetts
Reformed Church Military Schools		
1862	Riverview Military Academy	New York
1917	Massanutten Military Academy	Virginia
Moravian Church Military Schools		
1861	Nazareth Hall Military Academy	Pennsylvania
1918	Gettysburg Military Academy	Pennsylvania
Lutheran Church Military School		
1908	The Collegiate Institute	North Carolina

The Moravian Church had two military schools, both in Pennsylvania. The first was the Nazareth Hall Military Academy, a very old school established between 1743 and 1759. In 1861, the first year of the Civil War, the school adopted a military format. The school was said to develop "all that is best in discipline and training, while even teaching the doctrine of peace." The cadets were organized into two companies, and their drill was said to be to such "perfection . . . even the most ardent peace advocate could not help applauding the military proficiency and the manly, alert appearance." The school operated until at least 1929. Their second school was Gettysburg Military Academy in Pennsylvania and was in operation in 1918, but little more is known about it.[50]

Conclusion

The expansion of the military school into various dominations was a function of cultural attraction and similarities. Rev. Allan L. Burleson in 1896 referred to the military schools as "great moral agency for good." A little more than one hundred years later, the military school was identified as the vehicle for "enforcing support, and discipline, and to give children structure" in order to better blend two cultural elements of religion and academics. John Kraus supported that view in that the military school culture expressed the values of the church school in a modified form. In doing so the military school's culture was complementary to moral development. However, as will be seen in later chapters, this strong association did not prevent the impact of other political, cultural, and economic factors.[51]

★ 10 ★

Tested by War, Depression, and Fire

The period between the Spanish-American War and the Great Depression coincided with a peak in the popularity of the military school and the initial development of a significant opposition to both military schools and military training in schools. Much of the motivation of the latter grew from the high cost in human life of World War I. During that period American military schools passed through political controversies, adjustments in their mission, and legal and economic challenges, institutional fires, and war.

The Spirit of 1898

The Spanish-American War was a period of enthusiastic unification. It was a time when the divided nation came together for what was promoted by journalists as a noble cause. The mood in the United States was patriotic and supportive of the military school concept. The image of former Confederate general Joseph Wheeler, West Point graduate, and Gen. William R. Shafter, Union Medal of Honor recipient, recalled to active military service in Cuba and serving side-by-side, increased patriotic feelings for a nation seeking reconciliation.

As the war with Spain approached, the midshipmen of the US Naval Academy, who had graduated early in April 1898, and cadets from a Southern military college proclaimed they were "all sons of Johnny Rebs . . . ready to march in six hours." Several military colleges, including VMI and Virginia Tech, volunteered their service as VMI had done in the Civil War: "We hereby tender to the Governor of Virginia the services of the Virginia Polytechnic Institute [Virginia Tech] Corps of Cadets, consisting of four companies of infantry, one light battery of artillery, commissioned and noncommissioned staff and band for the defense of our country in the event of war."[1]

Despite the fact that both Virginia Tech's and VMI's corps of cadets were members of the state's active militia, the requests were turned down. Perhaps the memory of the deaths of young cadets of VMI on the field during the Battle of New Market played a part in the governor's decision. But back on campus, the atmosphere was affected: "The spirit of war was everywhere. There was a vague uncertainty, an indefinable something in the air. Among the cadets there was a listless longing for the end of the academic session. Many resigned from college and joined the armed forces." It was a ripe recruiting ground at military colleges. When Sergeant Loving recruited at Virginia Tech, he left with six enlistees for Company G, 2nd Virginia Infantry Regiment, and six more followed later.[2]

As the volunteers mobilized for war, the 2nd Virginia lacked its authorized band. In response, the band director at Virginia Tech resigned from the college and enlisted twenty-five men to fill the regiment's need. These men included fifteen Virginia Tech cadet band members from the classes of 1898 through 1903, five former members of the Glade Cornet Band, a town organization that supported the corps until the school organized its own band in 1893, and four alumni.[3] Although the 2nd Virginia Infantry Regiment did not see action, the cadets' service was reflective of the military spirit of the time and a legacy symbolized proudly more than one hundred years later in the Virginia Tech drum major's baldric.

As in the Civil War, the value of military leaders educated at military schools was reflected in the commanders of deployed combat formations in Cuba, Puerto Rico, and the Philippines. During the Spanish-American War and the associated Philippine Insurrection, deployed major formations included three corps and seven divisions. Review of the ten initial commanders of those organizations reveals that five were graduates of West Point. The remaining five commanders (three of whom were Medal of Honor recipients) all were volunteer officers in the Civil War who joined the regular army and became professional soldiers after the war. Despite the representation of combat-experienced veterans of the Civil War who had not attended one of the nation's military schools, West Point continued to carry a large percentage of the leadership burden at the higher echelons of command.

The spirit of 1898 extended well into the next decade. In 1916 the Mexican border crisis triggered by Pancho Villa resulted in an American incursion into Mexico as part of the Punitive Expedition under Gen. John

1st Squadron, 1st Vermont Cavalry Regiment, 1914. This unit, made up of cadets of Norwich University, was called to active duty in 1916. (Courtesy Norwich University Archives)

Pershing. On June 30 the 1st Squadron, 1st Vermont Calvary Regiment, began several weeks' training for deployment to the Mexican border. That training ended the summer vacations for Norwich cadets, as they were an integral part of the Vermont National Guard's 1st Squadron. Norwich's president, Col. Ira L. Reeves, was the regimental commander, and his cadets and members of the faculty were called to active duty. Despite Colonel Reeves's best efforts to send his regiment to the border, the War Department sent the cadets back to Norwich University, and they were demobilized.[4]

Peak of Military School Representation

The Spanish-American War brought on a patriotic resurgence in the United States, which helped expand the military school concept. In the years between 1898 and 1907, approximately 83 military schools were founded or existing nonmilitary schools were converted to the military format. Between 1903 and 1926, no fewer than 278 to 280 military schools operated in the United States. This was the peak for the number of military schools in the United States. It was a period when the military uniform and the military concept of school were favored throughout the country.[5]

The patriotic feelings of the period were even manifested with children dressed in a style called Dewey suits, which hailed Admiral Dewey, famous for his defeat of the Spanish fleet in the Philippines in 1898. One of the children dressed in the Dewey style was the son of Wesley Peacock, the headmaster of Peacock School for Boys, established in 1894 in San Antonio. Between 1894 and 1900, the Peacock School for Boys grew from twenty students to more than one hundred, and two new buildings were constructed. In 1900 Wesley Peacock began moving his school toward a military school format. Peacock added a military department, with VMI alumni on his staff, required the upper school students to wear uniforms, and officially changed the school name to the Peacock Military Academy (College).[6]

The 1900 presidential campaign included militarism as a major issue. The Democratic candidate, William Jennings Bryan, took a dim view of the fact that the United States, with no prior tradition of a large standing army, expanded the army from 25,000 soldiers in 1896 to 275,000 soldiers during the Spanish-American War. President McKinley, on the other hand, advocated military reforms and expansion of American influence, which required expansion of the regular army, greater control of the states' national guards, and pacification and protection of the gains the United States made in the Spanish-American War.[7] The mood of Americans helped defeat Bryan and ensured the election of McKinley along with his war-hero running mate, Theodore Roosevelt.

War Department Instructors

The limited assignment of active-duty army officers to land grant colleges under the 1866 Morrill Land Grant Act expanded. In 1888 this detailing was extended to schools and colleges that provided a stipulated level of enrollment and military training. The act also provided these schools with the arms and equipment needed to support the training. Whereas the 1866 act detailed twenty officers, in 1870 the number was increased to thirty, and through a series of increases the total in 1892 was one hundred army officers and ten navy officers. These officers joined faculties with full pay and allowances at no cost to the school. Later, in 1893, the schools that did not recruit an active-duty officer could request the assignment of a retired officer by assuming the responsibility to pay him the difference between active-duty pay and a retired pension.[8]

Further benefiting military schools that secured an officer, the officers could request to teach other subjects. One of the schools that took advantage of the military offer and transitioned to a military format was the Fairfield Seminary in New York. Established in Fairfield in 1885, the school was a coeducational, private, boarding, nonmilitary school with an enrollment of 207. As a private college preparatory school, it included pre-law and premedical courses. But with the expansion of public education in New York State, and with students' ability to stay at home and attend public school, enrollment suffered and placed the future of the school in doubt.[9]

The headmaster, Frank Warne, used his political connections to help the school secure the 1891 assignment of Capt. George R. Burnett, a wounded veteran of the Indian Wars, and later another retired officer, Lt. Warren R. Dunton, and the seminary became the Fairfield Seminary and Military Academy. The change gave the school's small faculty an addition to the staff with minimal expense and gave the school a unique feature over the public schools. In 1892 the school operated with fifty-one male cadets and forty-three female students, and it would continue until 1901.[10] The assignment of active-duty military personnel to military secondary schools would benefit those programs through the early 1970s.

Pacifist Movement

In his forty-page 1934–35 prospectus for Illinois Military School, Clyde R. Terry, the school's president, wrote: "In these days . . . pacifist propaganda attempts to undermine every institution with the name military attached." His words are a reflection of the time between 1923 and 1933 when the post–World War I pacifist movement began to have a negative impact on military schools and their reputation.[11]

The pacifist movement started just after the Spanish-American War, during the same time as the Philippine Insurrection from 1899 to 1902. The Anti-Imperialist League functioned from 1898 to 1921 and included among its thirty-thousand members former president of the United States Grover Cleveland, journalists, labor leaders, and educators. Their platform was opposition to the annexation of Hawaii and the Philippines as well as to increases in the power and size of the federal government, including the institutionalization of large standing armed forces.[12]

Among the educators in the pacifist movement was Paul B. Barringer, president of Virginia Tech from 1907 to 1913. Barringer was also the vice president of the Anti-Imperialist League of Virginia. While president of Virginia Tech, he recommended that military requirements at the school be abolished for juniors and seniors. Both recommendations were rejected, and he resigned and returned to his medical practice. However, the antimilitary mood grew as Americans watched the alarming European casualties in the first years of World War I. The Hill Military Academy in Oregon saw its enrollment drop to only ten cadets prior to the United States' entry into the war. A commonly heard slogan was, "I didn't raise my son to be a soldier."[13]

As World War I raged in Europe and calls for military preparedness in the United States increased, the recently organized American Union against Militarism became the strongest group in opposition. The union was led by liberal publishers, clergy, lawyers, and feminists and claimed six thousand members and fifty thousand sympathizers. A common theme was that universal military service and increased military preparedness were an avenue for "international financiers . . . to make large profits from military conflict. The cause was joined by the Women's Peace Party, which had three thousand attendees from at least nine women's political groups at its first annual meeting.[14]

Preparedness Movement

At the same time as various groups began supporting the pacifist cause and protesting changes to the country's military readiness profile, the Preparedness Movement began. This group of military reformers had concerns about the ability of the United States armed forces to defend the nation. The Preparedness Movement was founded on the positions taken by the important naval strategist Adm. Alfred Thayer Mahan, US Naval Academy class of 1858, and Elihu Root, secretary of war, 1899–1904. Its champions became President Theodore Roosevelt and Gen. Leonard Wood, advocates for an expanded military, reform and expansion of the militia, and training of a pool of civilians to lead a potential major expansion of the army if needed for war. Also a part of the Preparedness Movement's desires was universal military service, which the Progressive Party supported, although the movement's members from the Republican and Democratic Parties did not.[15]

General Wood, the army chief of staff, was vocal in his calls for action on the subject of preparedness, as was Theodore Roosevelt, who used his popularity, the Progressive Party, his status as a Spanish-American War hero, the war in Europe, the sinking of the Lusitania in 1915, and his position as associate editor of *The Outlook* magazine to argue the Preparedness Movement's cause. Roosevelt was able to publish his views in some of the most influential media sources of the day. His articles in *The Outlook* included "Chapters of a Possible Autobiography: The War of America the Unready" (1913), "The World War: Its Tragedies and Its Lessons" (1914), and "Admiral Mahan: A Great Public Servant" (1914). In the *New York*

Times he published "What America Should Learn from the War" (1914), "Preparedness Without Militarism" (1914), "The Navy as a Peacemaker" (1914), and "Criticism of the President's Message" (1915). *Metropolitan Magazine* published his "Peace Insurance by Preparedness against War" (1915), "Uncle Sam's Only Friend Is Uncle Sam" (1915), and "International Duty and Hyphenated Americanism."[16] These articles and his speeches, along with those of Leonard Wood, were in no small part responsible for an attendance of 135,000 people at the Citizen's Preparedness Parade in New York and 350,000 marchers in ten other cities.

In 1912, 1913, and 1914 the movement organized the "Plattsburgh" training camps, funded by both the army and contributions, and many of their trainees were young men who were enrolled at or had graduated from military schools. The total numbers grew from 225 in 1914 to thousands in 1915. The support for military training was a benefit to both military colleges and secondary schools. Leigh Gignilliat, in his book *Arms and the Boy: Military Training in Schools and Colleges,* published in 1916, skillfully attached the advantages of the military school with military training benefits as the system that meets the test of discipline, physical endurance, and organization. He illustrated this in the opening of the book with the story of the 1913 rescue of fourteen hundred citizens of Logansport, Indiana, from the flooding of the Eel and Wabash Rivers by cadets of Culver Military Academy. This was not the only instance in which a military school was called out to support the community during a natural disaster. In 1936 a devastating tornado struck Gainesville, Georgia, killing 162 and injuring 950. Martial law was established under the Georgia National Guard, and both the Riverside Military Academy cadets and North Georgia College cadets prevented looting and supported medical communications and even mortuary operations for several days.[17]

In 1916 the National Defense Act and Navy Act were passed after exhausting debate and hearings. In those acts the Preparedness Movement got most of its desires, including increases in the army and navy and incremental improvements in the National Guard's training and numbers. The funding for summer training camps and universal military service were not a part of the acts, but most important for military schools was the establishment of the Reserve Officer Training Corps (ROTC). By 1926 there were 118 colleges and a hundred secondary schools with ROTC units. Not all military schools had ROTC, but the program vastly expanded the military representation and funding from the federal government for those schools that did.[18]

World War I

World War I saw at least one case in which a military school's cadet corps was called upon for active service. On April 10, 1917, the Eddystone Ammunition Plant in Chester, Pennsylvania, experienced a series of explosions that killed 133 people, mostly female employees. A company of the Pennsylvania Military College deployed to the disaster and restored order with rifles, forced back crowds, and remained on duty there for five hours.[19]

Military schools again contributed military leadership to the war efforts of the United States in World War I. In the Civil War and Spanish-American War, the representation of volunteer senior officers, many of whom had little formal military training before their war experiences, was significant. In contrast, at least initially, the senior commanders deployed to Europe were overwhelmingly graduates of West Point. Seven corps and twenty-nine divisions reached Europe and entered combat. As these thirty-six organizations initially entered Europe, they were commanded by thirty-three different generals. Biographical information could be located on thirty of those officers. West Point graduates numbered twenty-six, and one other had attended West Point but not graduated.

The Marine Corps was assigned to army divisions in the war, and Gen. John Archer Lejeune, USMC, subsequently became commander of the US Army's 2nd Division. Lejeune had graduated from Louisiana State University at a time when it was a military school and afterward attended and graduated from the US Naval Academy in 1888.[20] Another army division commander from the Marine Corps was Gen. Eli Kelly, who commanded the 41st Division at the end of the war. Kelly was an 1888 Naval Academy graduate.

The army in World War I was dominated in the highest echelons by professional soldiers trained at West Point who had led a small professional army prior to the war. Although there were eleven National Guard divisions, they were formed under active-duty officers with the consolidation of state regiments, many dominated by military school graduates from various states. The second wave of divisions being prepared in the United States for deployment to France numbered twenty-two when the war ended. Two of those were commanded by Virginia Tech graduates, and likely many other military school graduates commanded the remainder.

The biggest contributions made by the military secondary and collegiate institutions to the war effort were in the officer ranks of lieutenant through colonel. The army in 1917 numbered only 121,000 in the regular army and 181,000 national guardsmen. They were led by 5,791 regular

Table 10.1. Military School Contributions to the World War I Effort

School	Location	Alumni in Service
Culver Military Academy	Indiana	2,853
Virginia Tech	Virginia	2,297
Texas A&M College	Texas	2,217
Clemson	South Carolina	1,549
Virginia Military Institute	Virginia	1,407
Pennsylvania Maritime Academy	Pennsylvania	800
New Mexico Military Institute	New Mexico	710
Augusta Military Academy	Virginia	500
Norwich University	Vermont	495
Gordon Military College	Georgia	450
Massachusetts Maritime Academy	Massachusetts	450
North Georgia College	Georgia	371
Pennsylvania Military College	Pennsylvania	350
The Citadel	South Carolina	316
New York Maritime Academy	New York	300
Wentworth Military Academy	Missouri	553
Columbia Military Academy	South Carolina	287
United States Coast Guard Academy	Connecticut	184
Linsey Military Academy (School)	North Carolina	106
Army and Navy Academy	California	72

officers (of whom 67 percent had a year or less of service) and about 6,000 National Guard officers. By the end of World War I, one year and five months later, the US Army in France numbered over 2,000,000 men.[21]

The military schools gave a trained cadre of officers for a military force that was required to be ready, quickly expanded, and sent to combat in France. Texas A&M provided the largest number with 1,233, followed by VMI with 1,176, Culver Military Academy with 1,166, Virginia Tech with 638, Norwich with 345, the Citadel with 276, Wentworth Military Academy with 216, and Pennsylvania Military College with 208. Particularly benefiting from these contributions were the National Guard's 28th Division from Pennsylvania; 29th Division from Virginia, Maryland, and

New Jersey; 30th Division from Tennessee, North Carolina, and South Carolina; 35th Division from Missouri and Kansas; and 36th Division from Texas and Oklahoma. Additionally, newly formed and recruited National Army divisions had the benefit of many military school alumni as well: 79th Division from Pennsylvania, Maryland, and Washington, DC; 80th Division from Virginia, West Virginia, and Pennsylvania; 81st Division from South Carolina, North Carolina, Florida, and Puerto Rico; 83rd Division from Pennsylvania and Ohio; and 90th Division from Texas and Oklahoma.

The Supreme Allied Commander during the war, Gen. Ferdinand Foch, visited the United States after the war and said there were two things he was most impressed with: the size of the country and the widespread network of military schools. Symbolic of that conclusion was the parade in his honor in Richmond in 1921. The parade's eight military units were led by the corps of cadets of Virginia's two military colleges, VMI and Virginia Tech, and the last two units were the cadets of two of the state's military high schools: Benedictine High School and John Marshall High School. Although no nationwide accounting of World War I military school contributions has been made, table 10.1 shows some examples.[22]

The Lost Generation

World War I, with its incredible toll in human life, prompted young Americans to rethink the patriotism of 1898. Young people of the post-war period had acquired a more independent point of view, in contrast to that of those raised in the 1890s. In a *Rotarian* article of 1932, E. E. Calkins wrote: "A potent factor in the disillusionment of youth was the World War." Christopher Nack described the post–World War I feelings in the United States as "a growing sense of disillusionment that would swell throughout the 1920s and . . . define public sentiment toward the Great War." The growing antimilitary sentiment was reflected in the literature of the Lost Generation, college students' protests, popular cinema, and congressional actions.[23]

The "Lost Generation" describes a generation "uprooted by World War I and subjected, as a result, to intense philosophic despair and disillusionment with traditional ideals and beliefs, especially the values of prewar middle-class America." Among the generation's authors who highlighted the futility of armed conflict were John Dos Passos, in his work *Three Soldiers*, published in 1921, and Ernest Hemingway in his 1929 novel *A Farewell to Arms*. These works and others provided an image for those who had not fought in the war: the ugly truth of war and a "catalyst for the public to develop "peace consciousness."[24]

In the late 1920s students at Midwestern civilian universities voted to abolish mandatory military training; similar protests occurred at eastern civilian colleges as well. In 1927 the American Federation of Youth, an umbrella for approximately fifty groups, called for resistance against being "drafted as cannon-fodder for future imperialistic wars." In 1934 a nation-wide student strike involved twenty-five thousand college students. Three years later a similar strike involved fifty thousand students. Further contributing factors to a growing antimilitary climate were films highlighted by the classic *All Quiet on the Western Front*, which won Academy Awards for best picture and best director as well as nominations for cinematography and script in 1930.[25]

Amid the antiwar feelings, the US government withdrew the US Marines from Nicaragua in 1927, and Congress passed the Kellogg-Briand Pact and the Neutrality Acts of 1935 and 1937. The Kellogg-Briand Pact, ratified by the United States and sixty-one other nations by 1928, called for outlawing war. The Neutrality Acts addressed prohibitions against arms merchants and the munitions industry that supported belligerent nations.[26]

Public Debate over Military Schools

As a result of antimilitary feelings in the United States, military schools were viewed by many as part of a program to "hoodwink youth which must furnish the raw material for adult wars. . . . The soldier mentality is utterly inconsistent with the spirit of modern youth."[27] Among the many groups that challenged the military school movement was the Committee on Militarism in Education (CME), established in 1925: "[Its purpose was] to combat military training requirements at public schools and universities. The CME fought to remove military training, in the form of Reserve Officer Training Corps (ROTC), from high schools and to eliminate compulsory ROTC service at state universities. Throughout its fifteen years of existence, the CME endeavored to oppose militarism in all institutions dealing with youth."[28]

In 1926 Congress held hearings to consider the abolishment of compulsory military training at public schools and state colleges. The focus of the hearings was not military schools, but those schools that were not essentially military and had adopted a mandatory military training requirement. The danger that faced the military school movement was twofold. Part of the goal of the proponents was to limit the funding to provide active-duty army trainers for programs, particularly at the high school level. The second danger lay with attacks on the positive attributes that the military school movement claimed were provided by their format of education. Although a great deal of testimony included concern for the general militarization of public education, there were attacks on the very image of the military school. Among the accusations was that the War Department was "making goose steppers out of school boys." The CME characterized military training in school as both murderous and un-Christian: "An army exists to kill men, when ordered, in the nation's quarrel regardless of its justice; it should train men to that single end. If we object to any of our citizens, those specializing in murderous and un-Christian activities, we should abolish the Army. If we want an Army, we should recognize it for what it is. We should not tell lies about it being a school of citizenship or manual training, nor clutter up its drill grounds with disciplines of these irrelevant arts."[29]

Articles that questioned the positive elements of military school education became more frequent. In a 1928 edition of the *Rotarian* magazine, Rev. Ernest Title called for a nationwide protest against the expansion of military training and the philosophy of militarism. He saw the increase of military training, not only in military schools but also through the institution of ROTC, as "raising a generation . . . impregnated with the idea that war is inevitable." A 1930 article in *The Outlook* magazine was more focused on military schools. The author, Archibald Rutledge, was the head of the English department of the nonmilitary Mercersburg Academy and portrayed military schools as "summer camps . . . where much is made of uniforms, drills, discipline and horseback riding." His view of military school discipline was that it was "necessary of course for an army but it is abnormal for children." In summary, Rutledge thought the military school concept was an attempt to combine "true education" with military training, resulting in "mating beauty with the beast."[30]

In 1931 the Prevention of War Council called on the legislature of Oklahoma to end support of ROTC as a "costly waste." This action may have been inspired in part by the establishment of the Oklahoma Military Academy, which was supported by both state and federal funds and served initially as a secondary preparatory school that later expanded to include a junior college. In 1930 the academy numbered 289 cadets and sixty horses and was supported by eleven soldiers to train the cadets in cavalry skills. The fact that Oklahoma Military Academy was a public institution made it an important target for the Prevention of War Council.[31]

The Emergence of the College
Prep Requirement

The criticism of the relevance of military schools inspired administrators to look internally to ensure they offered a high-quality education. This further focused the military schools at the secondary level to strive to send their cadets on to college well prepared for both college and a career outside the military. An excellent example in 1927 was Culver Military Academy's Committee for Academic Improvement. Under the leadership of Leigh Gignilliat and Alexander F. Fleet, the committee helped improve Culver's academic administrative organization, established procedures for faculty selection, and called for stricter academic standards and a broadened curriculum.[32]

Final formation

Culver Military Academy final year formation, from the 1945 yearbook. The corps was organized into a regiment of a battalion, cavalry squadron, artillery battalion, and band. (Courtesy Culver Academies)

Also involved with the defense of the military format was Prof. Michael V. O'Shea, a well-known educator of the period. Despite encouragement from his associates to condemn the military school format, he assisted in the format's defense. He supported military schools for the education of boys as a student-focused effort rather than a method of crushing individuality. He saw the "military schooling setting as requiring the child to work in a group to achieve a social, rather than an individual goal [and providing a vehicle for] achievement into wider interest and achievement demanded by wider social life."[33]

Culver's adjustments to curriculum inspired and prepared 20 percent more graduates intending to pursue college. Fleet was instrumental in pushing the academy in the direction of a college prep school at a time when few Americans went to college. The academy invited experts from several distinguished colleges to provide their input for academic improvements. Over the course of several years, the academy, with William R. Gregory as dean, developed the doctrine that Culver's cadets for the most part were there for college. With the announcement of that doctrine in 1935, Culver Military Academy led the secondary military schools, after much study and reflection, in a direction that would be clearly college preparation.[34]

Impact of the Pacifist Movement

By 1937 the objections of the pacifist movement to military training in schools were focused in four areas. First, there was no legal requirement for the land grant colleges to have ROTC. Second, the militarism created by military training was not in line with American ideals of peace. Third, compulsory military training was alien to modern education, and its use implied a future pending conflict. After the devastation of World War I, this last fear was strong and reflected the hope that the war was the war to end all wars. Finally, the conduct of military training was conducive to hazing and student mistreatment.[35]

Prior to World War II the influence of the antimilitary movement can be measured in the decreased number of ROTC programs on college campuses. By 1937 that number had decreased from 191 to 122. In addition, seven colleges had ended their compulsory requirements for males for the ROTC program.[36] Although the movement's direct impact on military schools, as opposed to the ROTC, is harder to measure, the antimilitary movement was a significant factor in the end of the expansion of the number of military schools and surely also had a hand in declining enrollments.

The Great Depression and
Other Challenges

The Great Depression took its toll on the military school movement. Particularly hard hit were the smaller military schools that had little endowment to fall back on and, with declining enrollments, became insolvent and closed. The Great Depression started in 1929, reached its depth in 1933, and extended into the onset of World War II, 1939–41. During that period construction, which had peaked in 1925, had fallen 47 percent by 1932. By 1933 the consumer price index had fallen 18 percent from 1929 levels, and the gross national product had fallen by 29 percent. Unemployment rose from 3.2 percent to 24.9 percent at a time when only eight states had any type of unemployment compensation, and these were totally inadequate.[37]

Between 1929 and 1938, discounting the military schools that transitioned to nonmilitary schools, fifty-one military schools closed. Those schools that survived took a variety of actions to remain economically viable. Carlisle Military School in South Carolina had a steady decline in enrollment from 110 to about 50, despite cutting the tuition from $470 to $380 per year. The Methodist Church withdrew support and was forced to lease out the school. What steps James F. Risher took after leasing the school were not documented, but he took the school back to 100 cadets in 1936, and as the Depression ended there were 150 cadets for the school year 1938–39. Harry Abells, the superintendent at Morgan Park Academy in Illinois, took on the crisis in revenue by opening a junior college program in 1933 and conducting summer school (likely nonmilitary) that also enrolled girls.[38]

Enrollment at Culver Military Academy in Indiana dropped from 695 cadets in 1929 to a low of 335 in 1934. In an effort to encourage enrollment, tuition was decreased from $1,500 to $1,100 per year. There was a net profit of $30 per cadet in 1929, but by 1934 the school took a loss of $60 per student. In 1932 two teachers were offered free room and board as payment. From 1932 to 1938 the economic crisis caused the school to make a significant reduction in teaching staff. In 1932 and 1933 staff or faculty positions were reduced by ten; between 1934 and 1936, faculty decreased by an additional forty, leaving a total of seventy-three faculty. Before summer enrollment in 1935, the school tried to increase its income and reduce costs by reducing salaries. St. John's Military School of Kansas similarly reduced salaries. There, 50 percent reductions were taken by the superintendent, commandant, and entire faculty and staff. The alternative was clear, with enrollments from 1934 to 1936 about 30 cadets.[39]

The St. John's Military Academy of Wisconsin had an enrollment decline from 1929 to 1932 of 50 percent. The town's bank and the school, with a $51,000 debt, were so closely linked and on the brink of bankruptcy that negotiations that kept the two institutions open involved a loan of the needed money by one of the bank's officers with personal guarantees from every one of the academy's board members. Additionally, the school expanded summer camp to include girls, rented the facilities for the Chicago Bears football team's summer training, and expanded opportunities to include aviation training to attract additional cadets. Finally, the school was transferred from the ownership of the Smythe family to a nonprofit corporation.[40]

It is no wonder that so many small military schools were forced to close. From 1932 to 1933 private schools had among the smallest enrollment numbers in history. Despite this, administrators such as Colonel Gignilliat of Culver and James Risher of Carlisle Military School took the actions required to save their schools from financial ruin. The top twenty-five private military boarding schools in 1932 were still able to attract 8,069 cadets, with an average enrollment of 323.[41] Those that survived the Depression did so through determined fiscal management.

Legal Challenge

In 1922 there was a major legal challenge to private military schools. That year the State of Oregon Compulsory Education Act amended the state's constitution, passing by a vote of 115,000 to 101,000. This act required children ages eight to sixteen to attend public school. At the time, there were two private military schools in the state, Hill Military Academy and the Oregon Military Academy. The preliminary campaign to amend the state's constitution was undertaken by a committee of Scottish Rite Masons, who stressed the importance of "children being democratically trained in the common schools." The Ku Klux Klan was an open supporter of the act and could produce fourteen thousand members in the state. Fighting against the act were the Catholics, Lutherans, Seventh-Day Adventists, Episcopalians, and at least twenty well-known Presbyterian ministers. The law required ten thousand private and parochial school students to transfer to public schools. Such a law could have become a challenge to the majority of nationwide military schools as other states adopted similar laws.[42]

The Sisters of the Holy Names of Jesus and Mary, representing their elementary and high schools, along with Hill Military Academy challenged the law in court. By the time the court challenge reached the US Supreme Court in 1926, the Oregon Military Academy had closed. The sisters argued that the act was a violation of "parents' choice of school, rights of schools and teachers rights to engage in a useful business profession." Hill Military Academy claimed the act "violated the Fourteenth Amendment of the United States Constitution against deprivation of their property without due process." The Supreme Court sided against the state and in favor of Hill Military Academy and the Sisters of the Holy Names.[43]

Public Education

Between 1879 and 1886 enrollment in private schools plummeted from 73 percent to 32 percent; in 1889 it was down to 18 percent, and 11 percent in 1910.[44] This was due to expansion of the availability of public education at the primary and secondary levels. The expansion of public education did not incorporate, for the most part, the military school format. Many smaller military boarding schools and most nonboarding military schools were faced with challenges in enrollment as their cadets and potential cadets were incorporated into free public education. This forced many military schools to close, and a number of their administrators took positions in public school district leadership. City University's founder, Prof. Edward Wyman of St. Louis School, became president of the board of directors of public schools, and St. John's Academy's Rev. Richard L. Carney became superintendent of public schools in Alexandria, Virginia.

The dramatic decline of private schools, along with an expansion of public education, came at a time of heightened post–Civil War patriotic spirit. The advent of public education resulted in the closure of many smaller nonboarding military schools, with their functions absorbed into public education. At the same time, the patriotic feelings of the period helped increase private boarding military schools, and as a result military schools occupied a larger portion of the total private school population. A few years before World War I, military schools and preparatory departments of private colleges accounted for 73 percent of private school enrollment.[45]

Fire

Clyde R. Terry, an Ohio Wesleyan graduate and former World War I army chaplain, opened Kansas Military Academy in 1919. On the night of January 10, 1927, a fire at the academy trapped Cadet J. Kenneth Lum, an eight-year-old boy, on the second floor. When the fire chief ordered his men not to enter the second-floor room, Colonel Terry climbed the ladder to the window and entered the room. The dense smoke and burning ceiling and walls made it impossible to see. Crawling into the room, Terry felt his way to the bed. Carrying Cadet Lum as far as the window, the colonel passed out as he tried to push the boy out the window. Observing the attempted rescue, the firefighters pulled the two from the burning building. They were unable to revive the young cadet, but they saved Terry, who had suffered serious burns.[46]

Fire, particularly in the early twentieth century, was a real danger for boarding schools. Unlike most schools that suffered major fires, the Kansas Military Academy did not close. Terry took five of his teachers and forty-four cadets by train to Aledo, Illinois, and with the financial help of Aledo businesses, they reopened the school under the name Illinois Military Academy. Within seven years the school grew to include first through twelfth grades, a thirteen-member faculty that included a US Naval Academy graduate as commandant, enrollment of one hundred cadets, and a small junior college.[47]

Terry's good fortune in finding support in his newfound home in Illinois demonstrates the importance of strong community support for private military schools. In September 1896 a fire at another military school, Missouri Military Academy, closed the school. All eighty-six cadets escaped, and in October the founder, Col. Alexander F. Fleet, with five staff and seventy-two cadets, traveled by train to Culver Military Academy in Indiana, where there was dormitory space. Although Colonel Fleet never returned to Missouri, the business association of Mexico, Missouri, along with the superintendent from the closed Alabama Military Institute (also due to fire), rebuilt Missouri Military Academy and reopened it in 1900. Other schools that survived their fires and rebuilt included Rectory School in 1867, California Military Academy in 1873, Greenbrier Military Academy in 1905, Peekskill Military Academy in 1909, Chamberlain Military Institute in 1911, Horner Military Academy in 1914, Randolph Macon Academy (Front Royal, Virginia) in 1927, and Valley Forge Military Academy, Pennsylvania, in 1929.[48]

A tragic fire in 1906 burned North Hall, Milner Hall, and the school annex of Kenyon Military Academy in Ohio. The early-morning fire occurred while eighty-five cadets slept. At the roll call formation, several cadets were missing, and the bodies of three cadets were found in the debris: Cadets James Fuller, eighteen years of age; Everett Henderson, eighteen; and Winfield Kunkel, just fifteen. The Kenyon Military Academy closed the next year.[49]

Table 10.2. Military School Closings Because of Fire

Year of Fire	School	Location
1855	Maryland Military Institute	Maryland
1895	Pullman Military Academy	Washington
1898	Worthington Military Academy	Nebraska
1900	Alabama Military Institute	Alabama
1900	Clinton Liberal Institute	New York
1902	Claverack College and Hudson River Institute	New York
1904	Rockland Military School in New Hampshire	New Hampshire
1908	Betts Military Academy	Connecticut
1908	Nebraska Military Academy	Nebraska
1912	Wilson Military Academy	New York
1915	Rock River Military Academy	Illinois
1915	Tupelo Military Institute	Alabama

Samuel Rogal, in his book on American precollege military prep schools, named fire as the "traditional demon of military schools." The examples shown in table 10.2 of military schools that closed as a result of fires in the early twentieth century confirm Rogal's statement.[50]

After 1908 the number of fires decreased partly because of safety measures enacted following the major tragedy at the Lakeview Elementary School in Collinswood, Ohio. On March 4, 1908, the public elementary school caught fire due to an overheated steam pipe and killed 172 children and two teachers. National reaction resulted in reforms that included outward opening doors, fire inspections, and stricter laws associated with safety and fire prevention. As a consequence, there was a new awareness about safety issues, and new construction for military schools included fireproof barracks. Examples of schools with new improved construction after the 1908 tragedy are Staunton Military Academy, Blee's Military Academy, Culver Military Academy, Northwestern Military and Naval Academy, Tennessee Military Institute, Texas Military Institute in San Antonio, and Augusta Military Academy. Missouri Military Academy advertised that their buildings destroyed by fire were replaced by fireproof ones. New York Military Academy claimed in 1912 to have "the most complete fire-proof military school in the United States."[51]

One Hundred Twenty Years of Growth
in Military Schools Ends

Between 1879 and 1910 the impact of public education decreased enrollment in private education by 61 percent. This shift to elementary and
secondary public education for the most part did not carry with it the
military format. However, despite the reversal the popularity of military
schools emerged with a growing number of military boarding schools.
Public education did not end the peak of military school representation in
America but limited that peak.

From 1898 to 1926 there was a period in the educational history of the
United States in which the military school format was generally accepted
as a positive force for the education of young boys and men. For 116
years, from 1802 to 1926, the number of military schools had grown and
peaked. However, the political environment changed, and with it came a
growing questioning of the military nature of schools that slowed growth.

The economic challenge of the Great Depression, beginning in 1929
and extending into the period 1939–41, ended the growth in number of
military schools. With military schools numerically centered on private
education, fifty-one schools closed, and most others were threatened. It
was the Depression above all other factors that ended the rise of the military school in the United States.

Military school educational leadership found ways to address the Depression and protect particularly the larger schools. They adjusted construction to safeguard cadets from school fires. They adjusted educational
goals and standards to help military secondary schools stay relevant against
the attacks of the pacifist movement and counter the attitude of the Lost
Generation. As World War II approached, the number of military schools
in the United States had peaked, but it remained strong and ready to again
support the military leadership needs of a nation at war.

★ 11 ★

World War II through
the 1950s

During the period between 1903 and 1926, the number of military schools in operation was approximately 278 to 280. After the impact of the Great Depression, by 1939 that number of military schools had decreased to 228. Despite the demise of more than 50 schools, military school enrollment reached its highest point prior to the entry of the United States into World War II. The reason for this was threefold: Cadets enrolled in the land grant colleges grew to even outnumber the federal military academies' enrollment. The surviving military schools' enrollments were increasing with a recovering economy. Finally, the need for a strong military, highlighted by the approach of another world war, caused the public to embrace the military school concept again.

In 1935 enrollment at West Point was authorized to increase to 1,960 cadets, and the US Naval Academy's barracks capacity was 3,100. But seven years later, in 1942, cadet enrollment at each of the three land grant military colleges exceeded 2,000, Texas A&M having 6,543 cadets, Clemson 2,364, and Virginia Tech 2,490, for a total of 11,397 cadets, up from 9,044 in 1938. Since their establishment, the land grant military colleges had steadily increased their cadet enrollment. The Citadel had 1,980 cadets; VMI, 761; North Georgia College, 536; Norwich, 528; and Pennsylvania Military College, 232.

The popularity of military education at the college level was indicated by its growth. In 1923, when Virginia Tech made participation in the corps voluntary for juniors and seniors, the vast majority of returning junior and senior cadets remained with their classmates in the barracks and with their cadet company. Their experience in the corps produced cadets who were devoted and felt "close camaraderie fused with self-discipline, personal honor, respect and fidelity." Virginia Tech's president, Paul Barringer, contrasted that esprit de corps with his experience as a University of Virginia graduate: "At the University [of Virginia] we had an academic like affection, but here it is like a personal matter if you touch the Virginia Polytechnic Institute [Virginia Tech]. It is like a loyalty I have never seen."[1]

From 1938 to 1942 there were eleven military junior colleges, six maritime academies, and several primarily secondary military schools offering one or two years of college. The military junior colleges functioning at the time were Castle Heights Military Academy; Gordon Military College in Georgia; Georgia Military College; Kemper Military Academy in Missouri; Marion Military Institute in Alabama; New Mexico Military Institute; North Texas Agricultural College; Valley Forge Military Academy and College in Pennsylvania; Wentworth Military Academy and College in Missouri; Schreiner Institute in Kerrville, Texas; and Texas Military College in Terrell. Together, these schools and colleges had approximately twenty-one hundred college-level cadets.[2]

In 1942 the six maritime academies, including the new federal US Maritime Academy, had a combined enrollment of approximately 2,500. This number includes the Maine Maritime Academy, established with just 28 cadets, and the other four state academies: Pennsylvania Maritime Academy, California Maritime Academy, New York Maritime Academy, and Massachusetts Maritime Academy. There were also schools that were primarily secondary in nature but offered one or two years of military programming. These colleges included Allen Military Academy (one year), Edwards Military Institute in North Carolina, and the Oklahoma Military Academy, which had a college enrollment of approximately 200 cadets.

With the 2,496 cadets authorized for West Point in 1942, a slightly larger number authorized for the US Naval Academy, and the Coast Guard Academy with 219 cadets, the total military school enrollment at the college level was approximately 26,300. In 1942 the United States had 172 military schools and military junior colleges with a combined enrollment of approximately 34,800 cadets at the primary and secondary school levels. The total enrollment of military schools in the United States in 1942 was approximately 61,100 cadets and midshipmen enrolled in primary, secondary, and college-level education.

As the United States involvement in World War II approached and the Great Depression receded, antimilitary sentiment by the pacifist movement weakened. The onset of World War II created a "common bond, a shared purpose, and a spirit that cemented relationships and concern for one another. There was conflict among individuals and groups as there always had been, but dedication to winning the war took precedence over all concerns." The changing social attitude toward the military and military schools was mirrored by Hollywood's themes.[3]

Hollywood's Promotion of the Military School Image

One of the first indications that Hollywood was reflecting a positive image of military schools was a series of films addressing the US Naval Academy. The first, in 1933, *Midshipman Jack*, was followed by three films in 1935: *Annapolis Farewell, Shipmates Forever,* and *Annapolis Salute.* Soon after the premier of these films, *Navy Blue and Gold* focused on the naval academy football team, and it, like the other movies, told a positive story of military school life. In fact, a New York film critic commented that he wondered how the academy recruited their midshipmen "before Hollywood took up its pious burden of glorifying it."[4]

In 1938 *Brother Rat,* with Ronald Reagan and Eddy Albert, was released. This film was about the Virginia Military Institute, and much of it was filmed on location. The theme emphasized the institute's honor code, its military heroes, and its ideals. In 1939, a positive image complete with the popular Jackie Cooper was set and filmed at Culver Military Academy: *The Spirit of Culver.* After the war started, *The Major and the Minor,* released in 1942 and starring Ginger Rogers, was set in a military secondary prep school and was heralded as "one of the year's freshest and funniest movies." Also in 1942, *Ten Gentlemen from West Point* was released, and it was a historical look at one of West Point's early classes. *We've Never Been Licked,* a movie with an espionage plot also made in 1942, was set and filmed on the Texas A&M College campus and featured the military traditions of the Aggie Corps.[5]

After World War II and through the 1950s, Hollywood movies continued to reflect the positive attitude in the United States regarding military schools. The *Spirit of West Point,* released in 1947, was a football story featuring two cadets who exemplified the academy's honor code and ethical and gentlemanly conduct. James Cagney and Doris Day starred in *The West Point Story,* a musical that featured a Broadway director trying to help West Point cadets put on their annual show.[6]

The 1950s saw the release of a number of films depicting military schools in a positive, light-hearted manner. In 1952 Donald O'Connor starred in a comedy film with a West Point theme, *Francis Goes to West Point*. In 1955, *The Private War of Major Benson* starred Charlton Heston as a hardened combat veteran who softens to the needs of his secondary military school cadets. The same year, *The Long Gray Line*, starring Tyrone Power and Maureen O'Hara, told the story of a positive relationship between an Irish sergeant athletic coach and his family. It featured the experiences of generations of cadets, including MacArthur, Bradley, and Eisenhower.[7]

An exception to the positive image was the movie *The Strange One* (1957), set in a Southern military college. Ben Gazzara starred as a senior cadet who uses his position to terrorize underclassmen. In the end the cadets unite against him with their corps leadership and physically remove him, placing him on a train out of town.[8] Despite this exception, the great majority of films produced in the 1930s through the 1950s supported the public positive sentiment toward the American military school.

Military School Associations

Between 1914 and 1950, two associations were formed among military schools. The first was formed in 1914 by military school owners and administrators. Educational associations were not a new concept at the time. The National Educational Association was established in 1861, and the Maine Educational Association in 1866. Patterson's American Educational Directory in 1918 listed 117 associations representing various types of schools, teachers, and administrators.

The Association of Military Colleges and Schools of the United States (AMCSUS) was formed through the efforts of Col. J. C. Woodward of Georgia Military College. Woodward contacted each of the military colleges, military junior colleges, and military secondary schools and invited them to attend a conference in December 1913 in Washington, DC. As motivation, the colonel proposed two questions: Were their institutions not performing a "far greater service for education in general, and training of our citizen soldiers in particular, than is generally known? Why have we so long labored without more definitely organizing effort?" Colonel Woodward alluded to the need for an organization to promote "national policy, relationship to the War Department, obtaining more assistance from the Government and gaining more public interest."[9] The meeting of AMCSUS was conducted in the Ebbitt House in Washington, DC. Attendees included representatives of Augusta Military Academy in Virginia, Culver Military Academy of Indiana, New Mexico Military Institute, VMI, and Virginia Tech.[10] As a result of the meeting, the association was established the next year in 1914, just in time to make a significant long-term impact for not only military schools, but also military training in schools in general.

In 1916 the House Committee on Military Affairs held hearings on a bill to "increase the efficiency of the Military Establishment of the United States." At that time Europe was deep into World War I, and defense was paramount in the minds of the government. Sebastian C. Jones, the second president of the AMCSUS, led the delegation to the hearings. Jones, the superintendent of New York Military Institute, described the association as having grown to a membership of forty-two military schools with 10,000 cadets, several colleges with enrollments of 1,000, private schools with enrollments of 300 to 450, and several other schools with enrollments of 150 to 175. The hearings included testimony not only from Jones, but also from representatives of several association military schools: Columbia Military Academy of Tennessee, Culver Military Academy of Indiana, Georgia Military Academy, St. John's College of Maryland, and VMI. The hearings and the resulting congressional action established the ROTC programs, and the act included advantageous support and recognition for military colleges and schools.[11]

The second military association established was the National Association of Military Schools in 1950. At its height of membership, it represented 126 schools. The founder was Col. W. C. Tommy Atkinson, the longtime superintendent of the Army and Navy Academy in California. The original members of the association were Army and Navy Academy, Colorado Military School, McCallie School in Tennessee, and Roosevelt Military Academy in Illinois.[12] Beginning in 1956, membership, which had been limited exclusively to military schools, was expanded to include both private and public schools and public school districts that had Junior ROTC or National Defense Cadet Corps programs. The National Defense Cadet Corps is an organization that is similar to JROTC but does not receive funding for instructors. The National Association of Military Schools was comparable to AMCSUS in that its goal was to provide a united voice for obtaining federal support for its membership.[13]

The National Association of Military Schools was successful in its mission, but with declining enrollments of military schools in the 1960s and many schools transitioning to a civilian format, the association encouraged amalgamation with AMCSUS. The president of the National Association of Military Schools, Col. Keith Duckers, from St. John's Military School in Kansas, and Col. Charles Stribling of Missouri Military Academy, president of AMCSUS, were personal friends, which helped to facilitate combining organizations. In 1972, the same year that Colonel Atkinson, the founder of AMCSUS, retired from the leadership of the Army and Navy Academy, the National Association of Military Schools merged into AMCSUS.[14]

AMCSUS maintained its membership as military schools rather than incorporate those nonmilitary schools with JROTC from the National Association of Military Schools. The membership consisted of military colleges and universities, including maritime academies, junior military colleges, and private military schools. In 2011 the membership included twenty-six private military schools, five junior colleges, and nine military colleges or universities, including the maritime academies. Oakland Military Institute College Preparatory Academy, a charter school, was the first public school to be admitted into the association. Since 1914, the association successfully advocated for a variety of issues associated with common themes of interest. Between 1923 and 1995 there were 133 topics addressed at the association's annual meetings. Examples of those topics include administration, advantages of military schools, alumni publications, asbestos problems, athletics, difficult cadets, business/financial management, coeducation in military schools, cooperation among military schools, development/fundraising, ethical relations among schools, faculty, federal military inspection ratings, the honor system, insurance, military training, private schools compared to military schools, and the value of military education. The topics reflect the challenges faced primarily by the private military secondary schools, which comprised the majority of the members.

The War Department on Campus

West Point's continuity was heavily impacted during World War I, with the class of 1918 graduating in August 1917 and the class of 1919 in June 1918. To make matters worse, the losses among officers in France resulted in the commissioning of the sophomore and junior classes (classes of 1920 and 1921) as lieutenants for service in France on November 1, 1918, leaving only the plebes who had arrived in June.[15] West Point, like all military colleges, depended on a class system that ran smoothly, and war disrupted that systematic assumption of responsibility.

After the war, Brig. Gen. Douglas MacArthur, an alumnus of both West Point and Texas Military Institute (called West Texas Military Academy prior to 1926), became the superintendent of West Point and made positive changes that resulted in his being named the father of modern West Point. He moved to outlaw hazing, expand the sports program, establish senior privileges, and modernize both academic and military training.[16]

In the army's quest to avoid the mistakes of the World War I disruption, West Point, at the onset of World War II, established a more orderly three-year curriculum. The class of 1943 graduated in January 1943, and the class of 1944 in June 1943. The traditional four-year curriculum would not be reestablished until September 1945. Likewise, the US Naval Academy accelerated its program, and the class of 1941 graduated in February 1941 and the class of 1942 in December 1941, and the school converted to a three-year curriculum with additional academies conducted in the summer that included 88 percent of the four-year curriculum's academic instruction.[17] The Coast Guard Academy graduated the class of 1942 in December 1941 and the classes of 1943, 1944, 1945, and 1946 each a year early.

The land grant military colleges were not offered an orderly alternative but were seen as an almost immediate source of soldiers in World War II. In 1918 most of the older cadets in the land grant military schools were enlisted in the army's Student Army Training Corps. These cadets became part of the enlisted military force of the army and navy and in 1919 were intended to feed officers into the services. The land grant military colleges and other military college campuses had an army unit that included both former cadets and a corps of cadets. Because the war ended in November 1918, the army quickly demobilized, because of the Spanish flu epidemic, and the war had little impact on the land grant military schools, with the exception of enlistments from the ranks of many of their faculty and cadets.

During World War II the profile for military colleges was very different from that of prior decades. After 1942, until the end of the war in 1945, many college-level military schools experienced a repeat of the situation that beset many Southern military schools during the Civil War. The enrollment of the cadets corps at the Citadel, Clemson, Pennsylvania Military College, Texas A&M, VMI, and Virginia Tech declined an average of 75 percent between 1942 and 1945. The cadet enrollment in 1945 was down to 427 at the Citadel, 757 at Clemson, 93 at North Georgia College, 134 at Norwich, and 27 at Pennsylvania Military College. There were 1,992 cadets at Texas A&M, 207 at VMI, and 211 at Virginia Tech.

Virginia Tech's experience was similar to that of the other land grant military colleges. In 1942 the corps of cadets at Virginia Tech was 2,640 strong. They were organized into a brigade of two regiments totaling five battalions, with each oriented toward a particular branch of the army: one infantry, one engineers, and three coast artillery. By 1945 the corps was reduced to a cadet battalion of two companies numbering only 211 cadets. Most of the underclassmen were eighteen years or younger, and many of the upperclassmen were either classified as 4F and rejected from military service or awaiting a reporting date for active military training.[18]

What transpired at Virginia Tech was similar to what happened at West Point during World War I. The curriculum was accelerated, and class loads were heavy. Seniors were called to active duty, followed by juniors.[19] In most cases these cadets were commissioned as officers. As the war accelerated, many cadets simply enlisted. By February 1943 the vast majority of senior and junior cadets had departed, and those remaining were assigned to the Army Specialized Training Program (ASTP), which had been activated on campus. Before the war ended, 3,387 men passed through the ASTP at Virginia Tech, the vast majority of them not former cadets.

The ASTP was established to train soldiers for technical skills associated with engineering, communications, medicine, dentistry, personnel psychology, and foreign languages. The commandant supervised active-duty organizations of the armed forces, including the ASTP, a Specialized Training and Reassignment (STAR) unit, an Army Specialized Reserve Training Program (ASRTP) unit composed of 296 seventeen-year-old high school graduates, and a naval preflight unit. Those still in the ASTP in February 1944, regardless of their training, were assigned as riflemen in the hard-hit infantry divisions in the European theater after heavy losses in Italy and France.[20]

The war reduced the collegiate level of military school enrollment outside the federal military academies and maritime academies by about 75 percent. Cadets were replaced by soldiers, and after a short time those soldiers were no longer former cadets but men who had no knowledge of the military traditions or military school ethos of those institutions. A small number of younger cadets and upperclassmen labeled unfit for active duty maintained the institutes' traditions as best they could. A greater challenge to those traditions would emerge after the war.

World War II

The contributions of military schools to the war efforts in World War II have never been completely documented. In 1944, Col. R. L. Jackson, superintendent of Western Military Academy, calculated the contribution of military secondary schools and military junior colleges to the war effort. Gen. Melton G. Baker of Valley Forge Military Academy in Pennsylvania estimated that in 1941 military schools provided 50,000 officers for the war effort in 1944.[21] Using these numbers as a starting point, and adding the officers of the maritime academies and federal military academies, along with officers produced by military schools late in the war, it can be estimated that the total number of officers contributed to World War II by military schools probably approached 95,500.

The influence of these individuals can be seen by examining the leaders of three echelons of command in the US Army. The first of these three echelons was the army (made up of multiple corps). Eleven armies were organized, ten of which saw combat. The next echelon was the corps, of which twenty-four were organized and twenty-two saw combat. The final echelon was the division, which numbered approximately sixteen thousand men and was organized as infantry, cavalry, armored, or airborne. Eighty-nine divisions were organized in World War II, and eighty-seven saw combat.

Twelve men commanded armies in combat. Of them, ten were products of military schools, West Point accounting for eight and VMI two. Thirty-four men led corps in combat; of them twenty-four had attended West Point, and of the remaining ten, three had attended military colleges—Mississippi A&M, Norwich, and VMI—and one a military high school: Shattuck Military School in Minnesota. At the corps level, twenty-eight of the commanders entering combat had attended one of America's military schools.[22]

A total of 155 men commanded the eighty-seven divisions in combat. Gary Wade selected 45 of them as a sample for his study of World War II leadership. A biographical survey of those men revealed that at least 29, or 64 percent, attended military schools: 22 were West Pointers, 3 were from VMI, 1 from Norwich, 1 from the US Naval Academy, 1 from Virginia Tech, and 1 from West Texas Military Academy.[23]

For the US Navy, the US Naval Academy's domination of flag officer ranks was virtually complete. Between the Spanish-American War and World War II, every active-duty admiral was a graduate of the naval academy. Although that did not remain the case, the overwhelming number during the war remained naval academy products, including all four of the fleet admirals (five stars). Coast Guard Academy graduates dominated that service's flag officers. The US and state merchant marine academies provided much of the leadership for a maritime force that lost 1,554 merchant

Texas A&M College Corps of Cadets in 1949 numbered 3,768 cadets organized into six regiments (two or three battalions each) and the band. (Cushing Memorial Library and Archives, Texas A&M University)

ships and 5,662 to 8,300 merchant seamen and officers in World War II—a sacrifice that remains largely forgotten.[24]

As in World War I, the largest contribution made by the military schools, both secondary and collegiate, outside the federal military academies was in field-grade (major to colonel) and company-grade (second lieutenant to captain) officers. Although VMI provided fifty generals, it also gave a total of 3,159 officers; Texas A&M had two generals but also provided 14,123 officers; Clemson, Norwich, and Virginia Tech provided sixteen generals and 4,142, 1,270, and 4,472 officers, respectively. Pennsylvania Military College and Wentworth Military Academy provided six generals and 495 and 670 officers, respectively. The value militarily of the nation's military schools is clear. Gen. Douglas MacArthur reflected on the sports fields of his beloved West Point: "On the fields of friendly strife are sown the seeds that on other days, on other fields will bear the fruits of victory." When leaders were needed, the military schools answered the call, as shown in table 11.1.[25]

Korean War

A similar look at the Korean War's leadership reveals the continuing pattern: military school alumni took the majority of leadership positions in those critical combat formations that fought and determined the outcome of the

Table 11.1. Examples of Military School Contributions to the World War II Effort

School	Alumni in Service	Alumni Killed
Augusta Military Academy (Virginia)		52
California Maritime Academy		11
Clemson (South Carolina)	6,984	301
Columbia Military Academy (Tennessee)		35
Culver Military Academy (Indiana)	6,008	302
Fork Union Military Academy (Virginia)	429	19
Kentucky Military Institute		76
La Salle Military Academy (New York)		31
Maine Maritime Academy	384	
Massachusetts Maritime Academy	1,100	21
New Mexico Military Institute	3,000	150
North Georgia College	2,953	100
Norwich University (Vermont)	1,711	861
Oak Ridge Military Academy (North Carolina)		42
Oklahoma Military Academy		67
Pennsylvania Military College	918	37
Texas A&M College	20,229	953
Texas Military Institute (West Texas Military Academy)	433	29
The Citadel (South Carolina)	6,300	200
United States Coast Guard Academy	786	20
United States Maritime Academy	5,600	210*
United States Military Academy	9,000	616
United States Naval Academy	10,819	
Virginia Military Institute	4,102	182
Virginia Tech	7,285	323
Wentworth Military Academy (Missouri)	813	80

*This number includes 142 cadets who died in service with the maritime forces during the war.

conflict. The Korean War saw three corps and nine divisions deployed to combat. The initial commanders included three from West Point, one from St. John's College (Maryland), one from VMI, and one from West Texas Military Academy. The last had been commissioned out of that secondary military school in 1912.[26]

There were twenty-seven US Air Force general officers in command positions in the Far East Air Force. The air force had only been stood up as an independent armed service in 1949 and did not yet have its own academy. Of the twenty-seven, twenty-four biographies were found, showing that seventeen, or 71 percent, had a military school background. Of the seventeen, sixteen had graduated from West Point (one of whom had attended Norwich before) and one had graduated from New York Military Academy (a secondary military school). The remaining general was a graduate of Texas A&M.[27]

The Marine Corps had two major commands engaged in the Korean War: the 1st Marine Division and the 1st Marine Air Wing. Of the eight generals who commanded these organizations during the war, two were US Naval Academy graduates, one attended VMI, and one had been commissioned right out of St. John's Military Academy of Wisconsin during World War I.

Despite the widening of sources of commissions with the establishment of ROTC in 1916, the contribution in senior leadership was significant from the nation's military schools. Fifty percent of the army's corps and division commanders in the conflict were products of military schools, and especially West Point. Seventy-one percent of air force generals in command positions in the Korean theater of operations were military school graduates, with a majority West Pointers. And 50 percent of the men who commanded the marine division or the marine air wing were military school products, half of whom had graduated from the US Naval Academy. Furthermore, among the generals included in this survey, two had been commissioned as officers straight from a secondary military school and went on to command an army corps and a marine division in a modern conflict. That speaks volumes about military schools and their historical military contributions.

Military Schools Fewer but Stronger

The numbers of military schools operating during the 1950s, as compared to the 1940s, had decreased by approximately 7 percent, but those remaining were stronger. The 1950s were a period of growth in enrollment for both military secondary schools and military colleges. Samuel Rogal's study of precollege military schools provides an extensive list of enrollments of military secondary schools. A review of them finds that sixty-five schools reflect enrollment figures from the 1940s, early 1950s (1950–56), and late 1950s through early 1960s (1957–63).[28] These figures indicate that the average growth in enrollment of secondary military schools between the 1940s and as late as 1956 was almost 18 percent. Among the sample schools the average size of the schools was 246 cadets in the 1940s and 290 cadets between 1950 and 1957. That growth continued into the early 1960s. The growth in enrollment between the period 1950–56 and 1957–63 was 27 percent. The average enrollment among the sixty-five military secondary schools had reached a healthy 369 cadets by the early 1960s.

When comparing those schools' enrollments in the 1940s with the late 1950s through early 1960s, the growth rate was 33 percent. Only six (9 percent) of the schools showed a decline in enrollment. Two of the six schools had one hundred or fewer students, and the decline in their enrollments averaged -38 percent. The average decline in enrollment of the remaining three larger military schools was only -6 percent. The overall health of the secondary military schools during this period and up to the Vietnam War was excellent and growing.

A good example of a school in this period of institution expansion is Castle Heights Military Academy. Castle Heights was established in 1902 as Castle Heights School in Lebanon, Tennessee, with ninety-six boarding students. The school transitioned to a military format during the first year of the United States' participation in World War I. The school quickly grew to some 415 cadets on its 150-acre campus by 1919. In 1929, Bernarr Macfadden, nationally known physical fitness advocate, millionaire, and publisher, purchased the school and would influence the institution to stress character development and physical development. Macfadden was orphaned at eleven years of age and worked on a farm. Lacking the education he desired, at twenty-one he became a coach and physical trainer at Bunker Hill Military Academy in Illinois, working for tuition and perhaps board. There he introduced a program of boxing, wrestling, and diet management that he exported to Castle Heights through Harry L. Armstrong, who he hired and who ran the school. Armstrong was assisted by Daniel Taylor Ingram, VMI class of 1921, World War I Marine Corps veteran, and former assistant commandant of Fishburne Military School. Ingram was commandant of Castle Heights from 1929 until 1963, when he became superintendent until 1965.[29]

Castle Heights Military Academy traditional end-of-year graduation circle formation, 1963. (Courtesy Castle Heights Military Academy Alumni Association)

Cadets Duane (1961) and Gregg (1963) Allman, Castle Heights Military Academy (Courtesy Castle Heights Military Academy Alumni Association)

Although the Depression hit the school hard, with enrollment dropping to just 150, by the 1940–41 school year the cadet corps enrollment was 415, and a junior college was added. By the 1960–61 school year there were 583 cadets. Among the cadets during that period were the brothers Duane and Gregg Allman. The brothers formed a rock band at the school, called The Misfits, in 1963, and went on to form the Allman Brothers Band a few years later—a band that was central to the rock scene from the late 1960s through the 1970s. The young culture that so loved their music was to dreadfully impact the story of military schools, and will be addressed in the next chapter.[30]

As the Vietnam War approached, the political, economic, and cultural environment in the United States was positive for military schools. For the most part, challenges in the past presented by the economy and the pacifist movement were absent. Schools had growing enrollments, and Hollywood was generally positive. In many ways this period from World War II up to the Vietnam War was a golden age for military schools. Although the number of individual schools did not reach levels that had existed prior to the Great Depression, the schools were stronger and had a positive image. However, to paraphrase Bob Dylan, "The times they were a-changin'."[31]

☆ 12 ☆

Vietnam and the Decline
of the Military School

Events of the 1960s, along with the Vietnam War, changed the perception of military schools and the military in general. The idealism reflected in President John F. Kennedy's "ask not what your country can do for you" speech died a slow death after his assassination, Robert Kennedy's murder, Rev. Martin Luther King's violent death, and the Vietnam War protests. Replacing patriotic idealism was a youth counterculture associated with long hair, rebellious dress and behavior, and political activism. All these behaviors ran counter to the military school ethos.

In 1975, *Newsweek* magazine contrasted earlier decades: "The nation's military academies regularly turned out battalions of neat, generally well-disciplined lads, whose handsome uniforms were often the envy of their civilian peers, and whose patriotism and values were for the most part mirror images of those that had been imparted to them by their parents and their military instructors." The article went on to say that military schools fell from favor due to "national frustrations over the War in Vietnam." This was a period when soldiers returning from Vietnam would change from their uniform when they arrived at the airport, because military dress was greeted with insults and degradation. The increasingly antimilitary national mood was reflected in a "public anathema toward military education."[1]

Young people diminished the value of a military school education and questioned the most basic principles of the military school ethos. Wiley Lee Umphlett, who has written extensively about the film industry, proposes that movie themes and messages in the 1940s reflected sociocultural changes in social behavior and the mood of America. Writing in 1984, he states that "displaying . . . a code of honor and standard of behavior that are no longer credible, the spirit theme of movies like the *Spirit of West Point* is as old fashioned today as the patriotic mood that inspired it."[2] This comment says as much about that earlier period as it does about the post–Vietnam War period and its attitudes toward the military school ethos.

The 1960s and 1970s, and, to a lesser degree, even the early 1980s, were heavily influenced by the antiestablishment generation. Originating in the latter part of the 1950s, Beat or Beatnik culture and its influence inspired youthful rebellion against traditional values that characterized the military school culture. A series of events pushed the countercultural movement to expand and touch the lives of millions. This influence played heavily on high school and college-age students.

Secondary Military Schools
Case Their Colors

During the twelve-year period between 1966 and 1978, seventy-three military schools closed or transitioned to nonmilitary format. At least thirty-three of the schools that made the transition also dropped JROTC. For those schools still operating in 2014, there was little or no mention in their publications that they had functioned as military schools in the past. During the same twelve years, only five military schools were newly established—one located in Puerto Rico and two associated with the heavily Cuban-influenced population of Miami. The Caguas Military Academy, established in 1975, is one of seven military schools in Puerto Rico in operation in 2014. Both the Miami Aerospace Academy and the Inter-American Military Academy operated in Miami, Florida, during the 1970s and into the 1980s. The other two schools that opened during this period of low popularity of military schools were General MacArthur Military Academy in New Jersey, which operated from 1966 to 1975, and the Penn Military Academy, Hesperia, California, which was active from about 1968 until 1970. The State of California offers an excellent example of how the educational landscape changed for military schools in a very short period of time.

In 1963 there were twenty military schools operating in California. Fourteen years later, in 1979, California had only four military schools: Army and Navy Academy, California Maritime Academy, St. Catherine's Military Academy, and Southern California Military Academy. Only Army and Navy Academy was a secondary boarding military school.[3] The military schools in California that closed during that period had operated for an average of forty-two years. San Rafael Military Academy (Mount

Mount Tamalpais Military Academy Cadet Corps in 1902. The school's founder, Dr. Arthur Crosby, appears in civilian clothes. (Courtesy Anne T. Kent California Room, Marvin County Free Library)

Tamalpais Military Academy) closed after operating for seventy-nine years, from 1892 to 1971. San Rafael Military Academy was founded as the Mount Tamalpais Academy in 1890 (and became Mount Tamalpais Military Academy in 1892). The school enrollment varied, with 174 in 1921 and 246 in 1960. In its early years the school boasted a mounted cavalry unit and a mounted artillery unit. However, shortly before the school's closing, the KFRC Fantasy Fair and Magic Mountain Music Festival was held in 1967 a short distance from the school. This event reflected the rapid expansion of a youth counterculture that doomed the school.

Some military schools attempted to adjust their programs to survive. Examples of schools that were successful include Texas Military Institute (TMI) and St. John's College High School in Washington, DC. In 1972 TMI admitted females, then allowed them the option to be cadets. Two years later, TMI made its corps of cadets optional for male students. With a military school format that harbored dual student bodies, civilian and cadet, the school continued to be successful.[4]

One school that made adjustments unsuccessfully was Columbia Military Academy in Tennessee. Established in 1905, the school was a victim of three social changes: the antimilitary feeling associated with the Vietnam War, desegregation, and the changing attitudes of young students. The antimilitary feeling during the 1960s made it difficult to recruit cadets to enroll in a military boarding school. And with the desegregation of public schools in the early 1960s, local day academies opened across the country.

These schools were additional competition to military boarding schools, which were already finding it difficult to retain cadets. Columbia tried to transition from a military boarding school to a military day school. In 1969 they admitted females, who were offered the option of participating in the military program. By 1978, too few male students wanted to participate in the military program, indicating that the "discipline and rigor of military training . . . central to the success and pride" had lost its appeal, and the school closed.[5]

Apparently it was not only the students who no longer saw the utility of military schools, but the parents as well. Western Military Academy of Illinois was established in 1868 and transitioned to a military school in 1892. In 1972, declining enrollment caused the school to close. The closure was attributed to parents who did not want to send their children to a military school.[6]

Military Colleges Caught in the Storm

The 1960s and 1970s were an intense period of antimilitary feelings in the United States. Military secondary schools, as well as junior colleges and military colleges, suffered declining enrollments and closed. Three military colleges were particularly hard hit. Norwich University suffered the indirect effect of declining freshman classes as fewer young men chose to enroll in military colleges. Norwich, a private college, found itself faced with a financial crisis. Between 1966 and 1975, cadet enrollment at the university dropped from 1,292 to 914. The decline in enrollment contributed to the operating deficit of the university. Financial support decreased from $426,091 in 1970, to $81,856 in 1974. Had it not been for endowments and increased enrollment beginning in 1975, the university's future might have gone the way of so many other military secondary schools.[7]

The antimilitary movement also affected Pennsylvania Military College and Virginia Tech. Neither of those schools was protected by the isolation provided by a purely military campus. Pennsylvania Military College (PMC) was established in 1853 by Theodore Hyatt, who converted the prep school into the Delaware Military Academy in 1858. The school moved to Pennsylvania in 1862 and changed its name to Pennsylvania Military Academy.[8] Like Norwich, the school was privately owned. By the 1960s Chester, Pennsylvania, had become an urban environment, complete with flames of petroleum refineries and the Delaware River shipyards close by. Just five miles away from Chester was Swarthmore College, which had a pacifist tradition and therefore supported an active antiwar movement.

Between 1962 and 1966, PMC started an evolution from a military college dominated by a uniformed cadet corps into a college of two student bodies, with the civilian numbers overshadowing the military. The school enrolled veterans as civilian male students, but in 1966 the administration opened enrollment to other male students, including "students whose educational and career objectives do not coincide with cadet living."[9] In 1967 the school began enrolling female students.

With the urban environment, hostile students from Swarthmore College, and a growing nonmilitary population at PMC, the corps of cadets dwindled. According to the PMC president, this was because of "the antimilitary climate of the nation."[10] The Citadel, VMI, and North Georgia College began to experience modest recovery in enrollment in 1971, while PMC reached its lowest cadet enrollment. The three military colleges that saw gains were isolated from many of the direct social effects because of their rural environments and military campuses, unlike the situation at PMC.

In 1960 PMC had 650 cadets, who made up 65 percent of the total enrollment. As the environment at military colleges deteriorated through external and internal influences and administrative enrollment decisions, cadet enrollment decreased in 1966 to 590, to 450 in 1969, and to 331 in 1970. In 1971 cadet enrollment was 277, just 18 percent of total enrollment. Furthermore, the demise of the military college was also attributed to Philadelphia's influential Quaker pacifist tradition, which faced a cadet alumni population that was largely socioeconomically blue collar and did not have the political influence to counter the Quakers' desires.[11]

Declining cadet enrollment was not only a matter of smaller incoming freshman classes, but also the defection of upperclassmen cadets. In 1969, 22 percent of cadets chose not to return to the school. The 1969 yearbook described a civilian college and praised the college president, as he "had placed the past in proper perspective and had given emphasis to the to-day and tomorrow of Pennsylvania Military College." In 1972 the school changed its name to Widener College and disbanded its corps of cadets. The school followed the recommendation of a management study and buried its 113-year military tradition as a "good marketing decision."[12] The alumni and remaining cadets, now civilian students, became highly embittered.

Until 1970 Virginia Tech seemed to be isolated from the antiwar and countercultural movement because it was "located in the Appalachian Mountains of Southwest Virginia in a region known for its cultural isolation and social and political conservatism." The civilian student body was characterized by antiwar activists as "apathetic" and the school as a "lousy cow college . . . [where] nobody cares." There were few, if any, activities at Virginia Tech during the 1969 nationwide campus protests, known as the Moratorium to End the War in Vietnam. Like Texas A&M, where a handful of students conducted a short march, Virginia Tech was known as a conservative school focused on academics coupled with a strong military tradition. In 1970 a small group of civilian students at the university would threaten the campus tranquility.[13]

With the invasion of Cambodia in April 1970 and a growing dissatisfaction among some students with the school administration's conservative student life policies, the signs of an antiwar and counterculture movement began to emerge. In 1970 Virginia Tech had a student population of eleven thousand civilians and 973 cadets. The cadets were housed in military barracks and were uniformed for class in their gray class uniforms or blue dress uniforms in a style similar to the West Point cut.

On April 14, 1970, protesters gathered to disrupt Tuesday afternoon cadet corps drill as the corps marched off the upper quadrangle to the centrally located drill field. Company A, the lead unit, diverted from its destination in an effort to avoid physical confrontation. Company B and the other companies of the corps arrived at their drill location to see Company I surrounded by protesters. They were forced to march and push their way through the gauntlet of chants and attempts to humiliate them; the protestors had toy guns, an upside-down American flag, and a large plastic pig. Members of the civilian student government attempted to get the protestors to end their activities, to no avail. The last company marching off the drill field was surrounded, and only through the maintenance of cadet discipline and assistance from a few civilian students were student leaders and cadet senior staff officers able to march the company back to their barracks, avoiding violence.

The event resulted in a closing of ranks among the cadets, coupled with feelings of isolation in the Upper Quadrangle. The university administration acted quickly, suspending ten leaders of the student demonstration and obtaining a court injunction against further action. After a march of two hundred on the main administration building in support of the demonstration, the campus returned to normal.

But the following month at Kent State in Ohio, four students were shot and killed by National Guard soldiers. At Virginia Tech, demonstrators reacted by occupying Cowgill Hall, which disrupted classes, but that occupation ended quietly. Then Williams Hall was occupied by 168

Protest at Virginia Tech, April 14, 1970. (Digital Library and Archives, University Libraries, Virginia Tech)

demonstrators. The state police responded to administration requests, and demonstrators were forcibly removed; all were loaded on tractor trailers for their trip to jail. The 107 students involved were suspended. A Virginia Tech online history module describes the resulting atmosphere: "The campus climate, once serene, became polarizing. Students found themselves forced to make a stand on both campus and national events." The result was that enrollment in the corps declined from 541 in 1972 to 355 in 1977. The negative civilian image of the campus, along with a less conducive environment, made recruiting cadets a difficult task. It would take many years of concerted efforts and a united front of alumni, cadets, and the commandant's office to increase enrollment of the Virginia Tech Corps of Cadets.[14]

Hollywood Becomes an Adversary

Following the Vietnam War, Hollywood portrayed through their films a poor image of military education. Films released in the 1980s included three films portraying military schools in a bad light. In 1980 *The Lords of Discipline* depicted a military college as an institution of bigotry where an African American cadet had to be protected from death threats. *Taps*, released in 1981, was particularly demeaning, as it addressed the closing of a military school and the devoted cadets who threatened violence to sustain the institution. These films were followed in 1986 by a two-part miniseries made for television: *Dress Gray* depicted a military college where the murder of a freshman cadet is covered up by the administration. The 1980s films depicted military schools "as dehumanizing institutions, where nonconformity was severely punished."[15]

The antiwar movement and rebellious-youth culture, encouraged by Hollywood's representation of military education, helped to diminish the number of military schools to levels not seen since the 1840s. In 1980 only eighty-one military schools remained in the United States, and that level would further decrease until 1998, when there were only seventy-four.

The War in Vietnam

The Vietnam War weighed heavily on the nation's military schools. West Point would lose 273 alumni in the long conflict; Texas A&M, 160; the US Naval Academy, 130; the US Air Force Academy, 95; VMI, 43; Virginia Tech, 37; Norwich, 22; and Oklahoma Military Academy, 22. In the jungles of Vietnam the higher commands of the army were dominated by

West Point graduates. They included the commander of all American forces in Vietnam, Gen. William Westmoreland, who had attended the Citadel prior to West Point, as well as Gen. Creighton Abrams. A review of a sample of sixteen biographical profiles of the commanders of the army's three corps and seven divisions showed seven West Pointers, one of whom also attended New Mexico Military Institute, and one graduate of Oklahoma Military Academy.[16]

The Marine Expeditionary Corps and division commanders had few military school graduates, but among them was Maj. Gen. Bruno Hochmuth, commander of the 3rd Marine Division, a Texas A&M graduate who was killed in Vietnam.[17] He and Maj. Gen. George William Casey, a West Pointer and commander of the army's 1st Cavalry Division, were the highest ranking men killed in the war.

While the country was in strife, the nation's military colleges and federal military academies continued to support their county's call. Many of those officers faced multiple tours of combat duty in Vietnam and came home to a disapproving public and being placed on civil order duty in the city riots of 1967 and 1968. For the military and military schools, the Vietnam War was a most desperate trial.

But like Sgt. Gordon Douglas Yntema, Culver Military Academy class of 1963, the alumni and cadets of America's military schools did not give up. Sergeant Yntema, of the 5th Special Forces Group, while leading a Vietnamese force of two platoons in a blocking position, was attacked by a company of Viet Cong. Pinned down under fire from three sides and under enemy mortar fire, Sergeant Yntema organized the repulse of enemy attacks. He carried the wounded Vietnamese commander and a fellow Green Beret fifty meters under heavy fire to cover and returned to the fight. Out of ammunition, and offered surrender, he fought using his rifle as a club against approximately fifteen enemy soldiers and was killed in order to overcome his resistance.[18]

Damaged Reputation

Many parents had the perception that secondary military schools, particularly boarding secondary school, were like reform schools in nature and were associated with juvenile misconduct.[19] Further enforcing those feelings likely were the negative images in the movies set in military schools that were released in the 1980s. In addition, articles appearing in the *Washington Star*, *Washington Post*, and *New York Times* in the mid- and late 1970s reported welfare and juvenile court students being sent to secondary military boarding schools in Virginia, further damaging the once proud public image of the gray-clad cadet.

The appearance of the secondary boarding school as a reform school for juvenile delinquents was so prevalent in 1976 that the magazine *Town and Country* printed an article titled "Good Boys Go to Military Schools Too."[20] Still, that perception, coupled with the advent of military-style behavior modification programs, starting in the mid-1980s, and the nation's heightened awareness of the cultural problem of hazing continued to negatively impact military schools.

Military-style boot camps with behavior modification goals were easily confused as military schools by an ill-informed public. State-sponsored boot camps largely run in a uniformed and regimented fashion spread in the late 1980s through the early 1990s. The first such boot camp opened in Georgia in 1983. By 1995 there were seventy-five boot camps for adults and thirty for juveniles, and 18 boot camps operated in local jails. The justification for the creation of juvenile boot camps was an increase in youth crime, overburdened juvenile courts, and cost reduction. In addition to these government-sponsored programs were a small number of programs in the style of boot camps calling themselves military schools. These were privately run and addressed the drug abuse and general misconduct of teenagers. The value or merits of these behavior modification programs, both state and private, which helped degrade the reputation of secondary military schools, are not the subject of this study. But the perception of military secondary schools that they developed is undeserved and was a point of discussion during annual meetings of AMCSUS.[21]

Clearly, this public perception was a radical departure from the image the secondary boarding military school had prior to the Vietnam War. The image of Jackie Cooper as a cadet in the films *Ginky* (1935) and *The Spirit of Culver* (1939) or of the young cadets of *The Private War of Major Benson* (1955) are views of the secondary military school that were absent in the period of declining enrollments.

Another negative impact on the reputation of military schools, both secondary and at the college level, is hazing. William Trousdale's study of secondary military boarding schools found that parents frequently feel that hazing and military school are synonymous. According to Ronald Holmes, hazing is an epidemic problem in American culture—a finding backed with statistics from the University of Maine's national research. The problem is present in secondary education, colleges, athletics teams associated with those institutions, and professional athletics. Perhaps the most prevalent source of hazing since the mid-1920s has been college fraternities. Prior to this period, most often hazing events occurred between freshman and sophomore male students. The University of Maine study found that 55 percent of college students involved in extracurricular activities and 47 percent of all high school students are hazed.[22]

The problem of hazing at educational institutions is far from a new one. Plato's commentary of the savage conduct of young boys in 387 BC indicates that it certainly isn't a recent problem. Harvard College in 1657 may have seen the first dismissal for hazing in America when two members of the class of 1684 were punished. Between 1838 and 2012, 164 deaths associated with hazing occurred in 123 schools in the United States.[23] Statistics bear out the fact that hazing is a problem far from centered on military schools. Between 1900 and 1984 the deaths associated with military schools numbered 4, each at a different military collegiate institution. This reflects 2 percent of the deaths at 3 percent of the institutions.

The military school incident historically that gained national attention was the 1900 death of former cadet Oscar Booz. Booz left West Point after a short period during his plebe year and died of a respiratory disease. Although neither the army nor Congress found his death the fault of West Point, the congressional inquiry revealed to West Point's great embarrassment unacceptable practices in the cadets' initial training (Beast Barracks) that certainly were hazing at its worst.[24]

Cadet Douglas MacArthur was one of the plebes interviewed by Congress and would later return to West Point vowing to eliminate hazing. Both West Point superintendents General MacArthur, 1919–23, and Gen. John M. Schofield, 1876–81, and Adm. Henry Wilson at the US Naval Academy, 1921–25, went to great efforts to end the practice completely. None could say they succeeded in its permanent complete elimination. Gen. Hugh Scott, superintendent at West Point 1906–10, dismissed several cadets for hazing and found himself defending himself before Congress and the President as a result of the complaints of well-connected parents. He told President Theodore Roosevelt that if he reversed the dismissals it would "do greatest damage to the discipline of the Military Academy." The historical pattern for military schools, secondary or collegiate, has been that hazing is dealt with vigorously.[25]

There are many examples of military school administrative actions to prevent hazing both before General Schofield's efforts and up to recent years. In 1858 Francis Smith at VMI dismissed a sophomore (third-classman) for hazing. A few days later, fifteen of the sophomore's classmates protested the superintendent's actions and were also dismissed. In 1868 and 1869 cadet freshmen (fourth-classmen) were required to take an oath against hazing prior to becoming third-classmen at West Point. In 1891 at Virginia Tech an altercation between the freshman cadets and the sophomore cadets occurred when the Rats refused to oblige the sophomores again with a traditional bucking (paddling). The rebellion resulted in probation for all the involved sophomores. Cadets being loyal to their classmates made demands on the administration, which after extensive consideration dismissed one cadet, permitted five to withdraw, and dismissed six with eligibility to be reinstated the next year.[26]

In 1947 Texas A&M housed incoming fish (freshmen) in separate housing from their companies. After upperclassmen protested, eighty-seven cadet officers were reduced to privates.[27] In the last twenty-five years both Texas A&M and Norwich have disbanded companies or, in the case of Norwich, made the function (the artillery battery) an extra volunteer duty rather than an assigned barracks resident organization as a result of hazing.

Unlike public schools or universities, military colleges, federal military academies, and military secondary boarding schools, as William Trousdale reflected in his study of military secondary boarding schools, are where "the military school is responsible for a boy's [cadet's] well-being during every minute of the day."[28] It is not a question of what transpires at a frat house or at someone's home after classes. Cadets live in military barracks where the rules are clear, and those responsible are cadet officers and the military department. The military regulations further do not stop at the athletic field. This is not to say that hazing at military schools does not exist, but rules against such action are clear and not a point of debate.

Coeducation

Prior to the 1970s documentation of the existence of female members of corps of cadets at military schools is difficult to find and perhaps does not exist. Several schools' female students functioned as auxiliary organizations but did not appear to be under military requirements. There is some documentation that at two schools their auxiliary function may have included military-style uniforms and some elements of participation in the corps of cadets well before the 1970s. The Fairfield Seminary and Military Academy, a private secondary military school in New York in the late twentieth century, and the Atlantic Air Academy in New Hampshire, which functioned from 1945 until 1949 or 1953, may have had female cadets, but the research has yet to reach a conclusion.[29]

Military schools changed to coeducational form for two reasons, for the most part: adjustment to the advances of women in the military, and enrollment. Military colleges and federal military academies were influenced to do so by the advancement of women into traditionally male-dominated career fields. This was a progression that started during World War II as women took on traditional male jobs while the men of America were in uniform, and it continued with the feminist movement. Additionally, during the war some 360,000 women were in uniform, many serving in remote locations under challenging conditions, including 68 who were prisoners of war.[30]

Examples of the progress of women during the period 1970–93 include increases in the percentage of women gaining medical degrees from 8 percent to 38 percent and law degrees from 5 percent to 43 percent, and significant increases in females in the military. Between 1973 and 1980 women in uniform went from 1.6 percent to 8.5 percent of military personnel. During that period, specifically in 1975, the army disbanded the Women's Army Corps (WAC) and integrated its officers and women into the rest of the branches of the army.[31]

In 1972 Senator Jacob Jarvis of New York and Congressman Jack McDonald of Michigan sponsored a bill passed by the senate to authorize women to enroll in the federal service academies. The bill died in Congress, but hearings were held on the topic in 1974. The army pushed back against the concept, arguing that West Point's purpose was to provide combat arms officers.[32]

In the meantime, several all-male secondary military schools facing declining enrollments not only enrolled female students but also offered female membership in the schools' corps of cadets. Among the schools that opened their military programs to female cadets at the secondary school level were Lyman Ward Military Academy in Alabama in 1973, Texas Military Institute in 1974, New York Military Academy in 1975, San Marcos Baptist Academy in Texas in 1975, and Leonard Hall Junior Naval Academy in Maryland in 1976. At the college level in 1973 two military junior colleges, Kemper Military School in Missouri and Valley Forge Military Academy in Pennsylvania, began enrolling female cadets. The same year, the corps of cadets at Virginia Tech organized an all-female company, L Squadron. The following year, 1974, the US Merchant Marine Academy and Norwich University enrolled female cadets, as did Texas A&M with its organization of the all-female Company W-1. These changes did not receive a great deal of press coverage, but what press was received was positive in nature.

The congressional hearings in 1974 resulted in a steady attack on the armed forces' reasoning for not enrolling women in the federal academies. West Point's argument was weak, as West Point graduates were serving in all the branches of the army, not just the combat units. Further arguments—women's supposed problems with Spartan living standards, costs of modifying barracks, double physical standards, and the impact on morale—slowed but did not stop the forces of change. The House of Representative passed, 303 to 96, the Defense Authorization Bill of 1976, which authorized women at the federal academies. The Senate also approved the bill, and President Ford signed it into law on October 7, 1975.[33]

The federal military academies followed, with enrollment of female cadets and midshipmen at West Point, the US Naval Academy, the US Air Force Academy, and the US Coast Guard Academy in 1976. This change gained a great deal of positive attention in the press. On the other hand, the series of legal battles to keep two state-supported military colleges, the Citadel and VMI, all male did not reflect well on military schools in general. In 1989 VMI and in 1990 the Citadel rejected applications from females based on their all-male policy.[34]

The outcome of the efforts of the Citadel and VMI to remain all-male institutions was foreshadowed by the 1984 US Court of Appeals decision against the all-male Massachusetts Maritime Academy for violation of both the Civil Rights Act of 1964 and the Equal Protection Clause of the Fourteenth Amendment. In 1990 the Department of Justice took VMI's enrollment policy into court. The outcome, upon appeal, in 1993 was that the court ruled against the institute. However, the court provided for two remedies: VMI could "set up a separate but equal parallel program for women within the state or go private if they wanted to maintain V.M.I.'s admission policy without violating the Fourteenth Amendment."[35]

Both the Citadel and VMI, with heavy alumni support, fought the order in court and raised funds to create alternative, all-female military college programs in their respective states. Through VMI's efforts, the Virginia Women's Institute for Leadership was established in 1995 with forty-two female cadets at the all-female Mary Baldwin College.

Also in 1993 the Citadel was taken to court by female applicant Shannon Faulkner. The 1994 decision in federal district court ruled against the Citadel for violation of the Fourteenth Amendment and gave the college until the 1995–96 school year to facilitate enrollment of women.[36] The college chose not to wait and enrolled a single female cadet in the fall of 1994. With college personnel unprepared and the male corps of cadets hostile, the first experience failed. Four more female cadets were enrolled the next year, and the misconduct associated with harassment from underclassmen resulted in the departure of two of the women and disciplinary action against a male cadet.

Neither the Citadel-proposed alternative female military college nor Virginia Women's Institute for Leadership satisfied federal courts. The Supreme Court overruled the decision of the lower court, which provided for VMI's "separate but equal parallel program for women." Justice Ruth Bader Ginsburg stated that although Virginia "serves the state's sons, it makes no provision whatever for her daughters. That is not equal protection." The court found that neither VMI's goal of producing "citizen-soldiers" nor its physically and mentally grueling program made the school inherently unsuitable for women. The ruling offered a forceful rendering of the nation's efforts to reverse sex discrimination from the time women got the vote in 1920 to the present.[37]

VMI learned from the Citadel's mistakes, and according to Lt. Gen. Winfield Scott, former superintendent of the Air Force Academy, "no other military college had done so much to prepare for the arrival of women." VMI enrolled thirty women along with 430 male cadets in their 1997 incoming class. The college also obtained nine female upper-class cadets from Norwich University and Texas A&M. VMI's president, Gen. Josiah Bunting, felt that the larger number and the steps taken with the faculty would create "'a genuine cohort' to form a support system." This approach was successful, and that same year, the Citadel, learning from their prior experiences, enrolled twenty-seven female cadets to join the two female upperclassmen remaining from the prior year. Part of the success at the Citadel was due to the fact that one of the two remaining cadets was the daughter of the newly assigned commandant of cadets, Gen. Emory Mace.[38]

In 2010 the US Naval Academy had 16 percent female cadets; West Point, 15 percent; Norwich University's corps of cadets, 14.4 percent; Virginia Tech's corps of cadets, 14 percent; Texas A&M's corps of cadets, 11 percent; Virginia Military Institute, 9 percent; and the Citadel, 8 percent.

The Gulf War

The short duration of combat and the light casualties of the Gulf War have not encouraged a consolidated account of contributions from the nation's military schools. Texas A&M's assessment of three hundred participants includes twenty-one cadets who were called to duty as members of the reserves or National Guard. Three Texas A&M cadet alumni lost their lives. VMI and Virginia Tech each lost two alumni in the war.[39]

A review of the leadership of the major units committed to combat reveals some interesting trends. The US Army provided the overall commander of forces, Gen. H. Norman Schwarzkopf, who was a cadet at the age of ten at Bordentown Military Institute in New Jersey, later graduated from the secondary school Valley Forge Military Academy in Pennsylvania, and then went on to graduate from West Point. The army component of Central Command, Third Army, was under the command of another graduate of Valley Forge Military Academy, Gen. John J. Yeosock. Yeosock had two corps and seven army divisions under his command. The VII Corps was under Lt. Gen. Fred Franks, a graduate of West Point. The 1st Cavalry Division was under Maj. Gen. John Tilleli, a graduate of Pennsylvania Military College; the 101st Airborne Division (Air Assault) was under Maj. Gen. J. H. Binford Peay III, from VMI,[40] and the 24th Infantry Division (Mechanized) and 82nd Airborne Division were both commanded by West Pointers: Maj. Gen. Barry McCaffery and Maj. Gen. James Johnson, respectively. Of the ten key command positions for the army, seven were

General H. Norman Schwarzkopf (1934–2012), Bordentown Military Institute; Valley Forge Military Academy, class of 1952; US Military Academy, class of 1956. (US Army)

men prepared at various military schools of the nation. This was a higher percentage than that indicated by the World War II studies previously cited.

The maritime component was commanded by US Naval Academy graduate Adm. Hank Mauz. Of his five principal task force commanders, three were US Naval Academy graduates: Rear Adm. Daniel P. March, commander, Carrier Group Five, Battle Force Zulu (four carriers), Task Force 154; Rear Adm. Bob Sutton, TG 150.3, Naval Logistics Support Force; and Rear Adm. John B. LaPlante, TF 156, the amphibious force. The Marine Corps provided a corps-level force with two divisions and a Marine air wing. Of the four commanding generals, the three biographical profiles located included the commanding general of the I Marine Expeditionary Force, Lt. Gen. Walter E. Boomer, who had graduated secondary school from Randolph-Macon Academy in Virginia, and one US Naval Academy graduate, Maj. Gen. William M. Keys, commanding the 2nd Marine Division.

It had been 150 years since the number of military schools had been so few. Further, the establishment of ROTC programs throughout the nation provided wide opportunity for non–military school graduates' military careers. However, the continued representation of military school graduates in the key leadership positions in the Gulf War and the War in Vietnam demonstrated the important part military schools played in providing the nation's military leaders. It also reflected the resilience of commitment to the defense of the nation and the unique continued contributions of more than 200 years.[41]

☆ 13 ☆

Resurgence of an
Old Educational Tradition

During the 1960s and early 1970s, a period of strong antimilitary sentiment, the focus of education was on the social and emotional growth of students and equality of education. Society voiced concern with the quality of education and its impact on the preparation of graduates for careers. This was a period of increased federal influence and desegregation. In response to concerns about the quality of education, the magnet school concept developed. The first magnet schools emerged in the early 1970s with specialized curricular or instructional themes. To achieve racial balance, as well as meet the demands of urban parental pressure, federal grants were established to open magnet schools. By 1982 one-third of urban districts had magnet schools. Amid the growing dissatisfaction with public education and the search for alternatives, military public schools started to appear as magnet public schools with minimal notice from the public.[1]

Richmond's Public Military Schools

In 1980 the Richmond, Virginia, Public Schools opened Franklin Military School, an inner-city high school. Richmond was not without precedent, as the John Marshall High School, also in Richmond, had become a military school within a public high school in 1915. The influence of VMI and Virginia Tech graduates in the Richmond area was likely a factor. In 1915 male students were given the option to be members of the John Marshall High School Corps of Cadets, which operated far beyond its JROTC requirements. Cadets drilled and marched in conducted formations daily and were routinely uniformed in VMI-style uniforms. The strict military program at Marshall continued until 1971, when declining numbers and antimilitary feelings caused its demise. Franklin Military School, when it opened in 1980, received some attention but no public protest. The same relatively quiet opening occurred for Cleveland Junior Naval Academy in Saint Louis, Missouri, and Marine Academy of Science and Technology in Highlands, New Jersey, in 1982.

Educational Reform: Setting the Stage

The 1983 report from the National Commission on Excellence in Education began what Wayne Hoy and Cecil Miskel called the "first wave" in educational reform. The report, published under the Reagan administration, alarmed the public with a list of comparisons of US education with that of other countries. Statistics placed the effectiveness of the US educational system in question. Among the most alarming findings was that the functional illiteracy rate was approximately 13 percent for seventeen-year-old students and as high as 40 percent for minorities. Standardized test scores were the lowest in twenty-six years; the College Board's Scholastic Aptitude Tests (SATs) had been in a continual decline since 1963. Students in the United States, when compared to those of other industrialized nations, never placed first or second and were last in seven of nineteen academic categories. Among the recommendations from the 1983 report was the incorporation of standardized achievement tests to certify advancement, identify needs for remedial help, and determine eligibility for advanced courses.[2]

The political mood of the United States regarding school effectiveness was voiced by President Reagan and the National Governor's Association. In his 1984 inaugural address, the President called for more competition among schools, promoted by high academic standards, because without them there could be "no excellence in education." Despite these concerns, there were no new public military schools for almost ten years until 1993, when the Murray-Wright Junior Naval Academy opened in Detroit. Four years later, in 1997, the Kenosha Military Academy in Wisconsin opened with no opposition, as did the first public military middle school, the Magnet Military School, in North Charleston, South Carolina.[3]

However, as Murray-Wright Junior Naval Academy was opening as a public school, another expansion in education was in its early stages: the charter military schools. The first of these schools was the Willamette Leadership Academy, originally opened as the Pioneer Youth Corps School in Oregon. The charter school movement had begun just two years before, in 1991, in Minnesota. By 2000 there were 2,306 charter schools in thirty-four states. By 2007 that number had grown to 4,132 charter schools serving 1.1 million students.[4] Charter schools would become a significant factor in the resurgence of military schools.

The seven public and charter military schools established between 1980 and 1997 were important not because they started the resurgence of military schools but because they were the prelude to a movement in military public education that would accelerate dramatically beginning in 1999. Several prominent educators and politicians moved to the forefront as proponents of public military schools, and others as antimilitary educational groups opposing them.

Military School Proponents

There are five individuals whose efforts played an essential role in the increase of public military and charter schools. Four of them are associated with multiple school establishments: former mayor Richard M. Daley and Arne Duncan established schools in Chicago; Paul Gust Vallas established schools in Chicago, Philadelphia, and New Orleans; Bert L. Bershon, with the Charter Military School Development Corporation in Florida, helped found three military schools in that state; and Gov. Jerry Brown of California used his public prominence to wage the fight to establish the Oakland Military Institute College Preparatory Academy in Oakland while he was mayor there. Later, Brown would further influence state educational policy as governor.

The two most prominent figures, former Chicago mayor Richard Daley and Gov. Jerry Brown, have little in their backgrounds that would lead one to predict they would be champions of military schools. Both Mayor Daley and Governor Brown are attorneys, have served in their respective state senates, and are from prominent political families. Mayor Daley's father, Richard J. Daley, was the mayor of Chicago from 1955 until his death in 1976. The elder Daley was also head of the Daley political machine, which played a critical role in the election of John F. Kennedy to the presidency. Governor Brown is the son of Pat Brown, who served as governor of California from 1959 to 1967.

Richard M. Daley had no military background and did not attend a military school. In spite of this, he became a leader of the military school movement not only in Chicago, but also in California through vocal support of Governor Brown's efforts there. From 1999 until the end of his term as mayor in 2010, a total of six public military high schools were established in Chicago: Chicago Military Academy in 1999, George Washington

Carver Military Academy and Phoenix Military Academy in 2000, Admiral Hyman George Rickover Naval Academy in 2005, Marine Math and Science Academy in 2006, and the Air Force Academy High School in 2009. Chicago has the largest concentration of military schools, public or private, of any city in the United States.

Governor Brown was another unlikely candidate to be a proponent of military schools. As a California state senator and California secretary of state, he was a strong critic of the Vietnam War. He ran for the Democratic nomination for President against William Jefferson Clinton and was successful in the state primaries of Maine, Colorado, Vermont, Connecticut, Utah, and Nevada. Two of his biggest supporters were Jane Fonda and Tom Hayden, both well known for their antimilitary feelings. He is both the youngest California governor, having served as such from 1975 to 1983, and the oldest governor, elected at age seventy-two in 2010.[5]

From 1998 to 2008, while serving as mayor of Oakland, Brown became a strong military school proponent. His only military background was as a high school JROTC cadet in a Catholic high school. In Oakland, Mayor Brown unsuccessfully fought the Oakland School Board for the establishment of the Oakland Military Academy. He then went to the State of California Department of Education, which authorized the formation of Oakland Military Academy as a charter school.[6] The military school concept garnered support from Senator Diane Feinstein, who had been mayor of San Francisco "when it still had the reputation as a sort of urban theme park of the left," in the words of the Democratic Leadership Council (DLC). Senator Feinstein hailed Governor Brown's efforts to establish a charter military school as the "idea of the week" in December 2000. The Democratic governor of California at the time was Joseph "Gray" Davis, a Vietnam veteran and graduate of Hollywood Military Academy, a private military school that closed in 1969. Also publicly supporting Brown's efforts to establish military schools was Mayor Richard M. Daley.[7]

In 2010 Jerry Brown was elected as the thirty-ninth governor of California, having served previously as the thirty-fourth governor. On the platform at his victory speech were several cadets from Oakland Military Academy. On October 8, 2011, he signed California Senate Bill No. 537, California Cadet Corps. Among the provisions in the bill was:

> The Adjutant General may enter into a cooperative agreement with the governing board of a school district or a county office of education for the purpose of establishing, pursuant to existing statutory

authority in the Education Code, a military academy to be operated as a charter school, pursuant to Part 26.8 (commencing with Section 47600) of Division 4 of Title 2 of the Education Code, or as one of the existing alternative education options, available under the Education Code. The program would provide a structured, disciplined/ environment that would be conducive to learning in a college preparatory environment. In addition to academic skills, students would develop leadership, self-esteem, and a strong sense of community. An academy established pursuant to this section shall comply with the Education Code.[8]

This action led to the opening of two more military charter schools in California in addition to the five already operating in 2010.

Two professional educators also associated with pushing the military school movement into public education were Paul G. Vallas and Arne Duncan. Vallas, as Chicago superintendent of public schools, opened the Chicago Military Academy in 1999. This school signaled the beginning of a resurgence of military schools in the United States. During his tenure as school superintendent in Chicago, Vallas opened two additional public military high schools. He left Chicago in 2001 to become superintendent of public schools in Philadelphia, where among the many changes he made there, he established the Philadelphia Military Academy at Leeds in 2004 and the Philadelphia Military Academy at Elverson in 2005. In 2007, in the wake of hurricane Katrina, Vallas took over the Recovery School District of New Orleans. Before his departure from New Orleans in 2010, he laid the groundwork for a sixth military school, the New Orleans Military Maritime Academy, which opened in the fall of 2011 as a charter military secondary school. While Vallas had previously served as an officer in the Illinois National Guard for twelve years, his replacement in Chicago, Arne Duncan, had no experience in the military that might indicate continued support of the military school format.[9]

Arne Duncan, who would serve as US Secretary of Education from 2009 to 2015, replaced Vallas in Chicago in 2001. A Harvard graduate, Duncan had played professional basketball in Australia and worked for many years as the director of the Ariel Education Initiative, a nonprofit educational foundation. He served as the superintendent of Chicago Public Schools until 2008, becoming the longest serving big-city superintendent in the country. His actions were credited with an all-time high in elementary students' success at meeting or exceeding state reading and mathematics scores and high school students' gains on American College Testing (ACT), advance placement courses, and college scholarships. During his tenure another three military schools were opened.[10]

Moreso than any of the other proponents mentioned, Burt L. Bershon's background in the military most likely led him to champion the military school movement. Bershon was a graduate of Culver Military Academy, a prominent private military school in Indiana; he also graduated from the Wharton School of Finance and served in the air force. He was the founder and president of the Charter Military Schools Development Corporation, which successfully established the Sarasota Military Academy in Florida in 2003. In 2014 the academy had 1,052 cadets and the best academic indicators in Sarasota County. Bershon unsuccessfully tried to establish a charter military school in Cincinnati, Ohio, but he was thwarted by a coalition of antimilitary groups and the teachers' union. He shifted his attention back to Florida and helped establish a charter military school in Palm Beach. The school failed, but in 2008 he helped open the Francis Marion Military Academy in Ocala, which prospered and graduated their first senior class in the 2010–11 school year.[11] His plans for an additional eight schools ended with his death in 2011.

One of Bershon's goals was the admission of a charter military school into the Association of Military Colleges and Schools of the United States. Oakland Military Institute College Preparatory Academy joined the association in 2011, the year he died.[12] The success of Sarasota Military Academy surely was a factor in the opening of two public military schools in Florida: Summerlin Academy (2006) and Hollywood Hills Military Academy (2012).

The condition of public education in California and throughout the country led to a search for alternatives to the failing education system. That search led to the return of the American tradition of military schools, among other strategies. While serving as mayor of Oakland, Jerry Brown referred to his city's public schools as a "disaster." Only 1,660 of Oakland's 3,757 students graduated in the class of 2000, and only 377 were prepared for college.[13] Proponents of public and charter military schools in various parts of the United States declared three principal advantages to their offering the military format: a positive education environment, charter development, and expanded opportunities.

Jerry Brown focused on the establishment of a positive educational environment as his motivation for the establishment of the Oakland Military Academy. Brown believed that the military school environment is one that fosters an "environment where creativity, leadership and intellectual curiosity are deeply respected." Brown's experience in a Catholic high school with its JROTC program gave him the inspiration to seek a similar alternative. Recognizing that a religious public school could not be established because of the First Amendment of the Constitution, his alternative was a military school, "with its emphasis on ceremony, discipline, inspiration and leadership training." Brown saw the military school as a perfectly legal alternative to help concentrate on building character and discipline in order to create an environment to prepare students for college attendance.[14]

This similarity between the religious school's and military school's environments was also echoed in John D. Kraus's dissertation, "The Civilian Military Colleges in the Twentieth Century." According to Kraus, both religious and military schools are characterized as having "a strong commitment to purpose, values and traditions," and religious concepts and beliefs are similarly emphasized in military schools through "honor, integrity . . . responsibility, self–respect and physical well-being."[15]

The positive educational environment that Jerry Brown spoke of is visible in the structure of the Philadelphia Military Academy at Leeds. Cadets follow strict rules and wear cadet uniforms, student leaders control movement between classes, and the educational process is free from disruption or danger. Furthermore, the military system "fosters peer pressure to behave and positions the most accomplished upper-classmen as role models and authority figures." The secret to success, according to Hugh Price, is giving cadets a positive peer group, a strong focus on motivation and self-discipline, close supervision, accountability and consequences, structure and rewards, and a safe and secure environment.[16]

The second principal advantage espoused by public charter military school proponents is that the schools' function is character development of their cadets. The Western New York Maritime Academy, a charter school established in 2004, places that concept up front in their mission statement: "Leadership and character development are the inherent cornerstones of the school." Likewise, the Public Safety Academy of California, another charter military school, uses the terms *leadership* and *character development* in its mission statement, and the mission of the public Philadelphia Military Academy at Leeds contains the terms *leadership*, *education*, and *character*. Maryland's public Forestville Military Academy describes its graduates as "leader[s] of character committed to the values of truth, honor, knowledge and service." Other schools, such as California Military Institute and General John Vessey Leadership Academy in Minnesota, address the same concept through the identification of core values. Other military schools use the same concept by identifying specific character traits, such as integrity, ethics, self-discipline, citizenship, and moral ideas. This pattern is essentially the identification of those positive traits of character and reflects the central role character plays in current military schools.[17]

The final advantage espoused by proponents of charter military schools is the expansion of opportunities that fall into three themes: alternate educational philosophy, special training, and successful format. Jerry Brown approached the subject initially in broad terms: "The principle here is the freedom of parents to educate their children as they see fit." He cited the US Supreme Court case *Pierce v. Society of Sisters* (a case that involved the Hill Military Academy in 1925) as a guarantee that parents do not have to accept instruction from public teachers. His agreement then closed with, "These families who want an outstanding education for their children have every right to choose among a wide array of educational philosophies, including an academic school operated in collaboration with the California National Guard."[18]

The desire to provide special training oriented toward law enforcement or emergency services influenced the establishment of a few military schools. For example, Southeast Academy High School and Public Safety Academy in California state that their programs help cadets who are seeking "careers in police or military science or public safety" and "law enforcement, fire emergency medical services," respectively. Most public or charter military schools would rather take Jerry Brown's view: "The goal here is to become leaders in business, government, and the arts. It's not engaging debates on Star Wars, or whether to go to the Gulf War." This reflected the evolution secondary military schools made away from a perception associated with a military career and toward preparation for leadership in the civil sector.[19]

For the most part, the "expanded opportunity" claims avoid the controversial subject of military training and recruitment. Opportunities are framed in terms of greater academic success. In his congratulatory letter to Jerry Brown on the opening of the Oakland Military Academy in 2001, Richard Daley stated that his first graduating class from the Chicago Military Academy had a 40 percent higher citywide average in reading and a 30 percent higher than average score in mathematics. A few years later, similar claims were made on behalf of all military schools, citing the Chicago Military Academy again as having better performance than the rest of the city by 11 percent in reading and 32 percent in mathematics. In Philadelphia, their first military academy touted attendance rates 8 percent above the district average and teacher absentee rates 7 percent lower.[20] Opportunities are not framed by potential advancement upon enlistment in the military, but rather in terms of military school productivity.

Military School Opponents

The scale of opposition to military schools in public education grew quickly, beginning in 1999, when steps were taken to establish public military high schools in Chicago, and in 2000, when a charter military high school in Oakland, California, was being founded. Since then, the controversy has grown, and opposition to a military component in public schools has been adopted by various groups. Among the national opposition to military schools in public education are the American Civil Liberties Union (ACLU), Coalition for Alternatives to Militarism in our Schools (CAMS), Military Families Speak Out (MFSO), Iraq Veterans Against War, American Friends Service Committee–Quakers, Code Pink, and Gold Star Families. On a local level, groups such as Orange County Peace Coalition and Orange County Green Party in California; Committee for High School Options and Information on Careers, Education, and Self Improvement (CHOICES), in Washington, DC; the Chicago Teacher's Union; and local school boards have risen to dissuade the establishment of additional military schools and have been successful in several cases.[21]

Excellent sources to better understand the various objections to the military school concept are Brooke Johnson, "From School Ground to Battle Ground: A Qualitative Study of a Military-Style Charter School," and the American Civil Liberties Union, *Soldiers of Misfortune: Abusive U.S. Military Recruitment and Failure to Protect Child Soldiers*. Johnson's qualitative study was conducted in a California charter military school between 2004 and 2008. Johnson was active in the antiwar movement in the United States and abroad and became interested in a military charter high school when she was "shocked to learn that 11 and 12-year-olds were attending a military-style school that was publically funded."[22] The central theme of her analysis served as a guidepost to curb the influence and advancement of militarism in US public education and described four methods to fight militarism in public schools. Brooke Johnson called for "halting the march of militarism in public education and returning the schoolhouse to a place for growth and learning, as a way to reduce structural inequality in education, and level the unequal opportunities of the most vulnerable US citizens, youth."[23]

Johnson viewed the charter and public military schools in the context of the nexus of neoliberalism and militarism. According to Kenneth Saltman, neoliberalism's "economic and political doctrine insists upon the virtues of privatization and liberalization of trade, while concomitantly places its faith in the discipline [in education] for the resolution of all social and individual problems." He saw militarized public schooling "in terms of the enforcement of globalization through implementation of all the policies and reforms that are guided by neoliberal ideal." Brooke Johnson agreed and called "the military the literal strong-arm of U.S. global and domestic neoliberal policies focused on the logic of free markets and bigger profits." Saltman further noted that there is an overrepresentation of ethnic minorities within the US military due to poverty, joblessness, and increased militarism present in public schools—particularly public military high schools.[24]

Brooke Johnson argues that neoliberalism has encouraged school choices, including charter and magnet schools, and that accountability and efficiency brought by the No Child Left Behind Act have increased militarism in school structure and curriculum. According to Johnson, the "military-style schools . . . are a new and flourishing trend in public education," and "structural inequalities in the local school district and in the community push parents and students towards [military schools] . . . while discipline and uniforms pull in student enrollment." She claimed that military schools "socialize students into a culture of militarism and war . . . [resulting in] viewing the military as an equally beneficial choice as attending college." Finally, she described military school culture as dominated by masculine authoritarianism: "Hegemonic military masculinity is exemplified at the [military school studied] through condoning violence and the warrior hero archetype."[25]

Brooke Johnson illustrated the three most common objections to the establishment of military schools: military recruitment, a culture of violence and masculinity, and violation of the United Nations' Optional Protocol on the Involvement of Children in Armed Conflict as well as the Convention on the Rights of the Child. The military school environment with its uniforms, rituals, and vocabulary socializes cadets to accept militarism as both a "rite of passage" and as "beneficial to their life chances," according to Johnson. This vulnerability to the acceptance of military service has its opponents on two levels. First are the objections to US foreign policy and the military as a national institution and the military involvement in Iraq and Afghanistan. "The increasing militarism of the U.S. public education system needs to be understood in relation to the growing influence of neo-liberalism" and "the enforcement of global corporate imperatives as they expand markets through the material and symbolic violence of war and education."[26]

This point of view is not limited to the establishment of military charter and public schools. In 2006 the San Francisco School Board voted to ban JROTC from district schools. The board declared its opposition to military recruiting in 2005. In the words of the JROTC Must Go Coalition, "As this country enters its sixth year of the illegal occupation of Iraq and Afghanistan, it's time for the school board to go back to its original decision to kick the military out of our schools."[27]

The second level of objection to military recruitment through the establishment of military schools is that the military is targeting the "poor and working class youth of color," and "military-style schools are being established overwhelmingly in poor and working class communities of color." The location and cadet populations of at least ten of the thirty-five public or charter military schools in 2014 support the assumption that these programs were located in poor and working-class areas. This action is seen as a detriment, because young people are lured into service with the promise of money for college and career skills. In many cases, according to Brooke Johnson, minority veterans never use the money for college and gain nontransferable skills. The view of those opposed to military schools is, in many cases, that the military experience "thwarts life chances rather than expanding the options and opportunities for these disadvantaged youth."[28]

The next area of objections to charter and public military schools includes the military's hegemonic military masculine culture, which Johnson claims condones violence and marginalizes women. The military school culture with its uniforms, rituals, and warrior heroes are seen as legitimizing violence by the glorification of the military culture and, in some cases, as providing an introduction to the use of guns with marksmanship training. The military school culture is seen as emphasizing the military and marginalizing those cadets, particularly females, who do not take on a tomboy "female masculinity."[29]

The American Civil Liberties Union (ACLU) has become the leading opponent of military schools, citing the United Nations' Optional Protocol on the Involvement of Children in Armed Conflict and the Convention on the Rights of the Child. This treaty was ratified by the US Senate in 2002, and in 2005 the United States submitted a binding declaration of the minimum age for voluntary recruitment as seventeen (a year older than required by the treaty).[30]

The ACLU position is that the JROTC program is a violation of the recruitment prohibitions of the treaty. Their view of the program is that JROTC is a military recruiting tool that promotes "children's perceptions of a career in the military and enhances military recruiting efforts." The ACLU identified three Chicago public military schools where 18 percent of graduates enter the military. The point of contention with JROTC in military schools is that they expose children to recruitment (not induction) in violation of the treaty. Recruitment should be genuinely voluntary, with individuals fully informed of the duties involved in the military, which is not the case for young cadets, and the ACLU contends that this is a clear violation.[31]

Military Schools Are a Viable Alternative

Each of these objections described above deserves to be addressed in a qualitative manner if possible and through the views of the courts. Conclusions can be drawn from many sources that address each of the objectionable aspects of the military school culture. Objections against military recruitment centered on the unfair burden it places on minorities and the recruitment of ill-informed young people. Among the public military schools in Chicago, where minority enrollment was high, the military option was selected by 18 percent of graduates in 2004. The burdens of military service in 2012 can be illustrated by comparing it with Chicago's high murder rate. Military service in Afghanistan was nineteen times more likely to result in being killed.[32] But the casualty figures for the conflicts in Iraq and Afghanistan did not support the conclusion that military recruitment resulted in an unfair burden of national defense being placed on minorities. American military deaths in Iraq and Afghanistan through September 2010 numbered 5,670 of which 9.1 percent were African American and 9.9 percent were Hispanic. The proportion of casualties for African Americans was far less than their percentage of the population, at 15.4 percent, while that of casualties for Hispanic was very close to the percentage of population at 9.7 percent.[33]

In 2009 the legality of military recruitment in public schools was tested in federal court. The statutes of Arcata and Eureka, California, prohibiting military recruiters from contacting minors were overturned. Military schools, by exposing their cadets to the military culture with uniforms, titles, ceremonies, and vocabulary, provided eighteen-year-old graduating cadets the opportunity to make a better-informed decision regarding their future. Those who enlist after graduation from a military school benefit from JROTC, which provides incentives including, in the army, rapid promotion to the rank of corporal with a significant associated pay increase. Other benefits are intrinsic to military service and include job skills based on military specialty, job security, structured environment, the quality of work ethic desired by employers, and educational benefits that carry over to postservice opportunities. Military school cadets are better prepared to make an informed decision with regard to enlistment and benefit from the preparation provided to them in school.[34]

The objections to military schools as promoting a culture of violence and masculinity were not supported by Remi Hajjar's study of a public military high school. Rather than violence, Hajjar found an environment that promoted civility. In the military schools, "through comprehensive academy mores and codes, cadets learn and develop etiquette, trust, mutual respect, duty and tolerance for others that enhances their civility; an important life skill." This type of behavior was found to be stronger among those who had been at the academy for a longer period of time.[35]

Rather than marginalizing female cadets, toleration and good behavior— "treating others well and being polite, respectful"—was more apparent among cadets. The military school in which Hajjar conducted his study had as its cadet commander a female cadet who embraced her leadership role and was far from marginalized. In my personal experience in private secondary military school and a senior military college, I found that female cadets occupied critical leadership positions far beyond their proportion of the cadet population. Brooke Johnson's critical view of the military schools' impact on female cadet development was heavily influenced by a political agenda that viewed the military as an evil tool of the "strong arm of U.S. global and domestic neoliberal policies."[36]

Objections to military schools as violations of the United Nations' Optional Protocol on the Involvement of Children in Armed Conflict and the Convention on the Rights of the Child centered on the military school as a violator of the recruitment prohibitions of the treaty. The opponents of military schools see cadets' exposure to the military culture as recruitment of children and not voluntary, with adults fully informed of the duties involved in military service. But membership in a military school does not constitute membership in any of the armed forces of the United States. Further, the United States meets the requirements of the protocol, in that

> the minimum age at which the United States permits voluntary recruitment into the Armed Forces of the United States is 17 years of age; the United States has established safeguards to ensure that such recruitment is not forced or coerced, including a requirement in section 505(a) of title 10, United States Code, that no person under 18 years of age may be originally enlisted in the Armed Forces of the United States without the written consent of the person's parent or guardian, if the parent or guardian is entitled to the person's custody and control; each person recruited into the Armed Forces of the United States receives a comprehensive briefing and must sign an enlistment contract that, taken together, specify the duties involved in military service; and all persons recruited into the Armed Forces of the United States must provide reliable proof of age before their entry into military service.[37]

Remi Hajjar concluded that the "unique military subculture builds solidarity, which in turn develops students' propensity for education, discipline, civility, leadership skills and agency." Overall, students at a public military school "develop critical life skills and accrue forms of social capital, which should increase their chances of achieving upward mobility." This educational propensity was reflected by cadets' intent to attend college, with 85 percent of sophomores and juniors desiring to go on to college. Among the student population in which 74 percent are considered poor or at risk, based on fee or reduced school lunch programs, and 90 percent are from minority African American or Hispanic backgrounds, the propensity of military school graduates for advanced education was borne out by the 2006 statistics for Chicago high schools, in which graduates of the city's public schools averaged 48 percent continuing on to college, as opposed to 81.5 percent for the city's public military academies.[38]

Military school subculture builds leadership competencies, according to Hajjar. "Cadets acquire leadership schemata, repertoires, and skills through classroom lessons and firsthand experience." Robert Tarrant's findings support this conclusion; when he compares private military schools with public high schools, the military schools increase the student's leadership values.[39]

Hajjar defined agency as the capacity to transpose and extend schemas to new context. Agency entails an ability to coordinate one's actions with and against others; to form collective projects; and to persuade, coerce, and monitor the simultaneous efforts of one's own and others' activities. His conclusion that military schools enhanced these skill sets was also supported by Tarrant, who found that military schools supported the development of skills related to technical information such as having a broad outlook and well-rounded knowledge, managing groups, and seeing the larger picture in relation to one's own activity. Among Hajjar's examples of the military school subculture's facilitation of agency among cadets were several female cadets, who adopted the academy's codes, cultural repertoire, and embraced a women's potential. Examples in his study were in direct contradiction to a culture of violence and masculinity, which opponents of military schools gave as minimizing women.[40]

According to Hajjar, the military school subculture institutionalizes discipline; self-discipline thrives in its environment. Examples of performances that he cites as positively impacted by self-discipline are attendance, homework completion, and adherence to military codes. Attendance in the inner-city military school of his study was 93.3 percent, ranking fifth of eighty-four inner-city public schools. The public military school of his study was focused on correcting minor military uniform and conduct violations rather than on having to address "widespread drug distribution and possession, frequent and serious acts of violence, substantial criminal gang influences, and other illicit behaviors that plague many inner city schools." When viewed from the perspective of inner-city education, minor military violations of uniform and protocol standards are largely insignificant, yet to the cadets they are important. Among his conclusions was that the military school subculture yielded "positive behavioral changes for teenage students, especially given the inner-city backdrop of this particular school."[41]

The importance of discipline was illustrated by a study by the nonprofit group Public Agenda, which found that 97 percent of teachers and 78 percent of parents felt that in order for a school to flourish, discipline was needed. However, the experience of 85 percent of teachers and 73 percent of parents was that education suffered because of a few "chronic offenders." Seventy-seven percent of teachers admitted that their effectiveness suffered due to "disruptive students." The study further concluded that schools are effectively addressing serious behavior problems by having armed police on campus but not addressing violations of rules that disrupt class. Public Agenda also pointed out that "problems with student behavior appear to be more acute in urban schools and in schools with high concentrations of student poverty." Teachers from those schools are three times as likely to cite discipline, and twice as many quit teaching because of the lack of student discipline. For both teachers and parents, seven out of ten felt the cause of the behavior problems was "disrespect everywhere in our culture—students absorb it and bring it to school."[42]

Efficacy of Public and Charter Military Secondary Schools

The most important aspect of the debate over public education in a military format must be efficacy. Since the 1980s the national standards for the determination of efficacy have been attendance, dropout rates, graduation rates, and adequate yearly progress on state standardized tests. What follows are the results using comparative data from military schools and nonmilitary schools with student bodies of similar socioeconomic status.[43] The study was conducted using data from 19 public-funded charter or public military secondary schools from fifteen school districts in twelve states in the 2009–10 school year. The sample schools enrolled 7,707 cadets, or 78.6 percent of the total 9,815 cadets in the 28 publicly funded military high schools funded in that year. These 19 military schools were compared to 151 nonmilitary neighborhood schools, 27 nonmilitary magnet schools, and 19 nonmilitary charter schools identified as being in the same school district or a nearby location and of socioeconomic status similar to that of the associated military schools. The socioeconomic status of the schools was determined by the percentage of students receiving free or reduced meals. This was the common means used by all state reports. Military schools were matched with nonmilitary schools having no more than a 10 percent difference in socioeconomic status (see appendix D).

The study's conclusions must be tempered by the limited size of the sample and the single year of data, but the results provide the initial trend of military school performance. When compared with socioeconomically similar neighborhood secondary schools, military schools outperformed them in measures of attendance, dropout rates, and performance on mathematics and English tests. The difference in both mathematics achievement and English achievement was particularly notable. Military school performance, when compared with socioeconomically similar magnet and charter secondary schools, showed significant differences again for dropout rates. In fact, dropout rates, a national problem, were statistically lower for military schools versus the three other categories of schools (neighborhood, magnet, and charter). Although there were many cases of statistically insignificant differences, none of the neighborhood schools (as measured by graduation rates), magnet schools, and charter schools (as measured by attendance, graduation rates, and adequate yearly progress on state standardized tests) statistically outperformed their military counterparts in any of the five categories of comparison.

The objections to public military schools do not withstand close scrutiny from an educational perspective. But the need for a school culture that addresses the challenges to education of an undisciplined student body can be met by the military school format. Furthermore, the military school culture provides propensities for higher education, civility, leadership skills, and agency which increase students' chances for upward social and economic mobility in their future.

Statistically there is initial strong evidence to project advantages of military schools over neighborhood public secondary schools in attendance, dropout rates, and mathematics and English test achievement. Further, the statistics show that military secondary schools outperformed in dropout rates both charter and magnet schools. The military schools actually outperformed charter and magnet schools in respect to attendance, graduation rates, and adequate yearly progress on state standardized tests, although not significantly. All these further justify the military school format as another alternative in public secondary education.

Fifteen Years of Dramatic Growth

Starting in 1999 and continuing until at least 2014, the decline in the number of military schools and their enrollment ended, and these indicators have reversed. The number of military schools increased 29 percent between 1999 and 2014. In 1998 there were only seventy-five military schools, and the number in 2014 was ninety-seven. The growth in numbers of institutions is a direct result of the addition of thirty-five public and charter military schools.[44] The cadet and midshipman population in 2014 was approximately 52,900 students, with approximately 60 percent at the college or junior college level. This figure is approaching the 1942 peak of military school population of 61,100 students. This enrollment growth can be attributed to expanded numbers at the military colleges and over 11,000 attending public and charter secondary military schools. Examples of the latter include Maryland's Forestville Military Academy, a public military high school with 860 cadets, and two of Chicago's military high schools, with enrollments of more than 500. Furthermore, five of the thirteen charter military secondary schools had enrollments exceeding 500 cadets, and Florida's Sarasota Military Academy enrolled 1,052 cadets in 2014.

In 1972 the only public military school in the United States closed in Richmond, Virginia.[45] By 2014 a reorientation of the military school movement towards public education had risen, and thirty-five new public or charter military secondary schools had been established in the United States. These eighteen public and seventeen charter schools were located in seventeen states: California, Connecticut, Delaware, Louisiana, New Jersey, New Mexico, New York, North Carolina, Maryland, Minnesota, Missouri, Oregon, Pennsylvania, Utah, South Carolina, Virginia, and Wisconsin.

This increase has taken place in a period of increased popularity of the military. Gallup polling placed confidence in the military as an institution in the fiftieth percentile between 1975 and 1984, with a low of 50 percent in 1981. From 1985 until 2001, that figure increased only to the sixtieth percentile, with the exception of 1991 during the Gulf War, when it climbed to 85 percent. After the September 11, 2001, attacks on New York and the Pentagon, the confidence in the military as an institution increased to between 73 percent and 82 percent between 2002 and 2011. This, as compared to the next highest rated institutions, big business at 59 percent to 67 percent and the church from 45 percent to 53 percent, placed the military at the height of institutions in the Gallup polls.[46] The impact on the popularity of military schools benefited from these attitudes as well.

Although the nation had positive feelings toward the military as it entered the 2000s, Hollywood's enthusiasm for military schools did not repeat that of the middle of the previous century. Movies were divided on their treatment of the military school environment, with *The Substitute 4: Failure Is Not an Option* (2001) and *Igby Goes Down* (2002) presenting negative images, while *Cadet Kelly* (2002), with Hillary Duff starring, and *Field of Lost Shoes* (2014) reflecting positively. Of the four movies, *Cadet Kelly*, released by Disney, was the most successful. *Field of Lost Shoes* is well worth the readers' time with its treatment of Virginia Military Institute's participation in the Battle of New Market.

Television was largely negative in its treatment through single episodes of *The Sopranos* (2001), *Cold Case* (2004), *The Killing* (2013), *NCIS* (2015), and *KC Undercover* (2015). On the other hand, multiple episodes of *Malcolm in the Middle* (2000–2005) were positive. All the negative examples reinforced the image of secondary boarding schools as a place for bad kids or hazing. The supportive nature of the media seen in years past is likely never to return; however, this has not stopped but perhaps slowed the growth rate of military school enrollments.

Federal Military Academies

In 1954 there were five federal military academies with the establishment of the US Air Force Academy in Colorado. In 2010 the total enrollment of the five academies was 15,661 cadets. The three largest academies—the Air Force Academy, the Naval Academy, and West Point—each averaged 4,575 cadets or midshipmen, with the Coast Guard Academy and the Merchant Marine Academy enrolling 973 and 964 cadets, respectively. Each of these academies requires five years of postgraduate military service, with the exception of the Merchant Marine Academy, which requires maritime industry employment or military service.

The establishment of the newest federal academy is patterned after the oldest military schools associated with Alden B. Partridge and Sylvanus

US Air Force Academy's initial leadership: Brig Gen Robert Stillman, Gen Hubert Harmon, and Brig Gen Robert McDermott. (Courtesy US Air Force Academy McDermott Library)

Thayer. The Air Force Academy's first superintendent was Gen. Hubert Harmon, West Point class of 1915, who was recalled to active duty to lead the academy. The academy's first permanent dean of faculty was Brig. Gen. Robert McDermott, who was originally a member of Norwich University class of 1941, then graduated from West Point with the class of 1943. He was joined on the faculty by a large number of West Point alumni. The first commandant of cadets was another West Pointer, Brig. Gen. Robert Stillman, who also recruited other military college graduates, including an air officer commander and several air training officers. These men strived to make the honor code "a foundation for everything the academy hoped the cadets would be"[47] and ensured that the newest academy would adopt the standards founded 150 years before. The code was almost identical to those of the other established military academies and clearly established the centric role that honor has in the military school culture. This historical honor code prevails in what many believe is the most modern and technological of the federal military academies.

It is interesting to note that the academy that deals with perhaps the highest technological challenges in aerospace engineering and is orientated

US Air Force Academy parade about 1959. (Courtesy US Air Force Academy McDermott Library)

toward the most modern form of warfare had in its establishment the influence of the two oldest American military schools. Like the US Naval Academy's original commandant, Lt. James H. Ward, from Norwich, the US Air Force Academy had its influence through General McDermott. The heavy influence of West Point graduates ensured an adherence to the military school culture.

Military Colleges and Universities

In 2014 there were twelve military colleges and universities in the United States with enrollment of approximately 13,500 cadets and midshipmen. These institutions included two colleges that are essentially military campuses, two military colleges with the corps of cadets representing a large portion of a dual student body, five maritime academies, and two land grant universities with corps of cadets and one female corps of cadets at a private college.[48] Virginia Military Institute had 1,569 cadets, and the Citadel had 1,899 cadets. VMI and the Citadel are the only military colleges maintained to be exclusively military campuses with a cadet student body. Both the Citadel and VMI have Army, Navy, and Air Force ROTC as options.

The United States has seven maritime academies, of which only the Great Lakes Maritime Academy is a civilian institution. The US Merchant Marine Academy, previously discussed, is a federal military academy. The

Virginia Military Institute, 2015. (Courtesy Virginia Military Institute; photo by Kevin Remington)

remaining five academies have dual student bodies, with their corps of cadets numbering 4,400 cadets and midshipmen. The other academies include the State University of New York Maritime Academy (established as the New York Nautical School in 1874), with 1,154 cadets, and the Massachusetts Maritime Academy with 1,162 cadets. Located in California, New York, Maine, Massachusetts, and Texas, the maritime academies have proven to be the second most viable style of military school after the federal military academies. All of the five state maritime academies operating in 2014 are state supported and include Navy ROTC as an option.

Norwich University and North Georgia College, one of the land grant colleges, both have dual student bodies, civilian and military. In 2013 the Norwich University Corps of Cadets numbered approximately 1,386 cadets, about 66 percent of the undergraduate student body. Norwich, the only private military college in the United States, has Army, Navy, and Air Force ROTC as optional programs. North Georgia College includes about 750 cadets, or 14 percent of the undergraduate population. North Georgia is unique among the military colleges in that it offers only Army ROTC and has a large number of cadets who also serve as members of the state's National Guard.

Two land grant universities maintain a significant corps of cadets. In 2014, the Texas A&M Corps of Cadets numbered 2,450, or 5.4 percent of a

The Virginia Tech Cadet Regiment formed in front of the War Monument, 2011.
(Courtesy Virginia Tech; photo by Col. Rock Roszak '71)

student undergraduate population of 44,681. Virginia Tech's corps of cadets grew dramatically, from 769 in 2011 to 1,036 in 2014. In 2014, the military population at Virginia Tech was 4.5 percent of the 22,824 undergraduates. Both schools have Army, Navy, and Air Force ROTC as options.

The Virginia Women's Institute for Leadership (VWIL) of Mary Baldwin College was established in 1995. The program grew from the legal battles regarding the admission of women into the corps at VMI. Under the leadership of Brig. Gen. Michael Bissell, who became commandant in 1999, enrollment at the Virginia Women's Institute for Leadership expanded to between 120 and 125 cadets. The cadets' dress uniforms are an unusual forest green color with a West Point–style cut. Perhaps even more unusual is the fact that Virginia Women's Institute for Leadership is the only all-female military school in the world.[49]

Military Junior Colleges

In 2014 there were five military junior colleges, whose enrollment in 2011 was approximately 3,513 cadets in secondary and junior college programs. All five colleges offered a two-year Army ROTC program that awards an army commission. Wentworth Military Academy and College and Valley Forge Military Academy and College are both private; the remaining three colleges are state-supported. All except Marion Military Institute, which discontinued its secondary program and its private status and became a state-supported junior college in 2009, maintain a secondary military school program.[50]

Kemper Military School in Missouri maintained a junior college until 2000 and illustrates the fiscal vulnerability of the two remaining private military junior colleges. The school was established in 1844 by Frederick T. Kemper and successfully operated as a civilian male boarding school until Kemper's death in 1871. The leadership of the school then fell to Col. Thomas Johnston, a former student, who in 1864 had enlisted as a sixteen-year-old Confederate soldier. Johnston returned to the school in 1868 after graduating from the University of Missouri.[51] In 1885 he converted the school to a military format, and in 1899 he changed the name to Kemper Military; there were 502 cadets by 1918. In 1923 the school added a junior college and reached its peak enrollment as a military junior college and military secondary school in the mid-1960s. The school became a nonprofit in 1957.

When the Vietnam War took its toll on the nation's military schools, Kemper's enrollment dropped in 1976 to 89 cadets. In an effort to save the school, female cadets were enrolled in the 1970s, and the junior college's costly athletic program was dropped in 2000. By 1980 enrollment recovered to 224 cadets and climbed to 300 by 1999.[52] However, mounting debt and low enrollment placed the school in bankruptcy and finally resulted in its closure in 2002. The school met its demise by the most common cause of private military school failures: financial problems.

Private Military Schools

In 2014 the United States had forty private military schools enrolling approximately 9,300 cadets. The average number for military enrollment of these schools was approximately 230 cadets. The ten schools with the largest enrollment averaged 427 cadets. This number compares favorably to the largest military boarding schools in 1932, when the average was 323 cadets, indicating the well-being of private military schools in 2010. The vast majority, thirty-seven of the forty, of these private military schools are secondary schools. Seven of the schools have dual student bodies, several restrict the corps of cadets to the higher grades, and several have volunteer membership. Regionally, 68 percent of the schools are located in the South. This can be misleading, as seven of the thirty-two schools are actually located in Puerto Rico. Eight percent of the schools are located in the western states, and 15 percent are in the Midwest. Fifteen percent of the schools are located in the North.

Between 2006 and 2014, seven private military schools were confronted with financial crises. Five of them closed, including Chamberlain Hunt Academy, established in 1879, where enrollment dropped to thirty cadets

Kemper Military School about 1927 was organized into a battalion of four companies and a band with 360 cadets. (Courtesy Echo Company, Kemper Military School)

before it closed in August 2014. The Millersburg Military Institute in Kentucky, with an enrollment as low as fifty-five cadets, closed in 2006. The school had been established in 1893; however, the school went into debt as much as a million dollars.[53] In 2012 the Millersburg facilities were used to open Forest Hill Military Academy, which was enrolling between seventy-five and one hundred cadets in its initial years. The other two closures were small, recently established institutions. Ontario Christian Military School in California opened in 2006 with fifty-seven cadets but closed in 2009. Eagle Military School of South Carolina opened in 2002 but closed in 2008, when the economic downturn impacted its modest enrollment. In 2009, Low Country Military Academy opened in Ladson, South Carolina, but in 2011 it enrolled only fifteen cadets, and it appears to have quietly closed in 2012.

For the eight years between 2006 and 2014 there had been rumors of the closing of Howe Military Academy of Indiana. As recently as May 2014 published reports questioned whether the school would open in the fall because of low enrollment and a two-million-dollar debt. However, each year the school has managed to open its doors. From 2010 to 2011, both Oak Ridge Military Academy of North Carolina and New York Military Academy were on the verge of closing. Oak Ridge hired a basketball coach who was also a hedge-fund manager. He donated $471,000 to the school to jump start a plan to build a $2.5 million sports complex. In 2011 fraud allegations against the coach threw the school into a state of financial instability. New York Military Academy announced it was closing, but quick action by alumni, including a threat to sue the governing board, resulted in a takeover by the alumni. Each of these examples illustrates the vulnerability of well-established schools, as well as newly founded private military schools, to financial mismanagement, fraud, and economic downturns.[54]

Public and Charter Military Schools

In 2014 there were thirty-five military secondary schools operating within public education as public or charter schools. Regionally these schools were located throughout the United States, with nine schools in the Midwest, eleven schools in the West, ten schools in the South, and five in the North. Between 1915 and 1971 there was only one public military school. This Richmond, Virginia, school closed and was not replaced until the 1980 establishment of Franklin Military School.[55]

Of the eighteen public military schools operating in 2014, ten were located in urban environments in Chicago, Philadelphia, Richmond, and

Saint Louis. In 2014 enrollment in these schools included approximately 5,765 cadets, with 1,196 in Chicago. The seventeen charter military schools had a total enrollment of approximately 5,368 cadets. The charter schools were located in eight different states: seven in California, two in Florida, and one each in Delaware, Louisiana, Minnesota, New Mexico, New York, North Carolina, Oregon, and Utah.

Of the thirty-five public and charter military schools, only five were established before 1999. The most recently established are the Air Force Academy High School in Chicago; Stanislaus Military Academy in California, established in 2009; New Orleans Military Maritime Academy in Louisiana, opened in 2011; Hollywood Hills Military Academy, opened in Florida in 2012; Moreno Valley Military School, opened in California in 2012; Paul R. Brown Leadership Academy, opened in North Carolina in 2013; North Valley Military Institute in California, opened in 2013; and Utah Military Academy, opened in 2014.

⋆ **14** ⋆

Conclusions

This book has tracked the evolution of the military school in the United States from the need identified during the Revolutionary War through the creation of the federal military academies and to the 1999–2014 resurgence of military schools in the form of public educational institutions. Each major change in the number of military schools in the United States can be related clearly to a political, economic, or cultural trend.

Political and Cultural Factors

By far the greatest impact on military schools of the United States has been war. The Revolutionary War brought the need for a military academy to the attention of the Continental Congress. Political factors delayed the establishment of West Point for years, and its early years, from establishment through the administrations of Partridge and Thayer, were guided and affected by politics.

The Citadel pass-in-review, 2015. (Courtesy Russ Pace, the Citadel)

The first decline in the number of military schools in operation occurred in the period immediately following the Civil War, as many Southern schools failed to reopen or transitioned to nonmilitary format as a result of the war and the policies of the federal government's Reconstruction occupation. The war and the policies of the Union reduced the number of military schools in the South by at least 78 percent. Nationwide, the decline was from 171 military schools operating during 1855–65 to 147 operating during 1866–78.

The Spanish-American War's impact on military schools on the contrary was a positive effect. The mood of the nation, with the Southern states back in the union, was a positive one, which translated into a positive image for the military and military schools. This positive political environment was further fostered by the Preparedness Movement, and it helped lead to the peak in the number of military schools between 1903 and 1926, when 278–80 operated in the United States. This positive political environment continued through World War I despite the efforts of the Anti-Imperialist League; however, the war itself again changed that.

World War I and its incredible causalities helped generate political opposition and an antimilitary feeling. The pacifist movement, with groups like the American Federation of Youth and the Prevention of War Council, campaigned against military training, ROTC, and the value of military schools. The pacifist propaganda of the movement made the successful operation of military schools particularly challenging from 1923 to 1933. World War II ended the negative political environment for military schools; in fact, the war was responsible for a peak historical enrollment in military schools.

But once again, the Vietnam War and the military draft mobilized a political opposition that vilified the military and helped make the military uniform and military schools something less than an appropriate educational experience. The political impact of the Vietnam War, along with a cultural shift among young people, was responsible for a 65 percent reduction in the nation's military schools. This was the largest change in the history of the military-school movement.

The political environment, particularly since the attacks on the World Trade Center and Pentagon of September 11, 2001, has been favorable to the military and military schools. In this atmosphere, concerns over public education evolved into the adoption of military charter and public military schools as viable alternatives in public education and a resurgence of the military school concept after the declines of the 1970s. This resurgence

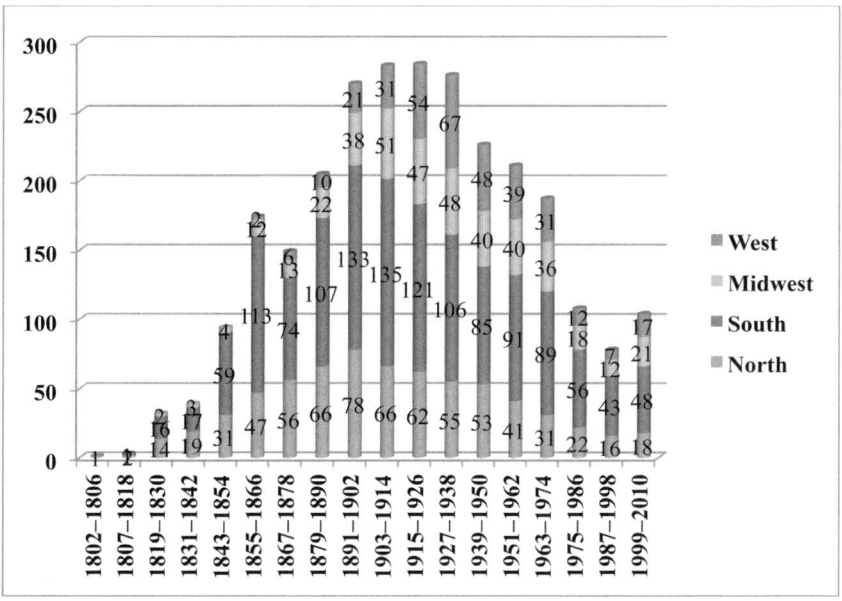

Fig. 14.1. Number of Military Schools in the United States, 1802–2010.

since 1999, and continuing through 2014, has been a significant factor in the history of military schools in the United States and a turn toward public funding in secondary education.

The Revolutionary War, the Spanish-American War, and World War II had positive effects on the military school movement. However, World War I and the Vietnam War mobilized political sentiment against the military educational format. War cannot be judged as either a multiplier or distracter of the military school historically. The political climate that emerges from the conflict is by far a more important factor than the conflict itself in predicting military school enrollments. The effect of each of these conflicts can be seen in the number of institutions in operation, as shown in Fig. 14.1. By far, political events have had the greatest impact on military schools, and this will continue in the future.

Cultural Factors

Three cultural changes affected military schools in the United States. One of those, in the South after the Civil War, was a positive factor. The other two surfaced around the effects of war and cultural disillusionment. The post–Civil War cultural conscience of the South focused on the Lost Cause. Confederate monuments, dedication ceremonies, literature, and the near worship of Gen. Robert E. Lee and Gen. Thomas "Stonewall" Jackson made military service the height of achievement for Southern manhood. The gray-uniformed cadets of postwar Southern military schools, coupled with the historical images of Civil War cadets in combat, made the military school a means to educate boys and young men in the martial virtues central to the Lost Cause. As a result, the South, after having its military schools drastically reduced in number, recovered by 1891–1902 and operated 133 military schools, or 49 percent of the nation's total of 270 military schools.

After World War I the Lost Generation became a cultural conscience centered on the disillusionment of war. It questioned the traditional values of the middle class and the military and of course the value of military sacrifice. Its leaders were influential authors, typified by Ernest Hemingway and others whose literature questioned patriotic notions. The popular media of the time furthered the disillusionment of the Lost Generation with the movie *All Quiet on the Western Front* (1930). As a result, military schools found it difficult to achieve needed enrollment, and political opponents of military training and military schools did not have to look far for supporters. This played a significant part in ending the peak in numbers of military schools in the 1930s.

The final cultural change was that of the antiestablishment generation. Although this cultural movement started with the Beat or Beatnik culture of the 1950s, it was the high school and college-age students of the 1960s who became the antithesis of the military school cadet. Through its message the counterculture turned several decades of American young people against the military school concept. The resulting closure of seventy-eight military schools between 1968 and 1978 was a combination of this cultural and political setting.

Economic Factors

The Great Depression was the largest factor that contributed to the end of the peak in the number of military schools established in the county. Between 1929 and 1938, discounting the military schools that transitioned to nonmilitary school formats, fifty-one military schools closed. Survival was a matter of aggressive financial management. Typically, the enrollment of both small and large military schools declined by 53 percent.[1] Staff reductions, tuition reductions, and sometimes radical salary adjustment were required for institutional survival. Those schools with small enrollment at the start of the Depression were more likely not to be able to adjust to the challenges.

Fathers of the American Military
School Movement

The military school movement has been heavily influence by economics, culture, and particularly politics, but the leadership of four men has been particularly important. Since 1802 in the United States at least 842 military schools have been established and have educated some of the country's most distinguished alumni (appendix E). The military school movement was modeled on West Point, which was developed as a unique school culture separate from that of the armed forces. This model was perfected under Sylvanus Thayer, who served as the superintendent of West Point from 1817 to 1833. At the same time, Alden B. Partridge expanded the military school concept outside the confines of West Point, first with the establishment of Norwich University and next through his personal involvement in the establishment of eighteen military schools as far south as Mississippi and as far west as Missouri. In a short thirty-seven years, by the time the Virginia Military Institute was opened in 1839, at least forty-three military schools had been established. Many, if not most, of those schools were founded by, were led by, or drew faculty from either West Point or Norwich alumni and were influenced by either Partridge or Thayer.

Among those alumni was Francis H. Smith, a former cadet who studied under Thayer at West Point. As superintendent of VMI from 1839 to 1890, Smith created a military school that became a cradle for the formation of educators in the South, particularly in Virginia. His influence, mentoring, and outright promotion of the skills of VMI graduates as educators influenced an explosion of military schools in the southern United States. Of the 138 military schools operating in the United States in the six years before the Civil War, approximately ninety-six were located in the South. A large factor in this growth and regional orientation was Smith and his VMI alumni, who founded or initially led at least twenty-six schools and provided faculty to many others. Finally, Stephen B. Luce, US Naval Academy graduate, whose midshipman years were influenced by both West Point and Norwich alumni, took the military school concept and transferred it firmly into the education of the merchant marine industry. His political lobbying and articles served as the catalyst for the merchant marine academy movement. His direct involvement in the formation of the New York Nautical School made the military school format the model for educating officers for the civilian maritime industry. The New York Nautical School (later named the New York Merchant Marine Academy) would involve the development of a series of merchant marine academies, including the US

Merchant Marine Academy. These military schools, which operate at the collegiate level, have become the second most enduring type of American military school outside the federal military academies.

Based on their contributions to the formation of a unique military school culture and the expansion of the military school movement throughout the United States, there are four men who should be considered fathers of the military school movement: Alden B. Partridge (1785–1854), Sylvanus Thayer (1785–1872), Francis H. Smith (1812–1890), and Stephen B. Luce (1827–1917). These men established the model and trained the school leaders and faculty in these founding military institutions that led to the spreading of a unique educational format throughout the United States.

Regional Historical Orientation of US Military Schools

The American military school is not an educational format restricted largely to the South or unique to Southern culture. Although Francis H. Smith, the alumni of the VMI, and, to a lesser degree, the alumni of the Citadel have contributed greatly to the popularity of the military school format in the southern United States, the historical representation of military schools throughout the country cannot be ignored. The image of the Southern cadet soldiers of the Civil War and the well-known private military secondary schools, particularly in the South and in Virginia, have created the misconception that military schools are a feature unique to Southern culture.

An examination of the number of military schools and their regional orientation for more than a century reveals that, prior to the establishment of VMI, the Northern states and Southern states were nearly equal in the number of military schools they contained. Since that time, the South has maintained a larger number of military schools than other regions of the country, as depicted in Fig. 14.2. After the South's recovery from the Civil War and Reconstruction, the number of military schools in operation from 1879 to 1890 was approximately 204. Of them, 107 were located in the southern United States, but 96 military schools, representing 47 percent of the total, were operating in other regions of the country. Sixty-five of these were in northern states, 21 in the Midwest, and 10 in the West.

Between 1903 and 1926 no fewer than 278 to 280 military schools operated in the United States, representing the peak in the number of military schools. The Southern states were home to approximately 46 percent of these

Fig. 14.2. Regional Distribution of Military Schools, 1802–2010.

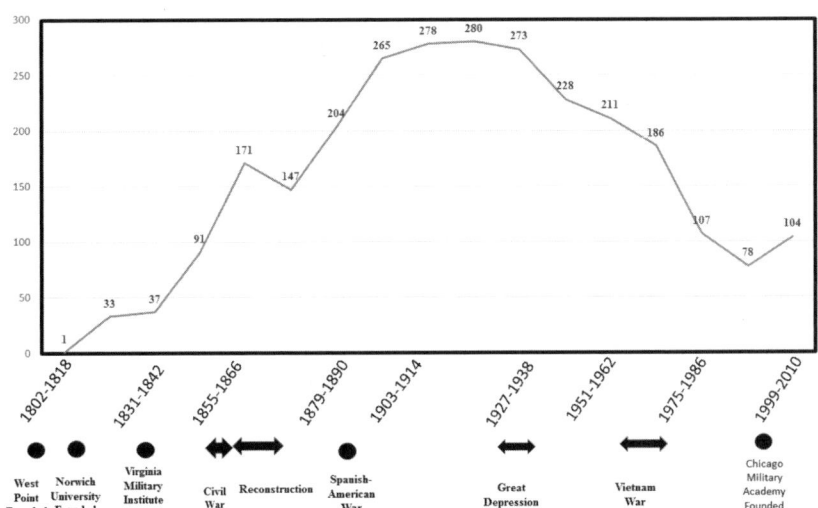

schools, while 23 percent were in the North, 18 percent in the Midwest, and 15 percent in the western United States. When looking at the number of military schools in operation in 2014 and discounting the seven military schools of Puerto Rico as part of the South, a clear picture is presented. The South (less Puerto Rico) accounts for only 39 percent of military schools in 2014. The majority of the schools are outside the Southern culture, and historically California is an excellent example of a non-Southern state where the military school format has had great appeal. California has had 93 military schools in its history, more than any other state. In 2014 it hosted 13 military schools, 14 percent of the total in operation.

The Future Challenge to Private Military Secondary Schools

The economic challenges of the 2008 recession have not significantly impacted financially shielded public secondary education or military colleges. Private military secondary schools have been highly vulnerable, however. The most susceptible are small boarding schools. In the 2011 and 2012 school years the majority of private military schools with enrollments less than 150 saw 2 percent to 8 percent declines in enrollment. During the same period the larger schools generally retained their enrollment or had minor growth.

The enrollment struggles of the private military secondary schools are dual in nature, as they are challenged both to retain students and maintain enrollment standards. The larger private secondary military boarding schools exceed many of the smaller schools by 30 percent and 40 percent in retention for two reasons: opportunity and comradery. The larger schools have more programs, more diversions, and better facilities. But more importantly, they can be more selective, so in turn the cadets desire to return to be with their friends. Cohesion and esprit de corps are strongest in a challenging environment with high standards and retention rates. When faced with enrollment challenges, smaller boarding schools are tempted to adjust their standards downward, but if they do, the few bad cadets that result can discourage four or five good cadets from continuing, and that unexpected drop in enrollment can result in a crisis. Further, many of the smaller schools have no endowment to help them get through challenging periods. The potential result could be very similar to that of the Great Depression, in which smaller and weaker schools closed.

The lessons of enrollment of less desirable cadets or midshipmen have been demonstrated historically by three school failures. Mississippi A&M was beset with problems throughout its forty years of operation as a military college. The dismal quality of enrollment, which resulted in the focus of the military program on control of behavior rather than character development, was incompatible for a military school's success. The use of military schools as a behavioral adjustment or reform school was similarly demonstrated by the rapid failure of two schools at the secondary levels. The Massachusetts state reform nautical training ship and the similar attempt in San Francisco met with disappointments in what the USS *Jamestown*'s captain in San Francisco called the "contaminating influences exerted by . . . wayward boys."[2]

The challenges of viability of private military schools can be questioned in the examination of the forty schools in operation in 2014. Puerto Rico is home to seven military schools established between 1959 and 2003. The fact that seven of those nine are still are in operation shows that they do not reflect the experience of military schools in the fifty United States. When the seven schools in Puerto Rico are removed from the list, thirty-three remain.[3]

The average date of establishment of the thirty-three schools is more than 115 years ago, in 1901. The only two schools established since 1965 are American Christian Military Academy, California (1999), and Forest

Hill Military Academy, Kentucky (2012). The future strength of American private military schools lies with those twenty-one schools that enroll two hundred or more cadets. The remainder must choose their presidents, headmasters, or superintendents with the greatest care and maintain boards that support healthy policies devoted to retention, careful enrollment selection, and fundraising to build solid schools.

Military Secondary Schools' Future

Public education is where the majority of military secondary schools will be in the future. Between 2009 and 2014, a period of five years, ten new military schools opened, and only two of them were private schools, with the remainder charter or public schools. During that same period four small private military schools closed, three of them after no more than five years. Also during that period the larger private military schools of 300 maintained their enrollment or showed growth, while the private military schools of 120, some with long histories, were struggling with enrollment.[4]

Of the eight public or charter military secondary schools, one, Utah Military Academy, opened with 329 cadets. Another, North Valley Military Institute in California, enrolled 224. While eight public schools were added, only two were closed. The trend for the secondary military school in the future appears to be with publicly funded charter and public schools. Symbolic of this was the addition of a military charter school, Oakland Military Institute–College Preparatory Academy of Oakland, California, to the Association of Military Colleges and Schools. Until that membership in 2012, the secondary schools of the association had been private for over ninety years.[5]

Political Challenge to the Military School Culture

Court rulings and military directives have had a direct impact on the role of religion in the military school model. In 1972 the US Court of Appeals ruled against mandatory church attendance at the federal military academies and state-supported military colleges and universities. The next federal court action was in 2003 against VMI's nondenominational and voluntary evening meal prayer. In 2005, Mike Weinstein, US Air Force Academy class of 1977 and founder of the Military Religious Freedom Foundation, brought suit against the US Air Force, demanding that it stop all members of the USAF, including chaplains and academy cadets, from evangelizing, proselytizing, or any actions "to involuntarily convert, pressure, exhort, or persuade a fellow member of the USAF to accept their own religious beliefs while on duty." This action was influenced by controversy at the US Air Force Academy about claimed religious intolerance and evangelical proselytizing. The deputy chief chaplain of the US Air Force stated the chaplains' position: "We will not proselytize, but reserve the right to evangelize the unchurched."[6] The importance to military schools that are publicly funded is that these actions and more that followed were, to some, an "attack upon the very existence of religious expression within the military context."[7]

A look at all four of the most important men in the military school movement, Alden B. Partridge, Sylvanus Thayer, Francis H. Smith, and Stephen B. Luce, shows their clear intent that religion and morality be of central importance in the military school culture. By 2006, however, both the air force and the navy issued guidance that "other than Divine/Religious Services, religious elements for a command function, absent extraordinary circumstances, should be non-sectarian in nature."[8] The Weinstein campaign continued with a series of lawsuits that included one in 2007 that removed Bibles with service insignia from sale on military installations.

In 2011, Air Force Chief of Staff General Norton Schwartz issued a directive that in part prohibited "use of [commanders'] position to promote their personal religious beliefs to their subordinates" and further warned, "Commanders . . . who engage in such behavior may cause members to doubt their impartiality and objectivity." That same month Walter Reed Military Hospital in Washington, DC prohibited visitors from leaving religious items such as Bibles. Weinstein met with the USAF lawyers in 2013, and that same year the House Armed Services Committee adopted an amendment to the National Defense Authorization Act in 2013, the objective of which was "to ensure protection of the rights of Armed Services members to hold, act upon, and practice freely their religious beliefs as long as they do not interfere with any constitutional liberties of others." The White House issued their position that senior presidential advisers would recommend a veto.[9]

In March 2014 a cadet at the US Air Force Academy posted Bible scripture on his personal whiteboard door card. In two hours the Military Religious Freedom Foundation had influenced the academy to have it removed. Following that, a number of cadets posted similar handwritten scriptures, including Qur'an text posted by some Muslim cadets showing their support, in what Weinstein describes as a "revolt." According to USAF regulations, leaders "must avoid the actual or apparent use of their position to promote their personal religious beliefs to their subordinates or to extend preferential treatment for any religion."[10] However, this circumstance did not appear to be a violation of USAF regulations, nor is it not in accordance with *Department of Defense Accommodation of Religious Practices Within the Military Services* update of January 22, 2014.

In September 2014 the US Air Force deleted the required "so help me God" from enlistment and commissioning oaths based on recommendation from the Pentagon's general counsel.[11] Two months later, in November, the Military Religious Freedom Foundation intervened again, and the army began to reconsider its policies, as a Christian college was requiring the hiring of an ROTC assistant military science professor to have the same qualification as that for other faculty there: being a Christian.[12]

Don Snider and Alexander Shine, in their US Army War College examination of character development and integrity, published in 2014, write: "We conclude from these examples that the institutional behavior of our military professions within the DoD [Department of Defense] manifests cultures that can fairly be described as increasingly hostile to personal moralities and their rightful expression, especially when based on religion." In 2012–13 I had many conversations with a chaplain of a secondary military boarding school. He had recently completed twenty-eight years of active duty with the army, five years with the National Guard, and two combat tours of duty. He inferred that by the time he retired, restrictions had limited his ability to accomplish his responsibilities as an army chaplain.[13]

An Assessment of the Military School Concept

The military school format encompasses three levels of education: primary, secondary, and collegiate. Student bodies may be all-male, all-female, or coed. The schools may require a military obligation or they may not. Some are sponsored by various religious denominations, some are funded privately, and some publicly. Despite the diverse nature of their enrollments, funding, and postgraduate requirements, a pattern emerges from the military school culture that can be assessed. Assessments have been made focused on postgraduation results.

There are only two military schools that enroll only grade school cadets: St. Catherine's Military Academy in California and the San Antonio

Academy in Texas. Both of these schools date back to the 1920s as military schools and no longer have boarding programs. San Antonio Academy is the premier grade school in the city. It attracts the best families and has a list of alumni that includes such men as Congressman Lamar Smith, who has served in the US House of Representatives since 1987. St. Catherine's is a Catholic school whose military program is robust despite the oldest cadets being in eighth grade. The religious nature of the school is complete, with both laypersons and nuns as teachers. Graduates of both schools go on to complete high school and on to some of the nation's best colleges.

The nation's military high schools are a mixture of publicly and privately funded schools. The alumni of the older private schools, many of which are boarding schools, are extremely impressive, and many of those schools boast of sending their graduates to federal military academies as well as some of the most competitive colleges. The military format at the high school level has passed the test of recent studies comparing schools of similar socioeconomic backgrounds in the critical areas of attendance rates, dropout rates, standardized test scores, and graduation rates in the public arena. Additionally, the effects of discipline affect self-esteem, college aspirations, and such basics as completion of homework.

Whereas many public school parents and teachers are disillusioned by the heavy impact of disorderly classes and student misconduct, the military schools in private and public education offer an effective alternative. Once more these schools impart a skill set through opportunities seldom offered in a nonmilitary school setting. These skills include interpersonal skills, leadership, and facilitation of agency that promote the ability to act in concert with, lead, and contribute to group efforts. Finally, these schools offer a character development setting that provides students with a moral code to adopt beyond that of obedience to public law and order. Military schools offer an abundance of teachable moments that are related to future adult success and ethical conduct.

Accolades for military junior colleges were recently quantified by the *CNN Money* ranking of community colleges. The list included three military junior colleges that ranked in the top fifteen. New Mexico Military Institute was ranked second, Georgia Military College ninth, and Marion Military Institute twelfth.[14] Another financial advantage is the opportunity these schools also offer in early commissioning after two years and postgraduate membership in the National Guard or reserves with the pay of a lieutenant.

The military colleges and federal military academies offer all the previously mentioned advantages. A look at a sampling of their academic performance is noteworthy. Each year Rhodes Scholars are selected nationwide among the finest college graduates. The list of the top fifteen historically includes West Point at number four, with eighty-five Rhodes Scholars, the Naval Academy at number ten with forty-three, and the Air Force Academy at number fourteen with thirty-five. Ranked alongside the academies were Harvard, Yale, Princeton, Stanford, Dartmouth, and MIT. The same academies took the top three ratings among the nation's high school guidance counselors. The widely used *U.S. News & World Report Best Colleges* ratings placed them at thirteen, seventeen, and twenty-seven, respectively. *Forbes* magazine's list of the top one hundred colleges list the federal military academies at nine, twenty-seven, and thirty-four, respectively.[15]

Perhaps most telling is the college rating system developed through the study of alumni outcomes and the understanding of alumni of the contribution their college made in their preparation. *The Alumni Factor* provides six ratings as follows: Top twenty liberal arts colleges included West Point, the US Naval Academy, the US Coast Guard Academy, and the US Air Force Academy. Top twenty national universities included Virginia Tech. Top regional colleges placed the Citadel at number one. The list of 227 colleges under the category of overall ratings included the Naval Academy at number three, the US Military Academy at number five, the Coast Guard Academy at number nine, the Citadel at number ten, the Air Force Academy at number eighteen, Virginia Tech at number thirty, and Texas A&M at number sixty. The top small colleges included the US Coast Guard Academy at number five and the Citadel at number six. The top medium-sized colleges included the US Naval Academy at number two, West Point at number four, and the Air Force Academy at number seven. The very large universities included Virginia Tech at number one and Texas A&M at number four.[16]

A clear comparison between the civilian and military school experiences in relation to results is reflected in a 2015 Gallup Poll study in which fourteen thousand Virginia Tech alumni, both civilians and cadets, examined their quality of life. The study linked educational experiences of graduates' "long term outcomes" in relation to "Purpose Well-Being, being motivated to achieve goals; Social Well-Being, having strong relationships and love in life; Financial Well-Being, having a secure economic status; Community

Gen. Douglas MacArthur visits the Alamo with a Texas Military Institute cadet escort, 1951. (Courtesy TMI—The Episcopal School of Texas)

Well-Being, engaging with and having pride in one's community; and Physical Well-Being, having good health and energy." The results are telling in relation to the influence of the military school experience. Virginia Tech results were highly favorable in comparison to national averages, but "leading the pack, former members the Corps of Cadets have the highest rate of overall well-being in all five areas."[17]

Those Who Took the Path Less Traveled

Over the 212-year history of military schools, their alumni have been giants of politics, business, literature, sports, education, and entertainment. The ranks of the cadets and midshipmen of our military schools have no list that could do their distinguished collective group justice. However, appendix C, Selected Military School Alumni, demonstrates that military schools provide their cadets unique skills that are transferable beyond military service. The educational environments facilitate learning and encourage reaching for incredible accomplishments along with honor, integrity, and moral courage. Gen. H. Norman Schwarzkopf said of his military high school experience, "Valley Forge Military Academy taught me to be proud to be an American . . . choosing the harder right rather than the easier wrong. . . . West Point prepared me for the military," he continued, "Valley Forge prepared me for life."[18]

Gen. Douglas MacArthur stated simply of his military high school, "This is where I started and I thank a merciful God that I am able to come back to the school again." Texas Military Institute prepared him for West Point, where

> they teach you to be proud and unbending in honest failure, but humble and gentle in success; not to substitute words for action; not to seek the path of comfort, but to face the stress and spur of difficulty and challenge; to learn to stand up in the storm, but to have compassion on those who fall; to master yourself before you seek to master others; to have a heart that is clean, a goal that is high; to learn to laugh, yet never forget how to weep; to reach into the future, yet never neglect the past; to be serious, yet never take yourself too seriously; to be modest so that you will remember the simplicity of true greatness; the open mind of true wisdom, the meekness of true strength. . . . The long gray line has never failed us. Were you to do so, a million ghosts in olive drab, in brown khaki, in blue and gray, would rise from their white crosses, thundering those magic words: Duty, Honor, Country.[19]

Appendix A

Military Schools of the United States,
1802–2014

Geographical Area	Number of Schools
Alabama	39
Arizona	3
Arkansas	7
California	93
Colorado	3
Connecticut	21
Delaware	5
District of Columbia	4
Florida	33
Georgia	25
Hawaii	2
Illinois	30
Indiana	3
Iowa	3
Kansas	4
Kentucky	20
Louisiana	8
Maine	3
Maryland	23
Massachusetts	13
Michigan	9
Minnesota	11
Mississippi	21
Missouri	28
Nebraska	4
New Hampshire	8
New Jersey	31
New Mexico	3
New York	78
North Carolina	52

Ohio	13
Oklahoma	3
Oregon	6
Pennsylvania	27
Puerto Rico	9
South Carolina	22
South Dakota	1
Tennessee	27
Texas	44
Utah	3
Vermont	7
Virginia	77
Washington	5
West Virginia	4
Wisconsin	6
Wyoming	1
TOTAL	842

Appendix B

Military Schools of the United States by Category, 2014

Legend

Data are in order of school; location; year of establishment (year transitioned to military school); denominational affiliation, if any; military program, if any; primary and secondary grade levels, if applicable; number of cadets in 2009–10 school year (unless established after, and data from, 2014).

Military Programs

AFJROTC: Air Force Junior Reserve Officer Training Corps
AJROTC: Army Junior Reserve Officer Training Corps
CCC: California Cadet Corps
MCJROTC: Marine Corps Junior Reserve Officer Training Corps
NDCC: National Defense Cadet Corps
NJROTC: Navy Junior Reserve Officer Training Corps
USNSCC: United States Naval Sea Cadet Corps

Federal Service Academies

United States Air Force Academy, Colorado Springs, CO, 1947, 4,619
United States Coast Guard Academy, New London, CT, 1877, 973
United States Merchant Marine Academy, King Point, NY, 1943, 964
United States Military Academy, West Point, NY, 1802, 4,553
United States Naval Academy, Annapolis, MD, 1845, 4,552

Military Colleges and Universities

California Maritime Academy, Vallejo, CA, 1929, 850
The Citadel, Charleston, SC, 1842, 1,866
Maine Maritime Academy, Castine, ME, 1941, 932
State University of New York Maritime College, Bronx, NY, 1874, 1,154
Massachusetts Maritime Academy, Buzzard Bay, MA, 1891, 1,162
North Georgia College and State University, Dahlonega, GA, 1873, 750
Norwich University, Northfield, VT, 1819, 1,123
Texas A&M University, College Station, TX, 1876, 2,177
Texas Maritime Academy, Galveston, TX, 1962, 300

Virginia Military Institute, Lexington, VA, 1839, 1,569
Virginia Polytechnic Institute and State University, Blacksburg, VA, 1872, 769
Virginia Women's Institute for Leadership/Mary Baldwin College, Staunton, VA,
    1842 (1995), 120[1]

Military Junior Colleges

Georgia Military College, Milledgeville, GA, 1879, 775[2]
Marion Military Institute Marion, AL, 1842 (1887), 399
New Mexico Military Institute, Roswell, NM, 1893, 962[2]
Valley Forge Military Academy and College, Wayne, PA, 1928, 451[2]
Wentworth Military Academy and College, Lexington, MO, 1880 (1882), 926[2]

Private Military Schools

Admiral Farragut Academy, Saint Petersburg, FL, 1945, NJROTC, 6–12, 348
American Christian Military Academy, San Bernardino County, CA, 1999,
    CCC, 6–12, 38
American Military Academy, Guaynabo, PR, 1963, AJROTC, 9–12, 50
Antilles Military Academy, Trujillo Alto, PR, 1959, AJROTC, 9–12, 100
Army and Navy Academy, Carlsbad, CA, 1910, AJROTC, 7–12, 320
Bayamon Military Academy, Bayamon, PR, 1975, AFJROTC, 7–12, 445
Benedictine High School, Richmond, VA, 1911, Catholic, AJROTC, 9–12,
    278[3]
Benedictine Military School, Savannah, GA, 1902, Catholic, AJROTC, 9–12,
    309
Camden Military Academy, Camden, SC, 1942, AJROTC, 7–12, 302
Caguas Military Academy, Caguas, PR, 1975, AJROTC, 1–12, 165
Carson Long Military Institute, New Bloomfield, PA, 1837 (1916), AJROTC,
    6–12, 207
Chamberlain-Hunt Military Academy, Port Gibson, MS, 1879 (1930), Presbyte-
    rian, 7–12, 112[4]
Christian Brothers Academy, Albany, NY, 1859 (1892), Catholic, 6–12, 506
Christian Military Academy, Vega Baja, PR, 2002, K–12, 100
Culver Military Academy, Culver, IN, 1894, 9–12, 455
Fishburne Military School, Waynesboro, VA, 1879 (1884), AJROTC, 7–12, 170
Florida Air Academy, Melbourne, FL, 1961, AFJROTC, 6–12, 350[5]
Forest Hill Military Academy, Millersburg, KY, 2012, 6–12, 75
Fork Union Military Academy, Fork Union, VA, 1897 (1903), Baptist, 6–13,
    535
Hargrave Military Academy, Chatham, VA, 1909, Baptist, AJROTC, 7–12, 340[6]
Howe Military School, Howe, IN, 1884, Episcopal, AJROTC, 7–12, 141
La Salle Institute, Troy, NY, 1850 (1891), Catholic, AJROTC, 9–12, 301
Leonard Hall School, Leonardtown, MD, 1909 (1941), 6–12, 51
Lyman Ward Military Academy, Camp Hill, AL, 1898, AJROTC, 7–12, 115
Marine Military Academy, Harlingen, TX, 1963, MCJROTC, 8–12, 267
Massanutten Military Academy, Woodstock, VA, 1899, AJROTC, 7–12, 170

Maita Luca Military Academy, PR, 1992, 4–12, 65

Missouri Military Academy, Mexico, MO, 1889, AJROTC, 6–12, 230

New York Military Academy, Cornwall-on-Hudson, NY, 1889, AJROTC, 7–12, 100

North Point Military Academy, Vega Baja, PR, 2003, 1–9, 45

Oak Ridge Military Academy, Oak Ridge, NC, 1852 (1866), AJROTC, 7–12, 77

Randolph-Macon Academy, Front Royal, VA, 1892, Methodist, AFJROTC, 6–12, 365

Riverside Military Academy, Gainesville, GA, 1907, AJROTC, 7–12, 350

San Antonio Academy, San Antonio, TX, 1886 (1920), 3–8, 217

St. Catherine's Military Academy, Anaheim, CA, 1889 (1923), Catholic, 4–8, 150

St. John's College High School, Washington, DC, 1851 (1915), Catholic, AJROTC, 9–12, 260

St. John's Military School, Salina, KS, 1887, AJROTC, Episcopal, 5–12, 193

St. John's Northwestern Military Academy, Delafield,WI, 1884 (1886), Episcopal, AJROTC, 7–12, 300

St. Thomas Academy, Mendota Heights, MN, 1885 (1905), Catholic, AJROTC, 9–12, 580[7]

Texas Military Institute, San Antonio, TX, 1893, Episcopal, AJROTC, 6–12, 108

Public Military Schools

Cleveland Junior Naval Academy, Saint Louis, MO, 1981, NJROTC, 9–12, 236

Franklin Military School, Richmond, VA, 1980, AJROTC, 9–12, 200

Marine Academy of Science and Technology, Highlands, NJ, 1981, NJROTC, 9–12, 279

Kenosha Military Academy, Kenosha, WI, 1997, AJROTC, 9–12, 132

Magnet Military School, North Charleston, SC, 1997, AJROTC, 6–12, 541

Chicago Military Academy–Bronzeville, Chicago, IL, 1999, AJROTC, 9–12, 549

George Washington Carver Military Academy, Chicago, IL, 2000, AJROTC, 9–12, 520

Southeast Academy High School, Norwalk, CA, 2000, 11–12, 158[8]

Phoenix Military Academy, Chicago, IL, 2000, AJROTC, 9–12, 404

Forestville Military Academy, Forestville, MD, 2002, AFJROTC, 9–12, 860

Philadelphia Military Academy at Leeds, Philadelphia, PA, 2004, AJROTC, 9–12, 249[9]

Philadelphia Military Academy at Elverson, Philadelphia, PA, 2005, AJROTC, 9–12, 3,199

Admiral Hyman George Rickover Naval Academy, Chicago, IL, 2005, 9–12, 397

Marine Math and Science Academy, Chicago, IL, 2006, MCJROTC, 9–12, 321

Summerlin Military Academy, Bartow, FL, 1851 (2006), AJROTC, 9–12, 451

Air Force Academy High School, Chicago, IL, 2009, AFJROTC, 9–10, 126

Stanislaus Military Academy, Turlock, CA, 2009, 9–12, 75

Hollywood Hills Military Academy, Hollywood Hills, FL, 2012, NJROTC, 9–12, 233

Bridgeport Military Academy, Bridgeport, CT, 2013, NJROTC, 9–12, 239[10]

Charter Military Schools

Willamette Leadership Academy, Veneta, OR, 1993, 6–12, 115

Public Safety Academy, San Bernardino, CA, 1999, CCC, 6–12, 562[10]

Oakland Military Institute–College Preparatory Academy, Oakland, CA, 2001, CCC, 6–12, 592

Sarasota Military Academy, Sarasota, FL, 2003, AJROTC, 9–12, 700

Delaware Military Academy, Wilmington, DE, 2003, NJROTC, 9–12, 538

California Military Institute, Perris, CA, 2003, CCC, 7–12, 545

La Sierra Military Academy, Visalia, CA, 2003, 7–12, 170

Western New York Maritime Academy, Buffalo, NY, 2004, NJROTC, 9–12, 325

General John Vessey Leadership Academy, Saint Paul, MN, 2004, AJROTC, 9–12, 120

Summit Leadership Academy, Hesperia, CA, 2004, 9–12, 197[10]

Bataan Military Academy, Albuquerque, NM, 2006, USNCC, 9–12, 120

Francis Marion Military Academy, Ocala, FL, 2008, AJROTC, 9–12, 130

New Orleans Military Maritime Academy, New Orleans, LA, 2011, MCJROTC, 9–12, 106

Riverside County Education Academy, Moreno Valley, CA, 2011, CCC, 9–12, 175

Paul R. Brown Leadership Academy, Elizabethtown, NC, 2013, 6–11, 103[11]

North Valley Military Institute, Granada Hills, CA, 2013, CCC, 6–12, 224

Utah Military Academy, Riverdale, UT, 2014, NDCC, 7–12, 329

Summary

Federal service academies, 5

Military colleges and universities, 12

Military junior colleges, 5

Private military schools, 40

Public military schools, 19

Charter military schools, 17

Appendix C

Military Schools of the United States,
1802–2014

ALABAMA

SCHOOL	Founded	First Year of Known Military Operation	Transitioned to Civilian	Last Year of Known Military Operation	Date Closed, or Still Open
Alabama Military Academy, Huntsville		1890		1891	
Alabama Military Institute, Anniston (1929: absorbed Anniston University School)	1904				1937
Alabama Military Institute, Eufala, Barbow County	1843			1849	
Alabama Military Institute, Tuskegee (1857–91: Park High School for Boys)	1857	1891			1900
Alabama Polytechnic Institute, Auburn[1] (1856–72: East Alabama Male College; 1872: East Alabama Male College in Auburn to the State of Alabama. The institution rechartered as the 1866–92 Agricultural and Mechanical College of Alabama; 1892–1960: Alabama Polytechnic Institute; 1960–: Auburn University)	1856	1872			1905
Alabama Presbyterian School for Boys, Anniston		1922			
Alabama Scientific & Military Institute, Tuskegee (also called Alabama Military Institute & Scientific; Macon County Military School; Military and Scientific Institute; Tuskegee Classical and Scientific Institute/Military Academy; Tuskegee Military Institute for Boys)	1846			1892	
Anniston University School, Anniston (also known as Alabama Presbyterian College; consolidated with Alabama Military Institute, 1929)	1905				1929
Central Military Institute, Selma		1853			
Glennville Collegiate and Military Institute, Glennville (also known as Major Goldsboro Military Institute)		1842		1846	

School					
Greenville Collegiate & Military Institute, Greenville (1846–58: Greenville Male Collegiate Institute)	1846	1858			1872
Howard College,² 1841–87: Marion; 1887–1913: Birmingham	1841	1871	1913		
Huntsville Military, Scientific & Classical School	1832			1843	
Hurt Military School, Montgomery		1943		1963	
LaGrange Military Academy, LaGrange, 1855–62: Florence (1855–62: Florence Wesleyan University)	1855	1858			1862
Livingston Military Academy, Livingston (1882–83: Cedar Grove Academy, also known as Cedar Grove Military Academy and Cedar Grove Academy)	1882	1883		1901	
Lyman Ward Military Academy, Camp Hill (1898–1954: Southern Industrial Institute)	1898	1954			Open
Marengo Military Institute, Demopolis		1890		1913	
Marion Military Institute, Marion (1842–87: Howard College)³	1842	1871			Open
Mobile Military Institute, Mobile		1880		1920	
Montgomery Military Academy (Institute), Montgomery		1932		1937	
Montgomery Military Academy, Montgomery		1860		1862	
Orville Military Academy, Dallas County (1852: Orville Institute)	1851	1852			
Rehobeth Male Academy, Rehobeth, Wilcox County (also known as Rehobeth Male Academy or Wilcox Male Institute/Military Academy)	1851			1856	

Note: Names and dates in parentheses in smaller type indicate actual name changes and variations on names and the years related to those differences, and in a few other cases they indicate a merger with another school.

School					
Scientific and Military Institute, DeKalb County (also called the Porter Military Academy [Benjamin F. Porter])		1854			
Selma Military Institute, Selma	1904				1908
Southern Military Academy, Fredonia Chambers County (also known as Gibson Hill's Academy, Wetumpka Military Academy)	1851				1864
Southern Military Academy, Greensboro	1919 or 1924				1935?
Southern Military Academy, Wetumpka, Elmore County (Wetumpka Military Academy)	1860				
Southern Polytechnic Institute, Eufala (also known as Alabama Polytechnic Institute)	1858				
Spring Hill College, Mobile	1918				1920
Starke University School, Montgomery	1887			1969	
Talladega Military Academy, Talladega		1894		1912	
University of Alabama, Tuscaloosa	1831	1860	1865		
University Military School, Clanton		1892	Prior to 1899[4]		
University Military School, Mobile	1893		1976		
Verner Military Institute, Tuscaloosa (may have been known as Edgar Military School)	1877			1913	
West Alabama Institute/Military Academy, Summerville	1851	1851		1855	
Wright's Military Academy/University Military School (Preparatory School), Mobile (also known as UMS-Wright's)	1893		1977		

ARIZONA

SCHOOL	Founded	First Year of Known Military Operation	Transitioned to Civilian	Last Year of Known Military Operation	Date Closed, or Still Open
Arizona Military Academy, Nogales, 1931–41: Tubac	1931			1941	
Brownmoor School, Phoenix (Boys' Department)	1931	1955		1957	
Southwestern Military Academy, Beaver Creek (Rimrock)	1924?		1963		

ARKANSAS

SCHOOL	Founded	First Year of Known Military Operation	Transitioned to Civilian	Last Year of Known Military Operation	Date Closed, or Still Open
Advance Military Academy, Fort Smith	1920			1927	
Arkansas Military Academy, Little Rock (1851: Little Rock Literary and Military Institute; also may have been known as Little Rock Military Academy)	1851			1910	
Arkansas Military Institute, Tulip, Dallas County (1849–50: George Alexander School)	1849	1850	1864		
Brown Military Academy of the Ozarks, 1937–before 1951: Sulphur Springs; 1951: Siloam Springs	1937	1951		1953	
Pine Bluff Military High School, Pine Bluff (also known as Jefferson High School)		1855		1860	
Speers-Langford Military Academy, Spearcy (1882–98: Searcy College)	1882	1898		1901	1924
St. John's College,[5] Little Rock, Knoxville		1859	1879 or 1882		

Note: Names and dates in parentheses in smaller type indicate actual name changes and variations on names and the years related to those differences, and in a few other cases they indicate a merger with another school.

CALIFORNIA

SCHOOL	Founded	First Year of Known Military Operation	Transitioned to Civilian	Last Year of Known Military Operation	Date Closed, or Still Open
Agualz Hall, Alta		1913			
American Christian Military Academy of Excellence, Rancho Cucamonga	1999				Open
American Military Academy, Long Beach	Prior to 1925			1930	
Anderson Military Academy, Irvington	1900				1914
Army and Navy Academy, 1910–36: Pacific Beach; 1936: Carlsbad-by-the-Sea (1910–36: San Diego Military Academy; 1936–38: Davis Military Academy; 1938–45: San Diego Army and Navy Academy)	1910				Open
Beaumont Military Academy, Beaumont (also known as Beaumont California Military Academy)	1965			1973	
Black-Foxe Military Institute, Hollywood	1928 or 1929		1961 or 1968		
Brentwood Military Academy, Los Angeles	1902				1972
Brown Military Academy,[6] 1937–58: San Diego; 1958–76: Glendora	1937				1977
Burbank Military Academy,[7] Burbank	1931 or 1932				1953
California Maritime Academy, 1931, 1933–40: California City; 1933: TS California State; 1940–43: TS Golden State; 1943–: Morrow Cove, Valejo (1931–39: California Nautical School)	1931				Open
California Military Academy, 1919–59: Palo Alto; 1959–74: San Jose	1919				After 1973

California Military Academy, Baldwin Hills, Los Angeles	1905			1964
California Military Academy, Oakland (1865–72: Oakland Military Academy, possibly known as Pacific Military Academy)	1865		1908	
California Military Institute, Coronado Beach		1906		1937
California Military Institute, Perris	2003			Open
California Preparatory School, Covina		1932	1953?	
California School for Boys, Los Angeles		1913		
Carlin Military Academy, Glendora	1932		1937	
Cheviot Hills Military Academy, Los Angeles (on grounds of old Pacific Military Academy)	1946 or 1947			1952
Coronado Military Academy, Coronado	1925			
Cromwell Military Academy, Los Angeles	1933		1936	
Del Monte Military Academy, Pacific Grove	1924		1930	
East Side Cadet Academy, San Jose	2000			2009
Elsinore Naval and Military School, Elsinore (founded as Lake Elsinore Military Academy)	1933			1977
Glen Turner (Taylor) Military School, Alameda	1925			1938?
Golden Gate Academy and Cadet School, Oakland		1878		
Golden Gate Military Academy, Puente	1930		1931	
Golden West Military Academy, Redondo Beach, Glendora	1929 or 1930			1935?
Great Western Military Academy, Van Nuys	1925			1933?

Note: Names and dates in parentheses in smaller type indicate actual name changes and variations on names and the years related to those differences, and in a few other cases they indicate a merger with another school.

School					
Harding Military Academy, Glendora	1934 or 1935				1972
Harvard Military School, North Hollywood/Los Angeles	1889				1890
Harvard Military School, North Hollywood/Los Angeles (1900–37: Bishop's School)	1900		1969		
Hitchcock Military School/Military Academy, San Rafael (1878: San Rafael College; 1888: Selborne School; 1907: Hitchcock Military Academy; 1925: Tamalpais School)	1878	1907	1925		1944
Hollywood Military Academy, Los Angeles	1912				
Hollywood Military Academy, Santa Monica	1923			1933	
King's Military Academy, 1923–26: Los Angeles; 1927–35: Upland	1923			1935	
La Monte Military Academy, Atascadero (1925: merged with and adopted name of Pasadena Glen School)	Prior to 1923		1925		
La Sierra Military Academy, Visalia	2003				Open
Laurel Hill Military Academy (1892: merged with Mount Tamalpais Military Academy)					1892
Long Beach Military Academy, 1919–36: Long Beach; 1936–41: Redondo Beach (1931–41: Blackwell Military Academy)	1919				1941?
Los Angeles Military Academy, Los Angeles	1895			1925	
Los Ceritos Military Academy, Long Beach	1926				1938?
Marcell Military Academy, West Altadena	circa 1936			1954	
Miramar Military Academy, Redondo Beach	1926				1936
Moreno Valley Military School, Moreno Valley	2011				Open
Mount Lowe Military Academy, Altadena	1937				1973

Mount Washington Military Academy, 1917: Los Angeles, Santa Monica (1921: absorbed West Lake Military Academy)		1917			1923?
Northridge Military Academy, Northridge		1954		1972	
North Valley Military Institute, Granada Hills (2004–13: North Valley Charter Academy)	2004	2013			Open
Oakland Military Institute College Preparatory Academy, Oakland	2001				Open
Oneonta Military Academy, South Pasadena	1915 or 1922				1942
Ontario Christian Military School, Ontario	2006				2009
Pacific Military Academy, Culver City	1922				1942, 1944, or 1949
Pacific Military Academy, Pacific Beach		1913		1916?	
Page Military Academy, Los Angeles	1908		1978 or 1982		
Palo Alto Military Academy, Palo Alto (1889: Manzanita Hall; 1919: Palo Alto Military Academy; 1972: merged with Harker)	1889 or 1893	1919	1968		
Park Military Academy, Menlo Park (1926–34: Pacific Coast Military Academy)	1926				1939?
Pasadena Army and Navy Academy, Pasadena	1917				1919?
Penn Military Academy, Hesperia		1968		1970	
Pershing Military Academy, Corona	1931				1935?
Public Safety Academy, San Bernardino	2000				Open
Raenford Military School, 1930–32: Puente; 1935–51: Sherman Oaks	1930?				1951
Ramsey Military School, Santa Monica	1932				1963?
Ridgewood Military Academy, Woodland Hills	1941		1976		

Note: Names and dates in parentheses in smaller type indicate actual name changes and variations on names and the years related to those differences, and in a few other cases they indicate a merger with another school.

School					
Riverside County Education Academy, Moreno Valley	2011				Open
Riverside Military Academy, Arlington	1928?		1933?		
Robert E. Lee Military Academy, El Monte	1928				1932?
Robling School, Redondo Beach (1930–31: Redondo Military Academy; 1931–33: Torrance Military Academy)	1930?				1939?
San Diego Military Academy, Del Mar	1958				1977
San Rafael Military Academy, San Rafael (1890–92: Mount Tamalpais Academy; 1892: merged with Laurel Hill Military Academy; 1892–1925: Mount Tamalpais Military Academy)	1890	1892			1971
Santa Monica Military Academy, Santa Monica (1919–25: Pasadena Military Academy, may also have been known as Pasadena Academy; 1917–20)	1905				1925
Seale Academy, Palo Alto (1924–29: West Coast Military Academy; 1929–33: Muldoon School)	1920			1933	
Sepulveda Military Academy, Sepulveda	1953?				1970?
Sierra Military Academy, Eagle Rock, Glendora	1924				1944
Southeast Academy Military & Law Enforcement High School, Norwalk	2009				Open
Southern California Military Academy, Long Beach (grade and junior high school associated with Brown Military Academy)	1924				1987
South Western Military Academy, Long Beach		1924		1925	
Southwestern Military Academy, San Mateo	1924		1969		
Southwest Military Institute Rock, Glendora	1932				1944
Stanislaus Military Academy, Turlock	2009				Open

School				
St. Augustine Military College, Benicia (also known as The College of St. Augustine)	1852	1867		1889
St. Catherine's Military Academy, Anaheim	1889	1923		Open
St. John's Military Academy, Los Angeles	1905			1968
St. Joseph's Military Academy, 1903–35: Rio Vista; 1935–52: Belmont	1903			1952
St. Matthew's Military School, Burlingame (1866–70: St. Matthew's School; also known as Brewer's Military Academy and College)	1866	1870		1915
Stonehurst Military & Naval Academy, Roscoe	1935			1937?
Summit Leadership Academy, Hesperia	2005			Open
Urban Military Academy, Los Angeles (1902–15: Urban School; 1915–40: Urban Military School; 1940–51: Urban Academy; 1951–59: Urban Military Academy)	1902		1959	
West Lake Military Academy, Santa Monica (1921: absorbed into Mount Washington Military Academy)	1915			1921
West Point Preparatory School, San Francisco		1937		
William Warren School, Menlo Park (1924: Menlo School)	1914		1924	
Yale School, Los Angeles	1899			1931?

Note: Names and dates in parentheses in smaller type indicate actual name changes and variations on names and the years related to those differences, and in a few other cases they indicate a merger with another school.

COLORADO

SCHOOL	Founded	First Year of Known Military Operation	Transitioned to Civilian	Last Year of Known Military Operation	Date Closed, or Still Open
Colorado Military Academy, Denver (1906–19: Hill School for Boys, Capitol Hill; 1919–23: Collegiate Military Academy; 1923–51: Colorado Military School; 1955: Colorado Academy)	1900		1955		
Jarvis Hall Military Academy, Montclair	1869				1904
United States Air Force Academy, Colorado Springs	1947				Open

CONNECTICUT

SCHOOL	Founded	First Year of Known Military Operation	Transitioned to Civilian	Last Year of Known Military Operation	Date Closed, or Still Open
Admiral Billard Academy, New London	1936				1953
American Literary, Scientific, and Military Academy (Norwich University), 1825–29, Middleton Connecticut: see under Vermont					
Betts Military Academy Stamford (1838–60: Stamford Classical Boarding School for Boys)	1838	1860			1908
Bridgeport Military Academy, Bridgeport	2013				Open
Brown Military Academy, New Haven	1859				
Cheshire Military Academy, Chester (1794–1862: Episcopal Academy of Connecticut; 1903: Cheshire School, also known as Cheshire Academy in the 1800s)	1794	1862	1903		

School					
Commercial and Military Institute, Bridgeport (also called Collegiate and Commercial Institute [School]; Penfield's Commercial and Military Institute; 1883: Connecticut Military Institute)	1862			1885	
Eastern Military Academy, Shippen Port, 1944–48: see Eastern Military Academy, New York					
Elm City Military Institute, New Haven	Prior to 1887				
Everest Military Academy, Hamden (1836–68: Rectory School; 1878: Atlantic Military Institute; 1886, 1888, 1889: Everest Rectory School)	Prior to 1836				1889
Jarvis Military School, Weston (1833–52: Weston Boarding School; in 1878 known as Weston Military Institute, Weston Commercial and Military Institute)	1833	1855			1888
Litchfield School for Young Boys, Litchfield	1922			1947	
Madison Military Academy, 1938–39: Long Island, New York; 1939–42: Old Lyme	1938				1942
Norwalk Military Institute, Norwalk (prior to 1901: Norwalk University School)		1888		1901	
Old Lyme Academy, Old Lyme	1942				1944
Overlook Military Academy, Norwalk (1905–: known as Connecticut Military Academy)	1903			1908	
Peck's Military Academy, Greenwich (1861–81: Peck's School)	1861	1881		1881	
Russell Military Academy (Institute), New Haven (New Haven Collegiate and Commercial Institute until 1840)		1840	1885		
Schenck's Military Academy, Danbury	Prior to 1850	1851			1859
South Norwalk Military Institute, Norwalk	1872	1897			

Note: Names and dates in parentheses in smaller type indicate actual name changes and variations on names and the years related to those differences, and in a few other cases they indicate a merger with another school.

School	Founded	First Year of Known Military Operation	Transitioned to Civilian	Last Year of Known Military Operation	Date Closed, or Still Open
Stamford Military Academy, Shippan Point, Stamford (1833–1913: Connecticut Literary Institution)	1833	1917	1937		
Suffield School, Suffield (1833–98: Baptist Connecticut Literary Institute)	1833	1916	1937		
United States Coast Guard Academy;[8] 1876–79: New Bedford, MA; 1897/1900–1910: Arundel Cove, MD; 1910–: New London (1876–1915: Revenue Cutter School of Instruction)	1876				Open

DELAWARE

School	Founded	First Year of Known Military Operation	Transitioned to Civilian	Last Year of Known Military Operation	Date Closed, or Still Open
Delaware Military Academy, Wilmington, 1859–65: see Pennsylvania Military College					
Delaware Military Academy, Wilmington	2003				Open
National Scientific and Military Collegiate Institute, 1853: Brandywine Springs; 1854: Bristol (became Bristol College)	1853		1854		
Rugby Military Academy, Wilmington	1872			1891	
Wilmington Literary, Scientific and Military Academy, Wilmington	1846				1848
Wilmington Military Academy, Wilmington		1900			1907?[29]

DISTRICT OF COLUMBIA

SCHOOL	Founded	First Year of Known Military Operation	Transitioned to Civilian	Last Year of Known Military Operation	Date Closed, or Still Open
American Institute		1828		1829	
John Holbrook's Military School	1828				1831?
St. John's College High School	1851	1915			Open
Washington United States Naval School		1811		1812	

FLORIDA

SCHOOL	Founded	First Year of Known Military Operation	Transitioned to Civilian	Last Year of Known Military Operation	Date Closed, or Still Open
Admiral Farragut Academy, Saint Petersburg	1945				Open
Alachua Military Academy, Newnansville, Alachua County	1853				
Bolles School, Jacksonville	1931		1963		
Crawford Hulvey Military Academy, Orlando	1925		1926		
East Florida Seminary, 1853–66: Ocala, Gainesville (1905: merged and became University of Florida)	1853	1881	1903		
Florida Agricultural College Lake City (1905: merged with University of Florida)	1884	1887	1905		

Note: Names and dates in parentheses in smaller type indicate actual name changes and variations on names and the years related to those differences; and in a few other cases they indicate a merger with another school.

Florida Air Academy, Junior Division, Plantation (also known as South Florida Air Academy)	1964				1979
Florida Air Academy, Melbourne (2015: Florida Preparatory Academy)	1961		2015		
Florida Military Academy, Fort Lauderdale	1953				1961?
Florida Military Academy, Green Grove Springs; prior to 1907–19: Jacksonville; 1919–25: Magnolia Springs; 1925–29: Dixie Cove Spring; 1929–53: Saint Petersburg (also known as 1913–20: Florida Military and Naval Academy; 1918: Florida (Hulvey) Military Academy; 1930: South Jackson Florida Military Academy)	1907				1953
Florida Military Academy, Plantation	1961				1966?
Florida Military Institute, Haines City	1930			1947?	
Florida Military School, Deland (1961–65: Florida Military College added)	1956				1971
Florida Naval Academy Junior College, 1939–40: Daytona Beach; 1940–53: Saint Augustine (1940: merged with Florida Preparatory School, established 1932)	1939				1953
Francis Marion Military Academy, Ocala	2007				Open
Gainesville Academy, Gainesville (1866: merged with East Florida Seminary)	1856		1866		
Hollywood Hills Military Academy, Hollywood Hills	2012				Open
Inter American Military Academy, Miami		1976		1992	
Miami Aerospace Academy, Miami		1971			1989
Miami Military Academy, Coral Gables	1923/1924				1974

(1924–29: Coral Gables Military Academy; 1929–32: Miami School for Boys Academy[10])
Note: This school may have been two separate schools owned by J. W. Williams.

School					
National Air Space Academy, Avon Park	1964			1965	
Palm Beach Military Academy, Palm Beach	2006				2007
Palm Beach Military Academy, West Palm Beach		1966		1971	
Quincy Military Academy, Quincy (1832–58: Quincy Male and Female Academies)	1832	1859			Circa 1861
Sanford Naval Academy, Sanford	1963				1976
Sarasota Military Academy, Sarasota	2003				Open
South Florida Military Academy, Palm Harbor, Miami	1923 or 1924				1974
South Florida Military and Educational Institute, Barstow (1895–1905: Florida Military College; merged with others to form University of Florida)	1890		1905		
St. Leo Military Academy/College,[11] Saint Leo	1880		1920		
Summerlin Academy, Bartow (1851–1970: Sumerlin Institute; 1970–2006: Bartow Senior High School; 2006–: commonly known as Summerlin Military Academy)	1851	2006			Open
Tampa Military Academy, Tampa	Prior to 1922 or 1925				1930?
West Florida Seminary/Florida Seminary, Tallahassee (1855–56: Florida Institute, also known as Florida Military & Collegiate Institute; later became Florida State University)	1851 or 1857	1859	1865		
Wyler Military School, Fort Lauderdale	1955			1969	

Note: Names and dates in parentheses in smaller type indicate actual name changes and variations on names and the years related to those differences, and in a few other cases they indicate a merger with another school.

GEORGIA

SCHOOL	Founded	First Year of Known Military Operation	Transitioned to Civilian	Last Year of Known Military Operation	Date Closed, or Still Open
Academy of Richmond, Augusta County[12]	1898		1959		
Benedictine Military School/High School, Savannah	1902				Open
Blakely Military Institute, Early County (1827–98: Blakely Academy)	1827	1898		1902	
Bowden Collegiate and Military Institute/College, Carroll County	1854		1861		
Elberton Military Academy, Elberton		1883		1885	
Erowah Military Institute, Canton		1895		1896	
Fannin Military Institute, La Grange (Established as Smith Academy; later known as Fannin Institute; also known as La Grange Military Academy)	Prior to 1837	Prior to 1861			1861
Georgia Military Academy, College Park (1886–90: Moreland Military Academy; 1890–1900: Georgia Military Institute; 1924–26: Chattahoochee Military School, also known as College Park Military Academy; 1966: became Woodward Academy)	1886		1966		
Georgia Military Academy, Greenville	1847			1850	
Georgia Military College, Milledgeville (1880–90: Middle Georgia Military & Agriculture College)	1880				Open
Georgia Military Institute, Marietta	1851				1865
Georgia Southern Military College, Bainbridge	1900 or earlier			1907	

School					
Gordon Military College (Institute), Barnesville (1852–61 and 1865–72: Barnesville Male & Female High School; 1861–65: Barnesville Masonic Female Seminary; 1872–1907: Gordon Institute; 1907–72: Gordon College)	1852	1890	1972		
Ingles Military High School, Atlanta	1889				
Locust Grove Institute, Locust Grove	1894	1919			1930
Marist Military School for Boys, Atlanta (also known as Atlanta Marist College)	1901		1974 or 1977		
Mountville Military Institute, Mountville	1895				1900
North Georgia College,[13] Dahlonega (1873–1929: North Georgia Agricultural College)	1873	1905			Open
Riverside Military Academy, Gainesville (1914: Riverside Military and Naval Academy)	1908				Open
Rome Military Institute,[14] Rome (also known as Magruder's Military Academy)	1855			1883	
Savannah Military Academy, Savannah (Chartered as Georgia Military Academy)	1883			1887	
Southern Military Academy, Manchester (later renamed College Park)	1911			1922	
South Georgia Military and Agricultural College,[15] Thomasville	1887	1914		1970	
Tugalo Military Institute, Carnesville (1900–1908: Tugalo Institute)	1900			1963	
University High School and Military Academy, Athens (also known as Athens Military High School)	1861				1865

Note: Names and dates in parentheses in smaller type indicate actual name changes and variations on names and the years related to those differences, and in a few other cases they indicate a merger with another school.

HAWAII

SCHOOL	Founded	First Year of Known Military Operation	Transitioned to Civilian	Last Year of Known Military Operation	Date Closed, or Still Open
Honolulu Military Academy	1911				1925
Kamehameha School for Boys (Military Institute)[16]	1887	1915 and 1951	1973		

ILLINOIS

SCHOOL	Founded	First Year of Known Military Operation	Transitioned to Civilian	Last Year of Known Military Operation	Date Closed, or Still Open
Admiral Hyman George Rickover Naval Academy, Chicago	2005				Open
Air Force Academy High School, Chicago	2009				Open
Bishop Quarter Junior Military Academy, Oak Park (1917–41: Bishop Quarter Academy)	1917	1936			1968
Bunker Hill Military Academy,[17] Bunker Hill (1859–83: Bunker Hill Academy)	1857	1883	1914		
Chaddock School for Boys,[18] Quincy (1853–74: Quincy English and German College; 1874–76: Johnson College; 1876–99: Chaddock College, also known as Chaddock Boys School)	1853	Prior to 1913	1960s		
Chicago Military Academy–Bronzeville, Chicago	1999				Open
Chicago Military Academy, Chicago	1902				1973
Dixon Military Institute, Dixon	1881	1882			1914

School					
George Washington Carver Military Academy, Chicago	2000				Open
Hansen Military Academy, Fulton	1903				1912
Irving Military Academy, Lake View	1876				1879
Junior Military Academy, 1932–1971: Chicago; 1971–73: Onarga (1971–73: Coe Military School)	1932 or 1934				1973
Lake View Military Academy, Lake View	1914				1917
Lithia Military Academy, Lithia	1912				1913
Marine Math and Science Academy, Chicago	2006				Open
Marmion Military Academy, Aurora (1933–35, 1994–: Marmion Academy, also known as Fox Valley Preparatory School)	1933	1935		1994	
Midwest Junior School, Knoxville	1929				1935
Morgan Park Academy,[19] Morgan Park; 1888: Highland Park (1873–77: Mount Vernon Military and Classical Academy; 1877–90: Morgan Park Military Academy; 1890–92: Illinois Military Academy; 1892: Morgan Park Academy, also called, 1892–1906, University School as part of University of Chicago)	1870 or 1873		1958		
North Shore Military Academy, Niles Center	1927				
Northwestern Military Academy and Naval Academy, Highland, 1888–1915: see under Wisconsin					
Onarga Military School, Onarga (1863–1917: Grand Prairie Seminary)	1863	1917			1971
Park Ridge Military Academy, 1930–45: Homewood; 1945–71: Park Ridge (1930–40: Chicago Military Academy; 1940–41: Homewood Military Academy; 1941–45: Barrington Military Academy; 1945–71: Park Ridge Military Academy)	1930				1971?

Note: Names and dates in parentheses in smaller type indicate actual name changes and variations on names and the years related to those differences, and in a few other cases they indicate a merger with another school.

	Founded	First Year of Known Military Operation	Transitioned to Civilian	Last Year of Known Military Operation	Date Closed, or Still Open
Phoenix Military Academy, Chicago	2001				Open
Rock River Military Academy, Dixon	1904				1916
St. Alban's Academy, Knoxville	1890				1917
St. Alban's School, Sycamore	1890			1932	
St. Joseph's Military Academy, LaGrange (absorbed by Nazareth Academy)	1909		1971		
Sylvan Dells School, Highland Park	1889				1908
Theodore Roosevelt Military School, 1918–24: Burlington, Kansas; 1924: Aledo, Illinois; 1932: Senior Branch located Abingdon (1919–24: Kansas Military Academy [1922: Kelly Military Academy merged]; 1924–31: Illinois Military Academy [School]; 1931: included junior college; 1931–70: Theodore Roosevelt Military School; 1931: Michigan Military Academy consolidated with Illinois Military School [1929–31])	1919				1973
Thorpe Academy for Boys, Lake Forest	1918				1933
Western Military Academy, Alton (1868–95: Wymann Institute)	1868 or 1879	1892			1971/1972

INDIANA

SCHOOL	Founded	First Year of Known Military Operation	Transitioned to Civilian	Last Year of Known Military Operation	Date Closed, or Still Open
Culver Military Academy, Culver	1894	1896			Open
Howe Military School, Howe (1884: Howe Grammar School; 1895: Howe Academy; 1940: Howe Military School)	1884	1895			Open
Le Mans Academy, 1955–68: Watertown, WI; 1968–2003: Rolling Prairie (1955–68: Sacred Heart Military Academy)	1955				2003

IOWA

SCHOOL	Founded	First Year of Known Military Operation	Transitioned to Civilian	Last Year of Known Military Operation	Date Closed, or Still Open
Epworth Military Academy, Epworth	1857				1928
Iowa Military Academy, Epworth	1928				1931?
Kemper Hall, Davenport	1886				1895

KANSAS

SCHOOL	Founded	First Year of Known Military Operation	Transitioned to Civilian	Last Year of Known Military Operation	Date Closed, or Still Open
Kansas Military Academy, Burlington, 1918–24: see Theodore Roosevelt Military School, Illinois					
Kelly Military Academy, Burlington (1922: merged with Kansas Military Academy)	1918				1922
Mount Barbara Military Academy, Salina	1893	1900			1903
St. John's Military School,[20] Salina (1887: Episcopal Military Institute)	1887				Open
St. Joseph's College and Military Academy, Hays (1908–32: Saint Joseph's College, Saint Joseph's College and Military Academy; 1952–70: Saint Joseph's Military Academy; became Thomas More Prep School)	1908	1932	1970		

Note: Names and dates in parentheses in smaller type indicate actual name changes and variations on names and the years related to those differences, and in a few other cases they indicate a merger with another school.

KENTUCKY

SCHOOL	Founded	First Year of Known Military Operation	Transitioned to Civilian	Last Year of Known Military Operation	Date Closed, or Still Open
Auburn Military Seminary, Auburn (may also have been known as Auburn Military Academy)	1892	Prior to 1902			1910
Bethel College, Russellville	1849	1918	1932		
Colonel Edgar's Military Academy,[21] Paris, Bourbon County (also called the Bourbon Military Academy)	1876				1885 or earlier
Daniel Boone Military Institute	1856				
Danville Classical and Military Institute, Danville (also known as Danville Classical and Military Academy)	1870				1880
(Major) Ferrell Military School, Hopkinsville	1873				1903
Forest Military Academy,[22] Anchorage (also known as Anchorage Classical and Military Institute and Anchorage Training School for Boys)	1855				Prior to 1885
Franklin Military Institute,[23] Franklin (1900–1901: also known as Franklin Military Academy)	1899				1908 or earlier
Hogsett Military Academy (School), Danville	1881	1895		1912	
Kentucky Military Institute,[24] Farmdale/Lyndon; 1932: Venice, Florida (1849: official name changed to Kentucky Collegiate and Military Institute; 1972: Kentucky Academy; 1849: absorbed Franklin Institute [Kentucky Collegiate & Military Institute; 1893: merged with Kentucky Training School])	1845	1895	1971		
Kentucky Training School, Mount Sterling (1893: merged with Kentucky Military Institute)	Early 1880s				1893

School	Founded	First Year of Known Military Operation	Transitioned to Civilian	Last Year of Known Military Operation	Date Closed, or Still Open
Louisville Military Academy, Louisville	1887				1889
Lynnland (Lynnlaud) Military Institute, Glendale	1874			1887	
(Colonel W. P.) Maury Military Academy, Mount Sterling		1905		1907	2006
Millersburg Military Institute, Millersburg (1893: Millersburg Training School)	1893				
Mount Sterling Literary, Scientific, and Military Academy, Old Fort Mason, Montgomery County	1847				1849
(L. S. Bauer) Mount Sterling Military Academy, Mount Sterling (may have also been known as Mount Sterling Military Academy)	1900				1901
Rugby Military Academy, Covington		1890		1904	
Warren County Military Institute, Warren County		1870			
Western Military Institute, 1847: Georgetown; 1850: Blue Springs; 1851: Drennan Springs; 1852: Nashville, Tennessee; 1854: Tyree Springs; 1855–62: Nashville, Tennessee (1855–62: Nashville Military College)	1847				1862

LOUISIANA

SCHOOL	Founded	First Year of Known Military Operation	Transitioned to Civilian	Last Year of Known Military Operation	Date Closed, or Still Open
Baton Rouge Military Institute, Baton Rouge		1854			
Clinton Military Academy, Clinton	1866				1869
Ferrell's Military Institute, New Orleans (Ferrell's School for Boys)	1873		1903		

Note: Names and dates in parentheses in smaller type indicate actual name changes and variations on names and the years related to those differences, and in a few other cases they indicate a merger with another school.

SCHOOL	Founded	First Year of Known Military Operation	Transitioned to Civilian	Last Year of Known Military Operation	Date Closed, or Still Open
Louisiana Seminary of Learning and Military Academy,[25] 1860–69: Alexandria; 1869: Baton Rouge, Rapides Parish (1870–74: Louisiana State University; 1874–77: Louisiana State University Agricultural & Mechanical College, Louisiana State A&M College, merged with Louisiana State, 1873)	1860		1877?		
Malavergne Military School, New Orleans		1848		1849?	
New Orleans Academy, New Orleans	1910	1931			1963 or 1987
New Orleans Military Maritime Academy, New Orleans	2011				Open
Rugby Military Academy, New Orleans (1894–97: Rugby School)	1894	1900			1963 or 1970

MAINE

SCHOOL	Founded	First Year of Known Military Operation	Transitioned to Civilian	Last Year of Known Military Operation	Date Closed, or Still Open
Forest City Military Academy, Portland (also known as Portland Military Academy or Portland Latin School)		1888			1893
Maine Maritime Academy	1941				Open
Whiting's Military and Classical School, Ellsworth (also known as Ellsworth Military Academy)	1841				1847

MARYLAND

SCHOOL	Founded	First Year of Known Military Operation	Transitioned to Civilian	Last Year of Known Military Operation	Date Closed, or Still Open
Aberdeen Academy, Aberdeen	1840				

School				
Avondale County School, Laurel	1927			1952
Baltimore School Ship USS *Ontario*	1857			1865
Bingham's Military School, Frederick	1827			1830?
Briarley Hall Military Academy, 1912–33: Poolesville; 1933–39: Beltsville	1912			1939
Cambridge Military Academy, Cambridge	1866		1868	
Catonsville Military Institute/Saint Timothy Hall Military Academy,[26] Catonsville (may have been known initially as Church Military School)	1845	1854		1868
Charlotte Hall School (Military Academy), Charlotte Hall	1774/1797	1850/1852		1976
Cochran-Bryan Preparatory School, Annapolis		1923	1953	
Forestville Military Academy, Forestville	2002			Open
Independent Academy, Carroll County	1856			
Landon Military Academy & Scientific Institute, Urbana (also known as Urbana Military Academy)	1846			1857?[27]
Leonard Hall Junior Naval Academy School, Leonardtown (1909–72: Leonard Hall School)	1909	1941		Open
Maryland Agricultural College,[28] College Park (later became University of Maryland)	1856	1907	1918	
Maryland Military Institute[29] (Military Academy), Oxford, Talbot County	1840			1856
Maryland Military Institute, Saint Dennis, Baltimore County; 1862: Catonsville (former Paradise Hotel)	1862			1867
Maryland Military and Naval Academy, Oxford, Talbot County	1880			1891

Note: Names and dates in parentheses in smaller type indicate actual name changes and variations on names and the years related to those differences, and in a few other cases they indicate a merger with another school.

School	Founded	First Year of Known Military Operation	Transitioned to Civilian	Last Year of Known Military Operation	Date Closed, or Still Open
McDonogh School, McDonogh, Owing Mills	1873	1905	1971		
Pikeville Literary, Scientific, and Military Academy, Baltimore County	1827				
Revenue Cutter School of Instruction, 1897/1900–1910, Arundel Cove: see US Coast Guard Academy, Connecticut					
St. James School, Rockville		1913			
St. John's College,[30] Annapolis	1696	18898	1923		
United States Naval Academy, Annapolis; 1861–65: Newport, Rhode Island (1845–50: US Naval School)	1845				Open
Western Maryland College, Westminster	1866	1920	1924		

MASSACHUSETTS

SCHOOL	Founded	First Year of Known Military Operation	Transitioned to Civilian	Last Year of Known Military Operation	Date Closed, or Still Open
Allen-Chalmers School, West Newton (also known as Allen Military School; Allen-Chalmers Military School; Allen Military School; Allen School for Boys)	1854	1917			1926
Ashfield Military Academy, Ashfield	Prior to 1931				
Boston United States Naval School, Boston		1815		1837	
Brig *Cleo*, School Ship, Nantucket	1828		1831		
Clover Leaf Seminary, Bershier Hills		1872		1875	
Collegiate and Military School, Boston		1862			
Highland Military Academy, Worcester	1856	1857/1858			1912

SCHOOL	Founded	First Year of Known Military Operation	Transitioned to Civilian	Last Year of Known Military Operation	Date Closed, or Still Open
Massachusetts Maritime Academy, Buzzard Bay (1891–1942: Massachusetts Nautical Training School)	1891				Open
Mitchell's Military School for Boys, Bellerica (1919–21: Mitchell Military School)	1870	1919			1930
New England Military School, Byfield	1965		1971		
Revenue Cutter School of Instruction, 1876–97/1900, New Bedford: see US Coast Guard Academy, Connecticut	Prior to 1870				
Springside Military Institute, Pittsfield	1922				
St. Clement Military School, Canton					1950
The Tabor Academy, Marion (naval school)	1876	1916	late 1940s		

MICHIGAN

SCHOOL	Founded	First Year of Known Military Operation	Transitioned to Civilian	Last Year of Known Military Operation	Date Closed, or Still Open
Barbour Hall Junior Military, Nazareth (1902–41: Barbour Hall)	1902	1951			1979
Carson Military and Naval Institute, Ann Arbor; 1922: Detroit; 1936: Ferndale (1920: known as Michigan Military and Naval Preparatory School)	1920		1941		
Garth Coast Military Academy, Detroit		1930			
Hall of the Divine Child, Monroe	1918				1968
Michigan Military Academy, Menominee (consolidated with Illinois Military Academy in 1931)	1929				1931

Note: Names and dates in parentheses in smaller type indicate actual name changes and variations on names and the years related to those differences, and in a few other cases they indicate a merger with another school.

SCHOOL	Founded	First Year of Known Military Operation	Transitioned to Civilian	Last Year of Known Military Operation	Date Closed, or Still Open
Michigan Military Academy, Orchard Lake	1877				1908
Murray-Wright Junior Naval Academy, Detroit (1965–90: Murray-Wright High School)	1965	1990			2007
Nazareth (Nasareth) Hall Military School, Grand Rapids (1928–35: Nasareth Hall)	1928	1935		1963	
Powell School for Boys, Grand Rapids	1883	Prior to 1913	Prior to 1917[31]		

MINNESOTA

SCHOOL	Founded	First Year of Known Military Operation	Transitioned to Civilian	Last Year of Known Military Operation	Date Closed, or Still Open
Col. Charles Young Military Academy, Saint Paul	2004				2005
College of St. Thomas, Saint Paul (1915: Saint Thomas Academy becomes separate from college)	1885	1906	1922		
Cretin (Military) High School, Saint Paul	1871	1917 or 1918	1982		
Grotin High School, Saint Paul		1963			
Military Academy, Frederick City	1829				
Pillsbury Military Academy, Owatonna (1877–86: Minnesota Academy; 1886–1920: Pillsbury Academy)	1877	1911 or 1920			1957
Shattuck Hall Military School, Faribault (1886–1913: Shattuck Hall School, also known as Shattuck School, incorporated as the Bishop Scabury Mission)	1886	1904	1971		
St. James School,[32] Faribault (merged with Shattuck School)	1901		circa 1964		

SCHOOL	Founded	First Year of Known Military Operation	Transitioned to Civilian	Last Year of Known Military Operation	Date Closed, or Still Open
St. Mary's Hall, Faribault (1858-94: Bishop Whipples School for Girls)	1858	1894	1915		
St. Paul Academy, Saint Paul	1900		1969		
St. Thomas Military Academy, Saint Paul (1885–1905: St. Thomas Academy part of College of St. Thomas until 1915)	1885	1905			Open

MISSISSIPPI

SCHOOL	Founded	First Year of Known Military Operation	Transitioned to Civilian	Last Year of Known Military Operation	Date Closed, or Still Open
Brandon Military School, Brandon (also known as Brandon State Military Institute)	1860				1863
Chalmers Institute,[33] Holly Springs	1850 or 1854				1861
Chamberlain-Hunt Academy,[34] Port Gibson	1879	1895			2014
French Camp Military Academy, French Camp (also known in 1878 as French Camp Military Institute)	1885			1913	
Greenville Military Academy, Greenville	1890			1926	
Gulf Coast Military Academy, Gulfport	1912				1976 or 1977
(John T.) Hairston Military Academy, Columbus		1859			
Houston Male Academy, Houston, Chickasaw County		1848			Prior to 1884
Jefferson Military College,[35] Washington (Jefferson College and Military Institute)	1811	1829			1964
Luka Military Academy, Luka	1858				1861
Meridian Military School and College, Meridian	1901		1914		

Note: Names and dates in parentheses in smaller type indicate actual name changes and variations on names and the years related to those differences, and in a few other cases they indicate a merger with another school.

School	Founded	First Year of Known Military Operation	Transitioned to Civilian	Last Year of Known Military Operation	Date Closed, or Still Open
Mississippi A&M College, Starkville (became Mississippi State University in 1932)	1880		1920		
Mississippi Military Academy, Pass Christian (known as Green's Military Academy for Boys and as Mississippi Military Institute)	1852				1861
Mississippi Military Institute, Aberdeen		1851		1878	
Mississippi Military Institute, 1848: Mississippi Springs; 1849: Fairchild's Well (Raymond, Hinds County)	1848				1854
Mississippi Military Institute, 1873–75: West Point; 1875–76?: Aberdeen; shortly after 1876–85: Pass Christian (also known as Captain Murfee's [Murphree's] Military Academy)	1873				1885
Sharon Male College, Madison	1838	1860			
Shubuta Institute and Military Academy, Shubuta		1899		1901	
Southern Mason Military Academy, Carrollton		1858			
St. Thomas Hall Military Academy,[36] Holly Springs	1838 or 1844	1849			1898
Tupelo Military Institute, Tupelo	1913		1947		

MISSOURI

SCHOOL	Founded	First Year of Known Military Operation	Transitioned to Civilian	Last Year of Known Military Operation	Date Closed, or Still Open
Arrow Rock Military Academy, Arrow Rock, Saline County	1839				1841
Breck's Military School, 1886–1916: Wilder; 1916: Saint Paul (1886–1918, 1959: Breck's School)	1886	1918	1959		

School					
Camden Point Military Institute, Camden Point (also known as Military Institute, Camden Point)	1894				1917
Christian Brothers College Military High School, Saint Louis	1850	1936?			1973
City University, Saint Louis (also called Wyman High School, Wyman Classical School, Wyman's School [semimilitary])	1861				1867
Cleveland Junior Military Academy, Saint Louis	1981				Open
De La Salle Military Academy, Kansas City	1910				1971
Jackson Military Academy, Jackson		1906		1917	
Kemper Military School[37] (Academy) (and College), Booneville (1844–1885/1899: Kemper Academy)	1844	1885			2002
Kirkwood Military Academy, Kirkwood	1882			1913	
Lafayette Military Academy, Lexington or Mayview, Lafayette County	1859			1861	
Marmadukes Military Academy, Sweet Springs	1891				1896
Military Institute, Camden Point		1905		1909	
Military School, Frederick City	1829				
Missouri Literary, Scientific, and Military College, Saint Louis (St. Louis Military Academy and may have been called Major Laws's Military Academy)	1844				1846
Missouri Military Academy,[38] Mexico	1889				Open
Missouri Military Institute, Lexington	1866				1871
Poole Military School, Pleasant Hill	1867				1883
Pruitt Military Academy, Saint Louis	1999				2006

Note: Names and dates in parentheses in smaller type indicate actual name changes and variations on names and the years related to those differences, and in a few other cases they indicate a merger with another school.

School	Founded	First Year of Known Military Operation	Transitioned to Civilian	Last Year of Known Military Operation	Date Closed, or Still Open
St. Charles Military College,[39] Saint Charles (1831–1901: St. Charles College; 1904–16: also called St. Charles Military Academy)	1831	1901			After 1916
St. Charles Military School, Saint Charles	1832				1861
St. James Military Academy, Macon (1899–1909/1910: Blee's Military Academy; 1890: Macon Military Academy)	1875				1909 or 1910
St. Joseph Military School, Saint Joseph		1858			
University City School, Saint Louis	1859	1861?			1867
University Military Academy, Columbia, Oak Hill (also called Welch's Military Academy)	1894				1918
Wentworth Military Academy, Lexington	1880	1881			Open
Western Military Academy, University City, Saint Louis (1859–92: Wyman's School/Institute, also known as Weyman's Military School [semimilitary; also spelled Wyman])	1859	1892			1971
Wyman's English and Classical High School, University City, Saint Louis (also known as Wyman's Military School, Wyman's School [also spelled Weyman])	1843				1853

NEBRASKA

SCHOOL	Founded	First Year of Known Military Operation	Transitioned to Civilian	Last Year of Known Military Operation	Date Closed, or Still Open
Agricultural School of the University of Nebraska, Lincoln	1914		1916		
Kearney Military Academy, Buffalo County (1892–98: Platte Valley Institute)	1892				1923

SCHOOL	Founded	Date Closed, or Still Open
Nebraska Military Academy, Lincoln	1908	1918
Worthington Military Academy, Lincoln	1892	1898

NEW HAMPSHIRE

SCHOOL	Founded	First Year of Known Military Operation	Transitioned to Civilian	Last Year of Known Military Operation	Date Closed, or Still Open
American Literary Seminary and Military Academy, Hanover		1821		1822	
Atlanta/Atlantic Air Academy, Rye Beach	1945				1949 or 1953?
Fairmount Military Academy, Nashua		1876			
Granite State Military and Collegiate Institute, Merrimack (1849–65: McGraw Normal Institute; 1872: McGraw Institute)	1849	1865	1872		
Gymnasium and Military Institute, Pembroke (also called Pembroke Military Academy)	1849 or 1850				1853
New Hampshire Military Academy, West Lebanon (1903: merged with Rockland Military Academy)	1898				1903
Rockland Military Academy, 1897–1903: Nyack on the Hudson, New York; 1903, 1905–14: West Lebanon; 1904: Hartford, Vermont (temporary, fire) (1903: merged with New Hampshire Military Academy)	1897				1914
Unity Scientific and Military Academy, Claremont	1836				1842

Note: Names and dates in parentheses in smaller type indicate actual name changes and variations on names and the years related to those differences, and in a few other cases they indicate a merger with another school.

NEW JERSEY

SCHOOL	Founded	First Year of Known Military Operation	Transitioned to Civilian	Last Year of Known Military Operation	Date Closed, or Still Open
Admiral Farragut Academy Pine Beach, Tom's River (also known as Farragut Academy)	1932				1994
Bordentown Military Institute, Bordentown	1881		1972		
Burlington Military (Academy) College, Burlington	1846		Prior to 1897		
De Vitte Military Academy;[40] Morganville	1917				1980s
Eagleswood (Englewood) Collegiate and Military Institute (School), Perth Amboy	1857			1872	
Edge Hill Military Academy, 1829–69: Princeton; 1869: Merchantville, Camden County	1829			1877	
Francis Military Academy, Stratford	1934	1937			1947
Freehold Military Academy (School), Freehold (also known as Freehold Military School 1920–22; 1912: merged with New Jersey Military Academy)	1901				1947
General MacArthur Military Academy, Mount Freedom (also known as General MacArthur Military Academy of New Jersey)	1966				1975?
Gerlack Academy, Brielle	1895			1913	
Kingsley School, Essex Falls	1910			1913	
Literary, Scientific, and Military Academy	1827				
Marine Academy of Science and Technology, Highlands	1981				Open

School					
Matawan Military Academy, Matawan (1834–57: Middletown Point Academy; 1857: Collegiate Institute of Middletown Point, later called Glenwood Collegiate Institute, Glenwood Military Academy; 1913: known as Matawan Institute)	1836	1857			1915
Montclair Military Academy,[41] Montclair (1837–91 and after 1925: Montclair Academy)	1837	1891	1925		
Montrose Classical and Military School, Orange	1871			1878	
New Jersey Institute, Orange	1828				1830
New Jersey Military Academy, Freehold (1912: merged with Freehold Military School)	1900				1916[42]
New Jersey Naval Academy, Beasley Point	1935			1939	
Newton Hill Military Academy, Newton (1913: known as Newton Academy)	1852	1900	1928		
Oakland Military Academy, 1934–50s: Oakland; 1950s–57: New Windsor, New York	1934		1957		
Orange Military Academy, Orange (also known as Adams and Prescott Military Academy)	Prior to 1878				After 1880
Paterson Military Academy, Paterson		1904			Prior to 1910
Pine Ridge Military Academy, Ocean Gate	1935				1944
Randolph Military Academy, Morristown	1912				1915
Roosevelt Military School, 1919–26: West Englewood; 1926–33: Nyack, New York (1932: Roosevelt Academy) (1919): merged with Sheldon School [nonmilitary, established 1915])	1919				1933
Sheldon School, West Englewood	1915	1917		1928	

Note: Names and dates in parentheses in smaller type indicate actual name changes and variations on names and the years related to those differences, and in a few other cases they indicate a merger with another school.

	Founded	First Year of Known Military Operation	Transitioned to Civilian	Last Year of Known Military Operation	Date Closed, or Still Open
St. John's Military Academy, Haddonville	1865			1897	
Stratford Military Academy, Camden	1935				1979
Wenonah Military Academy, Wenonah	1902				1933
Wilson Military Academy, Morristown	After 1908				1912

NEW MEXICO

SCHOOL	Founded	First Year of Known Military Operation	Transitioned to Civilian	Last Year of Known Military Operation	Date Closed, or Still Open
Bataan Military Academy, Albuquerque	2006				Open
Española Military Academy, Española	2004				2009
New Mexico Military Institute, Roswell (1891–93: Gross Military Institute)	1891				Open

NEW YORK

SCHOOL	Founded	First Year of Known Military Operation	Transitioned to Civilian	Last Year of Known Military Operation	Date Closed, or Still Open
Albany Military Institute, Albany (1813–70: Albany Academy)	1813	1870	2005		
Alexander Military Institute, White Plains (1844–57: Hamilton Military Institute; 1857–63: White Plains Military Academy, also known as Alexander Institute, 1863)	1844			1919	
Barnard Military School, New York City (also known as the Barnard School)	1886	1895 or 1905		1899	

School					
Bay View Institute, Babylon	1915	1878		1879	
Berkeley Military School for Boys, New York City					1916
Braden School, Cornwall-on-Hudson	1883				1965
Bronx Aerospace High School, Bronx	2002				2008
Bryant School, Long Island (1905: known as Long Island City High School; circa 1930: Long Island City High)		1887	Prior to 1905		
Calvary Military School, New York City	1926				1935?
Cardinal Farley Military Academy, Rhinecliff	1942				1970
Caswell Academy, Fishkill-on-the-Hudson	1897			1914	
Cayuga Lake Military Academy, Aurora (1799–1880/1882: Cayuga Academy)	1799	1880 or 1882	Shortly prior to 1900		
Chamberlain Military Institute, 1848–1913: Randolph; 1915: Perry (1915: known as New York Military and Naval School; 1916: Silver Lake Military and Naval Academy)	1848			1924	
Christian Brothers Academy, Albany	1854	1892			Open
Churchill Military Academy, Ossining on the Hudson (Sing Sing) (1869–1948: St. John's School)	1843				1948
Claverack College, Claverack (1830–54: Claverack Academy; 1854–1902: Claverack College and Hudson River Institute [The institute was for female students.])	1830	1854			1902
Clinton Liberal Institute and Military Academy, Clinton (The military school was restricted to the Boys' Department, also known as Fort Plain Military School)	1879	1891			1900
Col. (C. J.) Wright's Military Academy, Peekskill	1897	1897			

Note: Names and dates in parentheses in smaller type indicate actual name changes and variations on names and the years related to those differences, and in a few other cases they indicate a merger with another school.

Cook Academy, Montour Falls	1872	Prior to 1890	1922	1922	
Croton Military Academy, Croton Landing, Croton-on-Hudson (also known as Croton Military Institute)	1880			1893	
DeGraff Military and Collegiate Institute, Rochester (also known as DeGraff Military School)	1868				Prior to 1891
Deposit Military Academy, Deposit, Delaware County		1865		1866	
De Veaux School, Niagara Falls (also known as DeVeaux ["Devoe"] Military Academy; 1886–92, 1913: DeVeaux College [Orphanage])	1857		1950		
Dr. Holbrook's Military School, Ossining (Sing Sing)	1866				1914
Duff Military Academy, Cooperstown (also known as Cooperstown Classical and Military Academy)	1838				1843/1844
Eastern Military Academy, 1944–48: Shippen Port, Connecticut; Huntington, Old Springs, Long Island	1944				1979
Fairfield Seminary & Military Academy, Fairfield (1885–91: Fairfield Seminary)	1885	1891			1901
Franklin Military Academy, Franklin, Delaware County		1899			
Gerlack Academy, College Point (may also have been known as Furst's Military Academy)	1881				1895 or before
(North) Granville Military Academy, 1850–76: Stamford, Connecticut; 1876: North Granville; 1880: Mohegan Lake (1833–50: Stamford School for Boys; 1850–76: Stamford Military Institute; 1880, 1913, and 1932: Mohegan Military School; 1936: Mohegan Lake Academy [1953: Mohegan Lake School])	1833		Prior to 1940		
Harlem Literary and Scientific Academy, Harlem	1829			1831	

School					
Hudson River Military Academy, Finderne, New Jersey; Nyack-on-the-Hudson, New York (1897–98: known as Hudson River Military Institute)	1854	1856			1920?
Hudson River Naval Academy	1943				1953
Ithaca Military Band School, Ithaca (1922–29: Patrick Conway Military Band School)	1892				1931?
Jackson Military Institute, Tarrytown	1857		1878		
Kyle Military Institute, Flushing	1890				1935?
La Salle Institute, Troy, New York (1850–78: St. Joseph's Academy; 1878–91: St. Mary's Academy)	1850	1891			Open
La Salle Military Academy, New York, 1883–1906; Clason Point; 1906–26: Long Island (1883–1906: West Chester Institute [and Military Academy]; 1906–26: Clason Point Military Academy)	1883	1898			2001
Loyola School, New York City	1900			1956?	
Madison Military Academy, 1938–39, Long Island: see Madison Military Academy, Connecticut					
Marine Collegiate Military Academy, Stony Brook	1980				1983?
Mount Beacon Military Academy, Rochester	1880		1915		
Mount Pleasant Military Academy, Ossining (Sing Sing) (1814–45: Mount Pleasant Academy)	1814	1845			1925
Mount St. Joseph Semi-Military Academy, Newburgh (also known as St. Joseph Academy; 1953–55)	1924		1974		

Note: Names and dates in parentheses in smaller type indicate actual name changes and variations on names and the years related to those differences, and in a few other cases they indicate a merger with another school.

					Prior to 1891
Mount Vernon Military Academy,[43] 1852–75: Vernon; 1878–84: Rochester (1870–74: may have been located in Mount Vernon; 1852–53: Mount Vernon Boarding School; 1875: known as Mount Vernon English, Classical, and Military Academy)		1852			
Nazareth Hall Cadet School, Rochester		1953			1956
Newman's Military Academy, Tarrytown		1906			
New York Military Academy, Cornwall-on-Hudson (preparatory department known as Bard Hall)	1889				Open
New York School for the Deaf	1817		1952		
New York United States Naval School, New York City		1827		1844	
Nyack Military Academy, Nyack on the Hudson	1901			1903	
Peekskill Military Academy, Peekskill	1833	1857	1968		1968
Pine Forest Military Academy, Wingdale	Prior to 1919				
Poughkeepsie Military Institute, Poughkeepsie (also known as Warring's Military Boarding School and Warring's Military Institute)	1863			1903	
Riverview Military Academy, Poughkeepsie (1836–67: Poughkeepsie Collegiate School; also known in 1862 as College Hill Military Academy)	1836	1862			1920
Roosevelt Military School, 1926–33, Nyack: see New Jersey					
Rugby Military Academy, New York	1889			1903	
Rutherford Military Academy, Nyack (1858–62: Rutherford Academy)	1858	1862			1870
Sacred Heart Military Academy, Brooklyn?, Westchester	1883			1904	
Scientific and Military Academy, Whitesborough	1824			1828	

School					
Stamford Military Academy, Ossining	1917				1924?
Stanton Preparatory School (Academy), Cornwall-on-Hudson	1907	1937		1963	
Starr's Military Institute, 1858: Yonkers; 1874: Tarrytown; 1878 and 1883: Port Chester (1873: known as Port Chester Commercial, Collegiate, and Military Institute)	1854			1883	
State University of New York Maritime College, New York (also called New York Merchant Marine Academy; Maritime College at Fort Schuyler; New York Maritime College)	1874				Open
St. Austin's Military School, West New Brighton (1883: St. Austin's Episcopal School for Boys, also known as St. Austin's School)	1883	Prior to 1892; after 1888	Prior to 1917?		
Stella Niagara Cadet School, Stella Niagara	1908	1886	1971		
St. John's Military School, (Pebble Hill) Manlius (1869–81: The Manlius School, also called The Manlius Military School. Verbeck Hall was operated as the junior department from at least 1888 to 1914, then became Manlius–Pebble Hill School)	1869	1879	1969		
St. Matthew's Military School, 1898, 1901: Pocantico Hills; 1904/1905: Dobbs Ferry, Westchester County	Prior to 1898			1905	
St. Patrick's Military Academy, Harrison	Prior to 1945				1983 or 1986
St. Paul's Military School, Garden City, Long Island	1879				1895
Tarrytown Military Academy, Tarrytown		1863			
Trinity Military Institute, Tivoli (1847–78: Trinity School, also known as Clarks's Military Academy)	1847				1890s
Troy Military Academy, Troy		1893		1896	
United States Merchant Marine Academy, Kings Point	1943				Open

Note: Names and dates in parentheses in smaller type indicate actual name changes and variations on names and the years related to those differences, and in a few other cases they indicate a merger with another school.

	Founded	First Year of Known Military Operation	Transitioned to Civilian	Last Year of Known Military Operation	Date Closed, or Still Open
United States Military Academy, West Point	1802				Open
Vireum Military Academy, Ossining	1870				After 1878
Western Literary, Scientific, and Military Academy, Buffalo	1827				1846
Worrall Hall Military Academy, Peekskill (junior department of New York Military; merged with the New York Military Academy location in 1895)	1870				1924?
Xavier Military High School, New York City (1847–98: College of St. Francis, also called Xavier High School College of St. Francis)	1847	1898			1971
Yonkers Collegiate and Military Institute,[44] Yonkers (1861: Yonkers Military and Collegiate Institute, also known as Yonkers Military Institute, which may have been a follow-on new institute)	1852 or 1854			1891	

NORTH CAROLINA

SCHOOL	Founded	First Year of Known Military Operation	Transitioned to Civilian	Last Year of Known Military Operation	Date Closed, or Still Open
Asheville Military Academy, Asheville	1878			1890	
Atlantic Military Academy, Beaufort		1860			
Baird's School for Boys, Charlotte	1891			1936	
The Bingham School, 1844–64: Oaks, Orange County; 1864–91: Mebaneville; 1891–1922: Asheville	1815	1861			1927
The Bingham School at Mebane, Mebane	1877		1926		
Buie's Creek Academy, Buie's Creek (Buie's Creek Academy and Business College)	1897				1916?

School					
Bula Military School, Wilmington		1856	1861		
Cape Fear Military Academy, Cape Fear (1902–16: Cape Fear Academy)	1868		1902		
Carolina Military Academy, Maxton	1962				1973
Carolina Naval-Military Academy, Henderson	1919				1924
Catawba Military Academy, Rock Hill		1907		1910	
Charlotte Military Institute,[45] Charlotte (also known as Charlotte Military Academy and Carolina Military Institute)	1873			1910	
The Collegiate Institute, Mount Pleasant	1844 or 1854	1908			1933?[46]
Davis Military School, Winston, La Grange	1890				1893
The DeMeritte Military School, Jackson Springs	1916				1923?
Donaldson Military Academy, Fayetteville (1835–1914: Donaldson Academy)	1832	1911		1914	
Edwards Military Institute, Salemburg (1933–35: Edwards Memorial School; 1965–66: Southwood College)	1933	1935	1966		
Elizabeth City Academy, Elizabeth City	1820	1851			
Faison Military Academy, Faison (1876–1904/1905: Faison Male Academy)	Prior to 1876	Circa 1904		1908	
Fayetteville Military Academy, Fayetteville	1864			1865	
Franklin Military Institute, Duplin County (also known as Franklin Scientific and Military School)	1858	1861			1861
Franklinton Classical and Military Institute, Franklinton	1889			1891	
Henderson Military Academy, Henderson (1854: Henderson Male Academy; 1855: Bracey Military School)	1854	1855	1860		

Note: Names and dates in parentheses in smaller type indicate actual name changes and variations on names and the years related to those differences, and in a few other cases they indicate a merger with another school.

School					
Hillsborough Military Academy, Hillsborough,[47] (1867–82: NC Military and Polytechnic Academy)	1859				1882
Horner Military Academy, 1851–70: Oxford; 1870–76: Hillsborough; 1876–1914: Oxford; 1914–20: Charlotte (1851–70: Horner School; 1870–76: Horner and Graves School; 1876–79: Horner School, 1879–1920: Horner Military School)	1851	1879			1920
King's Mountain Military School,[48] Yorkville	1855				1886
Lenoir Academy, Kingston, Lenoir County (1904: Military Institute, Lenoir County; 1905–6: Kingston Military Academy)		1861		1906	
Linsey Military Academy		1970		1852	
Lovejoy Military Academy, Raleigh		1840s			
Moravian Falls Military Academy, Moravian Falls, Wilkes County (also known as Moravian Falls Academy)	1881		Prior to 1907		
New Bern Military Academy, New Bern	1904			1905	
North Carolina Classical, Literary, Scientific and Military Institute, 1830–31: Oxford; 1832–33: Raleigh (also known as Roanoke Classical Literary, Scientific, and Military; Oxford Military Academy; Lovejoy's Raleigh Male Academy)	1830				1833
North Carolina Classical, Literary, Scientific, and Military Institute, Fayetteville	1830				1831
North Carolina College of Agriculture and Mechanic Arts, Raleigh (later North Carolina State University)	1889	1894	1906		
North Carolina Literary, Scientific, and Military Academy, Raleigh (1840–44: North Carolina Classical, English, and Mathematical School, also known 1844–45, as Raleigh Institute)	1840	1844			1847

School					
North Carolina Military Academy, 1894–99: Fayetteville; 1899–1918: Red Springs (1894–99: Fayetteville Military Academy)	1894			1918	
North Carolina Military Institute, Charlotte (also known as Charlotte Military Academy)[49]	1859				1882
Oak Hill Military Academy, Granville County	1861				
Oak Ridge Military Academy[50] (Institute), Oak Ridge (1852–54: Oak Ridge Male Institute; 1854–1929: Oak Ridge Institute; 1929–: Oak Ridge Military Academy)	1852	1866 and 1917			Open
Paul R. Brown Leadership Academy, Elizabethtown	2013				Open
Raleigh Military Academy, Raleigh (also known as the Raleigh Classical, Mathematical, and Military Academy)	1844		1860		
Rhodes Military Institute, Kingston (1904: Military Institute, Lenoir; 1905–6: Kingston Military Institute)	1902			1913	
Rutherford Military Institute, Rutherfordton	Prior to 1890			1898	
Scotland Neck Military Academy, Scotland Neck (became Scotland Male Academy)		1887		1894	
Shelby Military Institute, Shelby		1890		1893	
Southern Military School, Williamsboro, Granville County, 1831: Raleigh	1829	1830			
Statesville Military Academy, Statesville (Iredell), Piedmont County (also called State Military Academy)	1856			1861	
Steven Lee's Military Academy, Chunn Cove, Ashville (also called Colonel Stephen Lee's [Military] School)	1846		1879		1879

Note: Names and dates in parentheses in smaller type indicate actual name changes and variations on names and the years related to those differences, and in a few other cases they indicate a merger with another school.

School	Founded	First Year of Known Military Operation	Transitioned to Civilian	Last Year of Known Military Operation	Date Closed, or Still Open
Thompson Military Academy, Siler City (also known as Thompson School)		1878		1900	
Turlington Military Institute (Academy), Smithfield (1886–before 1992: Smithfield Collegiate Institute, also known as Turlington Institute)		1892			1903
Wilmington Literary, Scientific, and Military Academy, Wilmington	1846				1848
Wilmington Military Academy, Wilson (also known as Colonel Ratcliff's Military School)	Prior to 1856				1861

OHIO

SCHOOL	Founded	First Year of Known Military Operation	Transitioned to Civilian	Last Year of Known Military Operation	Date Closed, or Still Open
Brook Military Academy, Cleveland	1874				1891
Fairmount Military Academy, Cincinnati		1862			1866 or earlier[51]
Kenyon Military Academy/College Preparatory School,[52] Gambier (1824–85: Kenyon Grammar School)	1858	1885?			1907
Miami Military Institute, Germantown	1894				1936
Nazareth Hall Military School, Grand Rapids	1875 or 1900/1928	1935			1982
Ohio Military Academy, Cincinnati	1832				1932
Ohio Military Institute, Cincinnati; 1893–94: Germantown (1833–90: Cary Academy [maybe Belmont College]; 1890–92: Farmer College)	1833	1890			1958?
Putnam Military Academy, Zanesville	1890 or 1891				1897

	Founded	First Year of Known Military Operation	Transitioned to Civilian	Last Year of Known Military Operation	Date Closed, or Still Open
Riser Military Academy, Cincinnati	1999				2000
St. Aloysius Academy and Military School (Cadet School), New Lexington (1915–55: St. Aloysius Academy)	1915				1969
St. Aloysius Military Academy, Fayetteville	1938				1982
Western Military Academy, Dayton	1879	1892			1971
Wilson Military Academy, Cleveland (also known as the Gen. Johnnie E Wilson Military Academy)	1999				2005

OKLAHOMA

SCHOOL	Founded	First Year of Known Military Operation	Transitioned to Civilian	Last Year of Known Military Operation	Date Closed, or Still Open
Oklahoma Military Academy, Claremore	1919 or 1920				1971
Ponca Military Academy, Enid	1940				1974
St. Joseph's Preparatory School, Muskogee	1903	Prior to 1932			1955

OREGON

SCHOOL	Founded	First Year of Known Military Operation	Transitioned to Civilian	Last Year of Known Military Operation	Date Closed, or Still Open
The Adams School, Portland (grades 1–6 of Hill Military Academy)	1938				1945
Bealey Military Academy, Troutdale	1924 or 1925			1949	

Note: Names and dates in parentheses in smaller type indicate actual name changes and variations on names and the years related to those differences, and in a few other cases they indicate a merger with another school.

SCHOOL	Founded	First Year of Known Military Operation	Transitioned to Civilian	Last Year of Known Military Operation	Date Closed, or Still Open
Bishop Scott Academy, Portland, Yamhill, Milwaukie, Portland (1852–70: Bishop Scott Grammar and Divinity School; 1870–78: Bishop Scott Grammar School)	1852	1887			1904
Hill Military Academy, Portland, 1931: Rocky Butte	1901				1962
Oregon Military Academy, Hillsboro (also known as Johnson's Military Academy)	1918 or 1919				1923 or 1924
Willamette Leadership Academy, Veneta (1993–2003: Pioneer Youth Corps School)	1993				Open

PENNSYLVANIA

SCHOOL	Founded	First Year of Known Military Operation	Transitioned to Civilian	Last Year of Known Military Operation	Date Closed, or Still Open
Allentown Collegiate Institute and Military Academy, Allentown (1848–67: Allentown Seminary; 1867–: Muhlenberg College)	1848	1864	1867		
American Classical and Military Lyceum, Mount Airy (1807–26: Collegiate Institute, also known as Mount Airy College, Mount Airy Military School)	1807	1828			1835
Carson Long Military Institute/Military School, New Bloomfield (Bloomfield Academy)	1836/1937	1915			Open
Cheltenham Military Academy, Ogontz	1871			1910	
Classical and Mathematical Academy, Bedford	1834				
Forest Hill Military School, Millersburg	2012				Open

					1954 or several years earlier[53]
Fort Washington Military Academy, Fort Washington (also known as Camp Hill Military Academy)	1946				
Gettysburg Military Academy, Gettysburg		1918			
Mantua Classical and Military Academy, Philadelphia		1827		1829	
Media Military Academy, Media (also known as Media Academy; Swithen C. Shortlidge Military Academy, Shortlidge Academy for Boys)	1874 or 1875	1889–1890	1894		
Mercersburg Academy, Mercersburg	1836				
Nazareth Hall Military Academy, Nazareth (1759–1861: Nazareth Hall)	1743 or 1759	1861			1929
Pennsylvania Literary, Scientific, and Military Academy, 1842–45: Bristow; 1845–48: Harrisburg (also known as Harrisburg Military Academy; Pennsylvania Military Institute, Bristol Military Institute)	1842				1848
Pennsylvania Military College, 1821–46: Wilmington, DE, 1862–68: West Chester; 1868–73: Chester (1821–46: Bullock School for Boys; 1846–53: Alsop School for Boys; 1853–59: Hyatt's Select School for Boys; 1859–62: Delaware Military Academy; 1862–92: Pennsylvania Military Academy)	1821	1858			1973
Pennsylvania Military Preparatory School, Chester[54]	1916				1956
Pennsylvania State Nautical School,[55] Philadelphia (1889–1919: Pennsylvania Nautical School at Philadelphia; 1940–47: Pennsylvania Maritime Academy)	1889				1947
Philadelphia Military Academy at Elverson, Philadelphia	2005				Open
Philadelphia Military Academy at Leeds, Philadelphia	2004				Open
Philadelphia United States Naval School, Philadelphia (also known as Philadelphia Naval Asylum School)		1839		1845	

Note: Names and dates in parentheses in smaller type indicate actual name changes and variations on names and the years related to those differences, and in a few other cases they indicate a merger with another school.

	Founded	First Year of Known Military Operation	Transitioned to Civilian	Last Year of Known Military Operation	Date Closed, or Still Open
Reading Boys' Military School, Reading (also known as Selwyn Hall Military School, The Diocesan School for Boys, The Diocesan Military School)	1875	1879			1895
Roth Military Academy (also known as Broad Street Academy)	1863			1892	
Saunders's Military Academy, West Philadelphia (also known as Professor E. Saunders's Military Academy, Courtland Saunders's Military Academy)	1852				1870
Scientific and Military Collegiate Institute, Reading	1850			1853	
St. Aloysius Academy, West Chester	1895				1949
St. Joseph Junior Military School, Pittsburgh	1937				1962
Stratford Academy, Philadelphia	1944				1945
Valley Forge Military Academy and Junior College, 1928–29: Devon; 1929: Wayne	1928				Open

PUERTO RICO

SCHOOL	Founded	First Year of Known Military Operation	Transitioned to Civilian	Last Year of Known Military Operation	Date Closed, or Still Open
American Military Academy, Guaynabo	1963				Open
Antilles Military Academy, Trujillo Alto	1959				Open
Bayamón Military Academy, Bayamón	1996				Open
Caguas Military Academy, Caguas	1975				Open
Christian Military Academy, Vega Baja	2009				Open

School	Founded	First Year of Known Military Operation	Transitioned to Civilian	Last Year of Known Military Operation	Date Closed, or Still Open
Lincoln Military Academy, San Juan	1963		1995		1995
Maita Luca Military Academy, Las Piedras	1992				Open
North Point Military Academy, Vega Baja	2009				Open
Roosevelt-Rhodes Military Academy, San Juan	1924				

SOUTH CAROLINA

SCHOOL	Founded	First Year of Known Military Operation	Transitioned to Civilian	Last Year of Known Military Operation	Date Closed, or Still Open
Aiken Military Academy, Aiken (Aiken Classical and Military Academy)	1842			1861	
Anderson Military Academy, Anderson		1860		1884	
Bailey Military Institute, Greenwood	1891				1936
Banks School for Boys, Columbia	1922				1924
Camden Military School, Camden (1949–58: Camden Academy; 1977: absorbed the Carlisle Military School)	1949	1958			Open
Carlisle Military School, Bamberg (1892–1914: Carlisle Fitting School of Wofford College until 1938; 1914–32: merged with Camden Military School)	1892	1932			1977
Catawba Military Academy, Rock Hill, York County	1903				Prior to 1908
Chick Springs Military Academy, Chick Springs	1916				1923
The Citadel,[56] Charleston (Military College of South Carolina; 1842–1910: South Carolina Military Academy; 1842–65: The Arsenal, Columbia, functioned as first-year cadets' school location)	1842				Open

Note: Names and dates in parentheses in smaller type indicate actual name changes and variations on names and the years related to those differences, and in a few other cases they indicate a merger with another school.

Clemson Agricultural College, Clemson (1964: Clemson University)	1893		1955		
Eagle Military Academy, Summerville	2002				2008
Greenville Military Institute, Greenville	1878			1887	
Kings Mountain Military School,[57] Yorkville	1855				1886
Low Country Military Academy, Ladson	2009				2012
Magnet Military School, North Charleston (also known as Toole Military Magnet School)	1997				Open
Mount Zion Academy, Winnsboro		1851	1858		
Patrick Military Institute, Greenville; 1887–1900: Anderson (1870–78: Greenville High School)	1887				1900
Porter Military Academy, Charleston (Military and Scientific Institute; 1867–86: Orphan Home and School Association of the Church of the Holy Communion; 1886–90: Holy Communion Church Institute; moved to Charleston, SC, before 1867)	1867	1891			1964
Rice Creek Springs Military Academy (School), Columbia (also called Rice Creek Springs School)	1827				1827
Richland Classical and Scientific Academy, Camden (also called Richland School)	1827				1831
St. John's Classical and Military School, Spartanburg (1854–56: St. John's College)	1854/1855	1856			1870s
Sumter Military Academy and Female Seminary, Sumter		1901		1903	

SOUTH DAKOTA

SCHOOL	Founded	First Year of Known Military Operation	Transitioned to Civilian	Last Year of Known Military Operation	Date Closed, or Still Open
South Dakota State College of Agriculture and Mechanic Arts, Brookings (later South Dakota State University)	1881		Circa 1919		

TENNESSEE

SCHOOL	Founded	First Year of Known Military Operation	Transitioned to Civilian	Last Year of Known Military Operation	Date Closed, or Still Open
Battle Ground Academy, Franklin (1889–1900: Wall and Mooney School; 1900–1924: Mooney School; 1924: Battle Ground Academy)	1889		1954		
Baylor School, Chattanooga	1893	1917	1971		
Bell Buckle Military Academy, Bell Buckle		1897			
Branham and Hughes Military Academy, Spring Hill (1892–1918: Branham and Hughes Academy)	1892	1918			1932
Brennan's Military Academy, Nashville (prior to transition to military was known as T. P. Brennan's School for Boys)		1891			
Brownsville Military School, Brownsville	1893				
Castle Heights Military School, Lebanon (1902–18: Castle Heights School)	1902	1918			1986
Chestnut Hill Military Academy, La Grange (also called La Grange Military Academy)	1857				1862

Note: Names and dates in parentheses in smaller type indicate actual name changes and variations on names and the years related to those differences, and in a few other cases they indicate a merger with another school.

School					
Columbia Military Academy,[58] Columbia (1978: Columbia Academy)	1905		1978 or 1979		
Dixon Academy, Shelbyville	1820	1885	Prior to 1890?		
Fayetteville Military Academy, Fayetteville	Circa 1825			1833?	
Fitzgerald and Clark Military Academy,[59] Trenton; 1911–: Tullahoma (1899–1906: Fitzgerald School; 1906–18: Fitzgerald and Clark School)	1899	1918			1921
Garrett Military Academy, Nashville (also known as the Military Academy of Nashville)	1884			1895	
Junior Military Academy, Bloomington Springs (1919–23: functioned as junior school for Castle Heights Military Academy; 1969–80: DeBerry Academy)	1919				1980
Lawrenceburg Military Academy, Lawrenceburg		1923		1925	
Massey Military School, Cornerstone; 1908–25: Pulaski (1903–17: Massey School, also known as Lawrenceburg Military Academy)	1903	1917			1924/1925
McCallie School, Chattanooga (also known as McCallie Military Academy; sometimes spelled MacCauley)	1905	1917		1970	
McMinn Military Academy, Rogersville (1802–50: McMinn Academy)	1802	1850		1913	
Memphis Military Institute, Memphis	Prior to 1894			1903	
Murfreesboro Military Academy,[60] Murfreesboro (also called Davis's Military School)	1858				1861
Nashville Military Academy, Nashville, 1855–62; see Western Military Institute, Kentucky					
Sewanee Military Academy, Sewanee (1868–1908: Sewanee Grammar School)	1868	1908			
Shelby Military Institute,[61] Germantown (also known as [Alexander McKinney] Rafter's Military School)	1858			1888	

School	Founded	First Year of Known Military Operation	Transitioned to Civilian	Last Year of Known Military Operation	Date Closed, or Still Open
Shelbyville Military Academy, Shelbyville, Bedford County (also known as Keiter's Military Academy)	1861				1862
Tennessee Military Institute, Nashville (formerly East Slide Academy)		1895			
Tennessee Military Institute, Sweetwater (1874–1902: Sweetwater Military College)	1874		1976		
University of Tennessee, Knoxville (1774–1807: Blount College; 1807–1869: East Tennessee College)	1774	1875	1887		
Western Military Institute, Nashville, 1852: see Western Military Institute, Kentucky					
West Tennessee Military School, Sommerville (1855–89: known as the Military School in Sommerville)		1855		1902	

TEXAS

School	Founded	First Year of Known Military Operation	Transitioned to Civilian	Last Year of Known Military Operation	Date Closed, or Still Open
Alamo Military Academy, San Antonio (also known as Alamo Military and Commercial Academy)	1883	1883			
Allen Military Academy, Bryan; 1896–99: Madisonville (1886–96: Madison Academy)	1883	1917			1976
Amarillo Military Academy, Amarillo (1910–17: Lowery-Phillips School)	1880	1912 or 1913			1917
Austin Military School, Austin	1852 or 1886	1917			1974
Austin Normal Military School, Austin	1869				1874

Note: Names and dates in parentheses in smaller type indicate actual name changes and variations on names and the years related to those differences, and in a few other cases they indicate a merger with another school.

School					
Bastrop Military Institute,[62] Bastrop; 1871–79: Austin (1851–57: Bastrop Academy; 1867–68: Bastrop Male and Female Academy; 1868–79: Texas Military Institute)	1851	1857			1879
Belton Academy (Wedemeyer Military Academy) 1886–1911: Belton, 1912–1915: Temple					1886/1915
Bishop Military Academy,[63] Corsicana	1857			1866	
Bryant School, Fort Worth	1912				1933
Burney Military Academy,[64] Itasca	After 1912				
Carlisle Military Academy, 1902–13: Arlington; 1913–17: Whitewright	1902				1913
Chilton Military Institute, Smith County	1867				1867
Cole's Classical and Military School, Dallas	1889				1902
El Paso Military Institute, El Paso	1907				1917
Forest Military School, Forest	1878				
Fort Worth Military Academy, Fort Worth	1907			1884?	
Garden Military Academy, San Antonio	1908				Prior to 1922[65]
Gideon's Military Academy, Corsicana	1865			1902	
Harding Military School, Dallas (also known as Hardin School for Boys)	1908			1921	
Harvard Military Academy	1898				
Lakeside Classical Institute, San Antonio	1905			1912	
Lancaster Military Academy, Lancaster (1903-4: Lancaster Academy)	1903	1904			1921
Luckin (Lukin) Military Academy, San Antonio	1917				1931?

School				
Marine Military Academy, Harlingen	1963			Open
Marshall College and High School, Marshall	1845	1857		1910[66]
Moye Military School (Academy), Castroville	1938		1959	
North Texas Agriculture College, Arlington (1913–17: Arlington Training School; 1917–23: Grubbs Vocation School; 1967: became University of Texas at Arlington)	1913		1954 or 1963	
North Texas Peacock Military Academy, Dallas	1930			1934
Peacock Military Academy, San Antonio (Peacock Military Academy [College]; 1894–1900: Peacock School for Boys)	1894	1900		1973
San Antonio Academy, San Antonio (1926: secondary school consolidated with West Texas Military Academy to form Texas Military Institute)	1886	1919		Open
San Marcos Baptist Academy, San Marcos	1907		2003	
Schreiner (Military) Institute, Kerrville (became Schreiner University)	1879	1923	1973	
St. Edward's Military Academy, Austin (also known as St. Edward's High School)	1878	1942	1946	
St. Mary's College, San Antonio	1852	1913[67]	1932	
Texas A&M University, College Station (1871–1963: Agricultural and Mechanical College of Texas)	1876			Open
Texas Maritime Academy, Galveston	1962			Open
Texas Military College,[68] Terrell	1906			1949
Texas Military Institute, 1854–56: Galveston; 1856–61: Rutersville	1854			1861
Texas Military Institute, Llano	1897			1898

Note: Names and dates in parentheses in smaller type indicate actual name changes and variations on names and the years related to those differences, and in a few other cases they indicate a merger with another school.

SCHOOL	Founded	First Year of Known Military Operation	Transitioned to Civilian	Last Year of Known Military Operation	Date Closed, or Still Open
Texas Military Institute, San Antonio	1893				Open
(1893–1926: West Texas Military Academy; 1926: secondary school of San Antonio Academy consolidated with West Texas Military Academy to form Texas Military Institute; also known as TMI, the Episcopal School of Texas)					
Texas Monumental and Military Institute, LaGrange	1856				1861
University Military School, Dallas	1898				1922
Waco Military Academy, Waco		1872			
West Texas Military Academy, Whitewright	1886				1904

UTAH

SCHOOL	Founded	First Year of Known Military Operation	Transitioned to Civilian	Last Year of Known Military Operation	Date Closed, or Still Open
All Hallows College, Salt Lake City	1886				1910
Ogden Military Academy, Ogden	1889				1896
Utah Military Academy, Salt Lake City	2014				Open

VERMONT

SCHOOL	Founded	First Year of Known Military Operation	Transitioned to Civilian	Last Year of Known Military Operation	Date Closed, or Still Open
Burnside Military School, Battleboro (before 1858: St. Michael's Boys School)	Prior to 1858	1858			1873
Jordan Hall Military School, Saint Albans	Prior to 1907			1913	

SCHOOL	Founded	First Year of Known Military Operation	Transitioned to Civilian	Last Year of Known Military Operation	Date Closed, or Still Open
Norwich University, Northfield; 1820–25: Norwich; 1825–29: Middleton, Connecticut; 1829–66: Norwich (1819–34: American Literary, Scientific, and Military Academy, Norwich)	1819 chartered; 1820 opened				Open
Pennock Military and Classical School, Rutland		1823		1826	
Rutland Military Institute, Rutland County (also known as Rutland Military Academy)		1874		1880	
Vermont Episcopal Institute, Burlington	1854/1860			1899	
Vermont Military Academy, Saxon River, Burlington (also known as Vermont Academy)	1876	1881	1908		

VIRGINIA

SCHOOL	Founded	First Year of Known Military Operation	Transitioned to Civilian	Last Year of Known Military Operation	Date Closed, or Still Open
Abingdon Military Academy (William King High School; Abingdon Academy until after 1897)	1849		1856		
Albemarle Military Institute, Charlottesville	1856	1857		1859	1859
Armstrong Military and Classical Academy, Upperville (also known as Upperville Academy or Upperville Military Academy)		Prior to 1852		1859	
Augusta Military Academy, Fort Defiance (1879–95: Augusta Male Classical Academy, also known as the Roller School)	1879	1895[69]			1983
Benedictine High School, Richmond (founded as Benedictine College, also known as Benedictine Military Institute)	1911				Open
Berkley Collegiate and Military Institute, Berkley, Norfolk County	1868				1963

Note: Names and dates in parentheses indicate actual name changes and variations on names and the years related to those differences, and in a few other cases they indicate a merger with another school.

School				
Bethel Military Academy, Warrenton (1867–69: Bethel Academy; 1869: merged with Evergreen Academy; also known as Bethel Classical and Military Academy)	1866	1869		1911[70]
Blackstone Military Academy,[71] Blackstone (1932: known as Blackstone School for Boys)	1912			1944
Cappahoosie Military Academy, Glouster County	1858		1860	
Capt. James D. Cobb's Military School, Chatham	1912		1930	
Chuckatuck Military Academy,[72] Chuckatuck (1857–59: Chuckatuck Female Institute; also known as Chuckatuck Male and Female Institute)	1854			1861
Classical and Military School, Petersburg (also known as the Classical, Mathematical, and Military School of William Maghee [McGee])	1843			Circa 1850
Confederate States Naval Academy, Drewry's Bluff	1863			1865
Craddockville Military Academy, Accomack County (may have been Margaret Academy Boys Department; Margaret Academy est. 1785, closed 1861, Joseph H. Hebard, VMI)	1853			1861
Culpeper Military Institute,[73] Culpeper (1865: Virginia High School; also called Col. Charles Lightfoot's Military School)	1857		1909	
Danville Military Institute, Danville (1820–59 and 1867–74: Danville Male Academy)	1817 or 1820	1859	1863	
Danville Military Institute, Danville (1908–18: Danville School for Boys; 1933–36: Virginia Presbyterian School[74])	1890			1939
Dinwiddie (Hatch) Military Academy,[75] Dinwiddie County (Winfield Academy)	1810–15	1828		1858
Elmington Classical and Military School, Nelson County	1872		1876	

School					
Fishburne Military School, Waynesboro (1879: Waynesboro High School; 1882–86: Fishburne School)	1879	1883/1884			Open
Fleetwood Academy, King and Queen County	1839				1860
Fork Union Military Academy, Fork Union (1897–1903: Fork Union Academy)	1897	1903			Open
Franklin Military Academy, Richmond	1980				Open
Franklin Military Institute, Franklin	1879			1910	
Frederick Military Academy, Portsmouth (1897–1903: Frederick Academy)	1897	1903			
Frederick Military College, Portsmouth (1961: Frederick Military Academy)	1958		*1961 (College)		1985 (Academy)
Greenwood Academy, Albemarle County	1881				1883
Hampton Military Academy, Hampton (1846–56: Hampton Academy; also called the Cary Institute)	1846	1856			1861
Hampton Normal and Agricultural Institute, Hampton (African American; later Hampton University)	1868	1872 or earlier		1901	
Hampton Roads Military Academy, Newport News (Hampton Academy; 1894–1901: Military Academy)	1894			1910	
Hargrave Military Academy, Chatham (before 1908: Warren Training School; 1909: Chatham Training School; 1909–25: Hargrave Academy)	1908	1925			Open
Harris Military Institute, Roanoke	1931		1934 or 1935?		
Hillsville Military Academy, Hillsville (1861: Southern Classical and Military Academy)	1861				

Note: Names and dates in parentheses in smaller type indicate actual name changes and variations on names and the years related to those differences, and in a few other cases they indicate a merger with another school.

Hinfield Academy, Danville	1930				1930
Hoge (Memorial) Military Academy, Blackstone	1893			1928	1913
Hoover Military Academy, Staunton (also known as Hoover's Select High School)	Prior to 1879				
Jefferson Military Academy, Tazewell (Jefferson Academy)		1849 or 1859			1910
John Marshall High School, Richmond (male voluntary corps of cadets)	1909	1915	1971		
Linton Hall Military School,[76] Bristow (1922–32: Linton Hall)	1922		1989		
Locust Dale Academy, Locust Dale	1858			1913	
Lynchburg Military College, Lynchburg (also known as Lynchburg College)	1855				1861
Maghee and Bryant Military Academy, Greensville or Sussex County	1922			1991	
Margaret Academy, Accomack County	1785	1856	1861		
Mary Baldwin College, Virginia Women's Institute, Staunton	1842	1995			Open
Massanutten Military Academy, Woodstock (1899–1917: Massanutten Academy)	1899	1917			Open
Military Academy at Middleburg, Loudon County (Middleburg Academy)		1827			
Miller School of Albemarle, Charlottesville	1827	1951	1957		
Newington Academy, Gloucester	1790	1823			
Newport News Military Academy, Newport News		1823			
Norfolk Military Institute, Norfolk (1746–1843: Norfolk Academy)	1746	1843	1854 or 1862		
Norfolk United States Naval School, Norfolk		1828		1837	
Petersburg Military Academy, Bath		1885			
Randolph-Macon Academy, Bedford	1890	1920			1934
Randolph-Macon Academy, Front Royal	1905	1910			Open

School					
Randolph-Macon College, Boydton (closed and reopened after war as civilian Randolph Macon College [University])	1830	1861			1863
Rappahannock Academy and Military Institute, Caroline County (1812–52: Rappahannock Academy; also called Rappahannock Military Academy)	1812	1848			1873
Ringgold Military Academy, Ringgold, Pittsylvania County	After 1848				1861
Rockingham Military Institute, Mount Crawford	1896			1915	
Rumford's Military Academy, King William County (Belmead Rock Castle; 1802–49: Rumford Academy)	1804	1849		1856	
Shenandoah Valley Academy,[77] Winchester (1832–65: Winchester Academy)	1832	1907			1939
Southern Military Institute (Academy), Blackstone	1859				
Southern Military Institute, Blackstone	1937?				1939?
Staunton Military Academy,[78] 1860–83: Charles Town, West Virginia; Staunton (1860–83: Charles Town Male Academy, also known as the Kable School)	1860	1886			1973
St. Emma Military School, Powhatan (1895–1945: St. Emma Industrial and Agricultural College, also known as Emma Industrial and Agricultural Institute; African American Catholic school)	1895				1972
St. John's Academy,[79] Alexandria (also known as St. John's Military Academy, Alexandria)	1833	1847 or earlier			1895
Suffolk Military Academy, Suffolk (also known as Nansemond Military Institute)	1875				1910
Virginia Collegiate Institute, Portsmouth (Virginia Literary, Scientific, and Military Academy; may be also Webster Military Academy, and Portsmouth Military Academy)		1828			

Note: Names and dates in parentheses in smaller type indicate actual name changes and variations on names and the years related to those differences, and in a few other cases they indicate a merger with another school.

School	Founded	First Year of Known Military Operation	Transitioned to Civilian	Last Year of Known Military Operation	Date Closed, or Still Open
Virginia Literary, Scientific, and Military Academy, Portsmouth (1849: became Virginia Collegiate Institute)	1839				1862
Virginia Military Institute, Lexington	1839				Open
Virginia Polytechnic Institute, Blacksburg (1872–96: Virginia Agriculture College; 1896–1944: Virginia Agricultural and Mechanical College and Polytechnic Institute; 1944: Virginia Polytechnic Institute; 1970–: Virginia Polytechnic Institute and State University; also known as Virginia Tech)	1872				Open
Washington Military Academy, Finney Mills, Amelia County (also known as Washington Academy, Old Field School)		1864			
Webster Collegiate Institute, Norfolk	1853				1862
Webster Military Institute, Norfolk	1867				1886
Westwood Military Academy, Lynchburg (1853–57: Westwood Academy)	1853	1857			1861
Williamsburg Military School, Williamsburg (Williamsburg Military Academy)		1852		1854	
Winchester Military School, Winchester (also may have been known as Winchester Academy)	1785	1856?			1861
Wytheville Military Academy, Wytheville	1881			1908	

WASHINGTON

School	Founded	First Year of Known Military Operation	Transitioned to Civilian	Last Year of Known Military Operation	Date Closed, or Still Open
Jefferson Military Academy, Pullman	1891 or 1893				1895
Latah Military Academy, Latah	1890 or 1891				1899
Marymount Military Academy, Tacoma (also known as St. Edward's Boys' School at Mary Mount)	1923				1971

	Founded	First Year of Known Military Operation	Transitioned to Civilian	Last Year of Known Military Operation	Date Closed, or Still Open
Puget Sound Naval Academy, Winslow (1950–51, 1953: Hill Naval Academy)	1937				1953
Washington State Nautical School, Seattle	1917				1920

WEST VIRGINIA

SCHOOL	Founded	First Year of Known Military Operation	Transitioned to Civilian	Last Year of Known Military Operation	Date Closed, or Still Open
Greenbrier Military School, Lewisburg (1812–90: Lewisburg Academy; 1890–96: Presbyterian School; 1896–1920: Lee Military Academy; also known in early years as the Brick Academy)	1808 or 1809	1890 or 1891			1972
Kanawha Military School, Charles Town	1880			1902	
Linsley (Linsley) Military Institute,[80] Wheeling (1814–61: Wheeling Lancastrian Academy; 1921–41: Linsley Academy; 1941: Linsley Military Institute; 1942–51: Linsley Institute; 1951–79: Linsley Military Institute)	1814	1877	1979		
Old Dominion Military Academy, Berkeley Springs	1898			1922	

WISCONSIN

SCHOOL	Founded	First Year of Known Military Operation	Transitioned to Civilian	Last Year of Known Military Operation	Date Closed, or Still Open
Danforth Military Academy, Oconomowoc	1949				1953
Northwestern Military and Naval Academy, 1888–1915: Highland Park, Illinois; 1914/1915–95: Lake Geneva (1995: merged with St. John's Military Academy)	1888				1995

Note: Names and dates in parentheses in smaller type indicate actual name changes and variations on names and the years related to those differences, and in a few other cases they indicate a merger with another school.

SCHOOL	Founded	First Year of Known Military Operation	Transitioned to Civilian	Last Year of Known Military Operation	Date Closed, or Still Open
Racine College,[81] Racine (also known as Racine Military Academy and Racine Military School and College; 1930–33: Racine Military Academy)	1852	1899			1933
Sacred Heart Military Academy, Watertown, 1968–2003: see Le Mans Academy, Indiana					
St. John's Northwestern Military Academy (1884–86: St. John's Academy; 1886–1995: St. John's Military Academy; 1995: merged with Northwestern Military and Naval Academy)	1884	1886			Open
Wisconsin Air Academy, New Berlin	2008				2009
Wyler Military School, Evansville (also known as Wyler School)	1928				1979

WYOMING

SCHOOL	Founded	First Year of Known Military Operation	Transitioned to Civilian	Last Year of Known Military Operation	Date Closed, or Still Open
S Bar H Ranch School for Boys	1924	1938			1943

Note: Names and dates in parentheses in smaller type indicate actual name changes and variations on names and the years related to those differences, and in a few other cases they indicate a merger with another school.

Appendix D

Number of Schools by Type for Each Military School Compared

Military School	Number of Neighborhood Schools	Number of Magnet Schools	Number of Charter Schools
1	63	11	8
2	63	11	8
3	11		
4	63	11	8
5	5	2	
6	20	5	
7	14		1
8	5		
9	10		
10	1		
11	2		2
12	10		1
13	2		2
14	9	3	2
15	4	5	3
16	1		
17	2		
18	1		1
19	20	5	1

Appendix E
Selected Military School Alumni

"This is where I started, and I thank a merciful God that I am able to come back to the school again."

—Remark made by Gen. Douglas MacArthur upon his return to Texas Military Institute, 1951

The following is a representation of alumni from a broad range of military schools. It is not intended to be complete in any of the categories, and additional categories, such as religious leaders, could be added. However, it does provide the reader a glimpse of the impact of military schools and a breadth well beyond the service in uniform.

Heads of State

Donald John Trump, President of the United States, 2017, New York Military Academy

James E. 'Jimmy' Carter, President of the United States, 1977–81, US Naval Academy

Dwight D. Eisenhower, President of the United States, 1953–61, US Military Academy

Ulysses S. Grant, President of the United States, 1869–77, US Military Academy

Jefferson Davis, President of the Confederate States of America, 1861–65, US Military Academy

León Febres Cordero, President of Ecuador, 1984–88, Charlotte Hall Military Academy

Martin Torrijos, President of Panama, 2004–9, St. John's Military Academy (Wisconsin)

Ricardo Martinelli, President of Panama, 2009–14, Staunton Military Academy

Manuel Antonio Noriega, military dictator of Panama 1983–1989, Peacock Military Academy

Luis Somoza, President of Nicaragua, 1956–67, La Salle Military Academy (New York)

Antastasio Somoza, President of Nicaragua, 1967–72, 1974–79, La Salle Military
    Academy (New York), US Military Academy
Fidel Ramos, President of the Philippines, 1992–98, US Military Academy
José Maria Figueres, President of Costa Rica, 1994–98, US Military Academy
Simeon Rylski, Tsar, 1943–46, and Prime Minister, 2003–5, of Bulgaria, Valley
    Forge Military Academy and College

Secretary of State of the United States

Alexander M. Haig, Jr., 1981–82, US Military Academy
George Catlett Marshall, 1945–49, Virginia Military Institute

United States Supreme Court Justice

Joseph Rucker Lamar, Associate Justice, 1911–16, Academy of Richmond

Generals and Admirals

Robert E. Lee, General in Chief, Confederate Armies, US Military Academy
Thomas J. "Stonewall" Jackson, Lieutenant General, Confederate Army, US
    Military Academy
John J. Pershing, Commander-in-Chief, Allied Expeditionary Forces, World War I,
    US Military Academy
Douglas MacArthur, Commander, Pacific, WWII and Korean War, Texas Military
    Institute, US Military Academy
Chester William Nimitz, Commander in Chief, Pacific Ocean Areas, US Naval
    Academy
William Frederick Halsey, South Pacific Area Command, US Naval Academy
George S. Patton Jr., Commander, Third Army, 1944–45, Virginia Military
    Institute, US Military Academy
William Westmoreland, Commander of US Forces, Vietnam, Citadel, US Military
    Academy
David McDonald, Chief of Naval Operations, 1963–67, Riverside Military
    Academy, US Naval Academy
H. Norman Schwarzkopf, Commander, Gulf War, Bordentown Military Institute,
    Valley Forge Military Academy, US Military Academy
Clifton Cates, Nineteenth Commandant of the US Marine Corps, Missouri
    Military Academy
Randolph McCall Pate, Twenty-first Commandant of the US Marine Corps,
    Virginia Military Institute
Alfred Richmond, Commandant, US Coast Guard, 1954–62, Massanutten
    Military Academy

State Governors

Albert Gallatin Brown, Mississippi, 1844–48, Jefferson College
Horatio Seymour, New York, 1852–54, 1862–64, Norwich University
Joshua Lawrence Chamberlain, Maine, 1866–71, Whitting's Military and Classical
    School
George Stoneman, California, 1883–87, US Military Academy

Joseph Kemp Toole, Montana, 1889–93, 1901–8, Western Military Institute (Kentucky)
Ernest W. Gibson Jr., Vermont, 1946–50, Norwich University
Daniel Walker, Illinios, 1973–77, US Naval Academy
Frank White, Arkansas, 1981–83, New Mexico Military Academy
Ernest F. "Fritz" Hollings, South Carolina, 1966–2005, the Citadel
Gerald L. Baliles, Virginia, 1986–90, Fishburne Military School
Joseph Graham Davis Jr., California, 1999–2003, Hollywood Military Academy, Harvard Military School
Rafael Hernández Colón, Puerto Rico, 1973–79, Valley Forge Military Academy
John Y. Brown Jr., Kentucky, 1979–83, Kentucky Military Institute
Carroll A. Campbell Jr., South Carolina, 1987–95, McCallie School
Matt Blunt, Missouri, 2005–9, US Naval Academy
Dave Heineman, Nebraska, 2005–, US Military Academy
Felix Gonzalez Canto, 2005–, Quintana Roo, Mexico, Riverside Military Academy
Homer A. Holt, West Virginia, 2010–, Greenbrier Military Academy
John Bel Edwards, Louisiana, 2016–, US Military Academy

Senators

Samuel Maxey, Texas, 1875–87, US Military Academy
Lawrence Tyson, Tennessee, 1925–29, US Military Academy
Richard Russell Jr., Georgia, 1933–71, Gordon Military College
Burnet Rhett Maybank, South Carolina, 1941–54, Porter Military Academy
Harry Pulliam Cain, Washington, 1946–53, Hill Military Academy
James P. Kem, Missouri, 1947–53, Blees Military Academy
Barry Goldwater, Arizona, 1953–65, 1969–87, Staunton Military Academy
Harry F. Byrd Jr., Virginia, 1965–83, Virginia Military Institute
Howard Baker Jr., Tennessee, 1967–85, McCallie School
Phil Gramm, Texas, 1985–2003, Georgia Military Academy
John McCain, Arizona, 1987–, US Naval Academy
Ike Skelton, Missouri, 1977–91, Wentworth Military Academy
Jack Reed, Rhode Island, 1997–, US Military Academy
Jack Webb, Virginia, 2007–13, US Naval Academy

Congressmen

Daniel Buck, Vermont, 1823–25, 1827–29, US Military Academy
Joseph Wheeler, Alabama, 1881–83, 1885–1900, US Military Academy
Carl Vinson, Georgia, 1914–65, Georgia Military College
Parker Corning, New York, 1923–37, Albany Academy
Carl G. Bachmann, West Virginia, 1925–1933, Linsley Military Institute
Frederick Charles Loofbourow, Utah, 1930–33, Ogden Military Academy
Edward C. Eicher, Iowa, 1933–38, Morgan Park Academy
Milton H. West, Texas, 1933–48, Texas Military Institute
John James Flynt Jr., Georgia, 1954–79, Georgia Military Academy
Alvin Paul Kitchin, North Carolina, 1957–63, Oakridge Military Academy
Barry Goldwater Jr., California, 1969–83, Staunton Military Academy

Ike Franklin Andrews, North Carolina, 1973–85, Fork Union Military Academy
James Kenneth Robinson, Virginia, 1971–85, Virginia Tech
Lamar Seeligson Smith, Texas, 1987–, San Antonio Academy, Texas Military
    Institute
Walter Beaman Jones Jr., North Carolina, 1995–, Hargrave Military Academy
Brett Guthrie, Kentucky, 2009–, US Military Academy
Michael Pompeo, Kansas, 2011–, US Military Academy
John Shimkus, Illinois, 2011–, US Military Academy
Martha McSally, Arizona, 2015–, US Air Force Academy
Roilo Golez, Republic of the Philippines, 1992–2001, 2003–13, US Naval
    Academy

United Nations Ambassador

John Bolton, 2005–6, McDonough School (Maryland)

Entertainment, Media and Sports

Cecil B. DeMille, Hollywood movie director, Pennsylvania Military College
Oliver Hardy, comic, Georgia Military College
Marlon Brando, actor, Shattuck Hall Military School
Jonathan Winters, comic, Norwich University
Jim Backus, actor, Kentucky Military Institute
Andy Rooney, Emmy winner, radio and television writer, Albany Academy
Dan Blocker, actor, Texas Military Institute
Hal Holbrook, actor, Culver Military Academy
Will Rogers, humorist, actor, Kemper Military School
Spencer Tracy, actor, Northwestern Military and Naval Academy
Jackie Coogan, actor, Urban Military School
Sylvester Stallone, actor, Charlotte Hall Military Academy
Casper Robert Van Dien, actor, Admiral Farragut Academy
Dale Adam Dye Jr., actor, Missouri Military Academy
Dean Martin, singer, Urban Military School
Owen Wilson, actor, New Mexico Military Institute
Marlin Perkins, television host, zoologist, Wentworth Military Academy
Greg and Duane Allman, rock musicians, Castle Heights Military Academy
Johnny Ramone, rock musician, Staunton Military Academy
Stephen Stills, rock musician, Admiral Farragut Academy
Dale Earnhardt Jr., stock car racing driver, Oakridge Military Academy
Pat Robertson, media head and Baptist minister, McCallie School
Ted Turner, media head, McCallie School
Howard Kalmenson, head, communications company, Riverside Military Academy
William S. Paley, chief executive officer, CBS, 1928–46, Western Military Academy
    (Illinois)
John B. Sias, president, ABC-TV, 1986–93, the Citadel
Pierson Mapes, president, NBC, 1982–94, Norwich University
Sam Donaldson, ABC newsman, New Mexico Military Institute

Walter O'Malley, owner, Brooklyn and Los Angeles Dodgers, 1950–79, Culver
   Military Academy
George Steinbrenner, owner, New York Yankees 1972–2010, Culver Military
   Academy
Glenn Davis, football player, Heisman Trophy winner, 1946, US Military Academy
Joe Bellino, football player, Heisman Trophy winner, 1960, US Naval Academy
Roger Staubach, football player, Heisman Trophy winner, 1963, New Mexico
   Military Institute, US Naval Academy
Eddie George, football player, Heisman Trophy winner, 1995, Fork Union
   Military Academy
Vinny Testaverde, football player, Heisman Trophy winner, 1986, Fork Union
   Military Academy
David Robinson, member, Basketball Hall of Fame, US Naval Academy

Business

Henry du Pont, president, Wilmington and Western Railroad, 1879–99, US
   Military Academy
Sewell Avery, Chairman of the Board, Montgomery Ward, 1930–54, Michigan
   Military Academy
James "Bud" Walton, cofounder, Wal-Mart, Wentworth Military Academy
Jack Eckerd, founder, Eckerd Pharmacy, Norwich University
Conrad Hilton, founder, Hilton Hotel chain, New Mexico Military Institute
Robert W. Woodruff, president, Coca-Cola Company, 1923–55, Georgia Military
   Academy
William Seawell, chairman and CEO, Pan American Airways, 1971–81, US
   Military Academy
Frank Borman, president, Eastern Airlines, 1975–86, US Military Academy
William R. "Bill" Tiefel, chairman of the board, CarMax, 2002–7, Ritz-Carlton,
   1998–2002, Valley Forge
Jim Kimsey, founder, America On Line (AOL), St John's College High School,
   US Military Academy
William Burnett Benton, CEO, Encyclopedia Britannica, 1943–73, Shattuck
   Military Academy
Daniel Frank Gerber, CEO, Gerber Baby Food 1952–74, St. John's Northwestern
   Military Academy
Bob McDonald, CEO, Procter & Gamble, 2009–13, US Military Academy
Robert Pamplin Sr., president, Georgia-Pacific, 1957–76, Virginia Tech
Harry Thayer, president or chairman of the board, AT&T, 1919–28, Norwich
   University
Donald Trump Sr., founder, Trump Organization, 1984–, New York Military
   Academy
Robert Benmosche, CEO, MetLife, 1998–2006, AIG, 2009–14, New York
   Military Academy
Charles Phillips, president, Oracle Corporation 2003–10, US Air Force Academy
Marshall Larson, president and chair, Goodrich Corporation, 2003–12, US
   Military Academy

Alex Gorsky, CEO, Johnson & Johnson, 2012–, US Military Academy

Leroy Raffel, Cofounder, Arby's Restaurants, Riverside Military Academy

Labor Leader

Joseph Lane Kirkland, president of the AFL-CIO, 1979–95, US Merchant
Marine Academy

Authors and Poets

Lyman Frank Baum, *The Wonderful Wizard of Oz,* Peekskill Military Academy

Edgar Rice Burroughs, *Tarzan,* Michigan Military Academy

Pat Conroy, *The Great Santini* and *The Lords of Discipline,* the Citadel

Stephen Crane, *Red Badge of Courage,* Claverack College

Paul Horgan, two-time Pulitzer Prize winner, New Mexico Military Institute

Edgar Allan Poe, *The Raven,* US Military Academy

J. D. Salinger, *The Catcher in the Rye,* Valley Forge Military Academy

Robert Heinlein, science fiction writers, US Naval Academy

Paul Henderson, Pulitzer Prize winner (investigative reporting), Wentworth
Military Academy

David Potter, Pulitzer Prize winner (history), Academy of Richmond

William Rose Benét, Pulitzer Prize (poetry), Albany Academy

Stephen Vincent Benét, Pulitzer Prize (poetry) twice, Hitchcock Military Academy,
Albany Academy

Winston Groom, *Forrest Gump,* and Pulitzer Prize (nonfiction), University
Military School

Nobel Prize for Physics Recipients

Albert Michelson, 1907, US Naval Academy

Robert Richardson, 1966, Virginia Tech

Henry Way Kendall, 1990, US Merchant Marine Academy

Martin Lewis Perl, 1995, United States Merchant Marine Academy

Nobel Peace Prize

George Catlett Marshall, 1953, Virginia Military Institute

James E. "Jimmy" Carter, 2002, US Naval Academy

Medal of Honor Recipients

US Military Academy, 84

US Naval Academy, 73

Texas A&M University, 8

Virginia Military Institute, 7

Virginia Tech, 7

Culver Military Academy, 5

Norwich University, 5

Whitting's Military and Classical School, Churchill Military Academy, Albany
Academy, 1 each

US Air Force Academy, Pennsylvania Military College, Fork Union Military
Academy, 1 each

Kentucky Military Institute, Kemper Military Academy, Columbia Military
   Academy, 1 each
North Texas Agricultural College, Edwards Military Institute (North Carolina),
   1 each
Saint Thomas Academy, Western Military Academy (Illinois), the Citadel, 1 each
New Mexico Military Institute, Georgia Military Academy, Academy of Richmond,
   1 each
Texas Military College, Texas Military Institute (West Texas Military Academy),
   1 each

Astronauts Who Have Walked on the Moon

Edwin "Buzz" Eugene Aldrin Jr., US Military Academy
Alan Bartlett Shepard Jr., Admiral Farragut Academy, US Naval Academy
David Randolph Scott, Texas Military Institute, US Military Academy
James Benson Irwin, US Naval Academy
Charles Moss Duke Jr., Admiral Farragut Academy, US Naval Academy

Notes

Preface

1. Lester Webb, "The Origin of Military Schools in the United States Founded in the Nineteenth Century." PhD diss., University of North Carolina, 1958, iv; Alvan Hadley, "The Association of Military Colleges and Schools of the United States (AMSCUS) and the Struggle for the Survival of the Military Preparatory Schools in America," PhD diss., University of Kentucky, 1999, 195–96.

2. Nancy Beadie, *Education and the Creation of Capital in the Early American Republic*, 109.

3. Hadley, "Association of Military Colleges and Schools," 124. Recent discovery indicates the additional fourth public military school was in its first year. Magnet Military School, North Charleston, South Carolina, started in 1997 as the first public military middle school and later added a high school.

4. One additional school, North Point Military Academy in Puerto Rico, had grades one through nine.

5. Association of Military Colleges and Schools of the United States (AMCSUS), *Association of Military Colleges and Schools of the United States Constitution*.

6. Chester Finn and Bruno Manno, *Charter Schools in Action: Renewing Public Education*, 17.

7. Chicago Public Schools, "Chicago Public Schools Website," accessed May 11, 2011, http://www.cps.edu/Schools/High_schools/Pages/Highschools.aspx.

8. Ibid.

9. US Department of the Army, *Senior Reserve Officers' Training Corps Program: Organization, Administration and Training, Army Regulation 145–1*, 6.

10. Ibid.

11. Greg Gallagher and Alan Nolan, eds. *The Myth of the Lost Cause and Civil War History*, 1.

12. The figure 833 was accurate until 2011. By 2014 the figure had increased to 842.

Chapter 1

1. William Firestone and Karen Louis, "Schools as Cultures," in *Handbook of Research on Educational Administration*, 2nd ed., ed. Joseph Murphy and Karen Louis, 297–323; Edgar Schein, *The Corporate Culture Survival Guide*, 29, 217–18.

2. Kim Hays, *Practicing Virtues: Moral Traditions at Quaker and Military Boarding Schools*, 219; Remi Hajjar, "The Public Military High School," *Armed Forces and Society*, October 2005, 54.

3. William Trousdale, *Military High Schools in America*, 96; Leigh Gignilliat, *Arms and the Boy: Military Training in Schools and Colleges*, 83–84; Conversations between the author and Harry Temple, 1997–2000. Temple is author of *The Bugle's Echo: The*

Chronology of Cadet Life at the Military College at Blacksburg, Virginia, 6 vols; Hajjar, "Public Military High School," 44–62.

4. Lewis Sorley, *Honor Bright: History and Origins of the West Point Honor Code and System*, 34; Texas Military Institute, *Bugle Notes* (San Antonio: Texas Military Institute, 2006).

5. Norwich University, *Cadet Handbook*, ii.

6. Trousdale, *Military High Schools*, 129; Sorley, *Honor Bright*, 47.

7. Sorley, *Honor Bright*, 17, 46; Lloyd Matthews, "Ideals, Military," in *The Oxford Companion to American Military History*, ed. John Chambers and Fred Anderson, 323.

8. Schien, *Corporate Culture Survival Guide*, 217–18; Abraham Maslow, "A Theory of Human Motivation," in *The Greatest Writings in Management and Organizational Behavior*, ed. Louis Boone and Donald Bowen, 115.

9. Army and Navy Academy, "History and Traditions," accessed April 11, 2011, http://www.armyandnavyacademy.org/about-ana-military-academy/school-history-traditions/.

10. William Ouchi, *Theory z: How American Business Can Meet the Japanese Challenge*, 244.

11. Virginia Military Institute Archives, "New Market Ceremony History," accessed January 15, 2016, http://www.vmi.edu/archives.aspx?id=21777; George Washington, "Sentiments on a Peace Establishment," in *Soldier-Statesmen of the Constitution*, ed. Robert Wright and Morris MacGregor, 199.

12. Firestone and Louis, "Schools as Cultures," 297–99; James Burke, "Military Culture," in *Encyclopedia of Violence, Peace, and Conflict*, ed. Lester Kurtz and Jennifer Turpin, 3:447–62; Schein, *Corporate Culture Survival Guide*, 29.

13. Burke, "Military Culture," 447; Frederick J. Manning, "Morale, Cohesion, and Esprit de Corps," in *Handbook of Military Psychology*, ed. Reuven Gal and David Mangelsdorff, 453–54; Rod Andrew, *Long Gray Lines: The Southern Military School Tradition, 1839–1915*, 67.

14. Burke, "Military Culture," 453.

15. Temple, *Bugle's Echo*, 4000–4001.

16. Ibid., 4012–13.

17. Theodore Crackel, *West Point: A Bicentennial History*, 88; John Kraus, "The Civilian Military Colleges in the Twentieth Century: Factors Influencing Their Survival," PhD diss., University of Iowa, 1978, 99–101.

18. South Carolina State Superintendent of Education, *Thirtieth Annual Report, 1898*, Reports and Resolutions of the General Assembly of the State of South Carolina, 1898, vol. 2 (1898): 145–63, https://babel.hathitrust.org/cgi/pt?id=chi.09622 7585;view=1up;seq=155. During this period the function of command of the cadet battalion was that of the commandant aided by the cadet adjutant.

19. Andrew, *Long Gray Lines*, 66; John Adams, *Keepers of the Spirit: The Corps of Cadets at Texas A&M*, 72. The founder of West Texas Military Academy (now Texas Military Institute), the Episcopal bishop of Texas, firmly endorsed the board's action in this matter of reenrollment.

20. Donald McKate, *Tradition: A History of the Presidency of Clemson University*, 135–37. Another cadet revolt over food occurred at the Michigan Military Academy in 1900. After painting "Down with the Dog" upon the door of the school, cadets demanded action on claims of food shortages and mistreatment. The faculty supported the cadets' grievances, and the end result was that the board supported the school's

superintendent, and two faculty member were fired. James Starbuck, "The Michigan Military Academy at Orchard Lake," *Michigan History Magazine* 50 (September 1966), reprint with additional notes by Brian Bohnett in possession of the author.

21. Trousdale, *Military High Schools*, 90; Karen Dunivin, "Military Culture: Change and Continuity," *Armed Forces & Society* 20, no. 4 (Summer 1994): 531–47; US Army, *The Soldier's Guide*, Field Manual 7–21.13:1.

22. South Carolina Department of Agriculture, *Handbook on South Carolina*, 36; Francis Smith, *The Virginia Military Institute: Its Building and Rebuilding*, 242.

23. Fork Union Military Academy, "A Little about Fork Union Military Academy," accessed May 11, 2010, http://www.forkunion.com/military-school-info/about-fork-union-military-academy.html; Chicago Military Academy, "Chicago Military Academy at Bronzeville," accessed May 12, 2010, http://www.chicagomilitaryacademy.org/; Trousdale, *Military High Schools*, 392

24. Edgar Schein, "Coming to a New Awareness of Organizational Culture," *Sloan Management Review*, Winter 1984, 3–16.

25. Lance Bethos, *Carved from Granite: West Point since 1902*, 9–11.

26. Ibid.

27. Sorley, *Honor Bright*, 47.

28. Howe Military Academy, "Howe Military Academy," accessed June 1, 2010, http://www.thehoweschool.org/wp/; Oakland Military Institute, "About," accessed May 24, 2010, http://oakmil.org/omiacademy/site/default.asp.

29. Culver Educational Foundation, "Culver Academies," http://www.culver.org/about/.

30. Maslow, "Theory of Human Motivation," 114–15; Trousdale, *Military High Schools*, 382.

31. Edward Mansfield, "The United States Military Academy at West Point," *American Journal of Education* 30 (March 1863): 17–47; *Anderson v. Laird*, 466 Court of Appeals, District of Columbia Circuit. 283 (1972), accessed April 5, 2006, http://law.justia.com/cases/federal/appellate-courts/F2/466/283/424650/; Associated Press, "4th Circuit Won't Review Ruling Striking Down VMI Prayers," First Amendment Center, August 14, 2003.

32. Kim Hays, *Practicing Virtues: Moral Traditions at Quaker and Military Boarding Schools*, 70.

33. Ibid., 1–4, 8.

Chapter 2

1. Lester A. Webb, "The Origin of Military Schools in the United States Founded in the Nineteenth Century," PhD diss., University of North Carolina, 1958, 174–75.

2. Jannus Langins, *Conserving the Enlightenment: French Military Engineering from Vauban to the Revolution*, 94–95.

3. The Visitor Center, "Tadeusz Kosciuszko," Lafayette Square, accessed May 14, 2008, http://www.the-visitor-center.com/pages/Lafayette-Square/slides/lafayette-square-050.htm.

4. United Kingdom Army, "The History of RMA Sandhurst," accessed January 19, 2009, http://www.army.mod.uk/documents/general/history_of_rmas.pdf.

5. Webb, "Origin of Military Schools," 3.

6. Ibid., 2.

7. Edward Boyington, *West Point and Its Military Importance during the American Revolution, and the Origin and Progress of the United States Military Academy*, 176.

8. Ibid., 177.

9. Webb, "Origin of Military Schools," 5; Theodore J. Crackel, *West Point: A Bicentennial History*, 36; Richard McMaster, "The Contribution of West Point to American Education," master's thesis, University of Texas, 1951, 4.

10. Webb, "Origin of Military Schools," 5.

11. Ibid., 8, 14.

12. Ibid., 9, 11–12; George Washington, "Sentiments on a Peace Establishment," in *Soldier-Statesmen of the Constitution*, by Robert Wright and Morris MacGregor, 211.

13. Stephen Ambrose, *Duty, Honor, Country: A History of West Point*, 11; David Myer, "West Point and Jefferson's Constitutionalism," in *Thomas Jefferson's Military Academy: Founding West Point*, edited by Robert McDonald, 55.

14. Myer, "West Point," 18; Edward Holden, *The Centennial of the United States Military Academy at West Point, New York: 1802–1902*, 26; Crackel, *West Point*, 35.

15. Webb, "Origin of Military Schools in the United States Founded in the Nineteenth Century," PhD diss., University of North Carolina, 1958, 25, accessed July 15, 2006, http://www.worldcat.org/title/origin-of-military-schools-in-the-united-states-founded-in-the-nineteenth-century/oclc/471930802; Boyington, *West Point and Its Importance*, 188.

16. Boyington, *West Point and Its Importance*, 188–90.

17. Ambrose, *Duty, Honor, Country*, 11.

18. Boyington, *West Point and Its Importance*, 192; Ambrose, *Duty, Honor, Country*, 12, 17–18.

19. Crackel, *West Point*, 46, 51.

Chapter 3

1. John Fredriksen, "Williams, Jonathan," in *The Encyclopedia of the War of 1812: A Political, Social, and Military History*, ed. Tucker Spencer, 1:783–84.

2. *Reminiscences of West Point in the Olden Time*, 19; Jonathan Shallat, *Structures in the Stream: Water, Science, and the Rise of the U.S. Army*, 82, 83.

3. Shallat, *Structures in the Stream*, 87.

4. Ibid.; Robert Cowley and Thomas Guinzburg, *West Point: Two Centuries of Honor and Tradition*, 23. Swift reported to West Point as a cadet in October 1801, the academy not having been officially established until March 1802. Having been commissioned in October 1802, he was considered the first graduate. According to Shallat, letters of protest from line officers helped block Williams's desires for command. These officers saw Williams as an academic with only two years' troop duty and already elevated to lieutenant colonel.

5. Ambrose, *Duty, Honor, Country*, 21.

6. Ibid., 96; Dean Baker, "The Partridge Connection: Alden Partridge and the Southern Military Education," PhD diss., University of North Carolina 1986, 2.

7. Baker, "Partridge Connection," 2.

8. Ibid., 5–30.

9. Ambrose, *Duty, Honor, Country*, 44.

10. Ibid., 34.

11. Cowley and Guinzburg, *West Point*, 26.

12. Larry Manning, "The Contribution of Sylvanus Thayer and the United States Military Academy to Engineering Programs in Higher Education in the United States," PhD diss., Texas A&M University, 2003, 103.

13. Larry Webb, "The Origin of Military Schools in the United States Founded in the Nineteenth Century," PhD diss., University of North Carolina, 1958, 11–12; Baker, "Partridge Connection," 77; École Polytechnique, "History and Heritage," accessed May 7, 2010, http://www.polytechnique.edu/home/about-ecole-polytech nique/history-and-heritage/.

14. Baker, "Partridge Connection," 2.

15. Ibid., 81.

16. Ibid., 83.

17. Ibid., 83, 89–91.

18. Manning, "Contribution of Sylvanus Thayer," 103.

19. Crackel, *West Point*, 84.

20. Baker, "Partridge Connection," 106–7.

21. Ibid., 94; William Ellis, *Norwich University, 1819–1911: Her History, Her Graduates, Her Roll of Honor*, 3.

22. American Literary, Scientific, and Military Academy, *Catalogue of the Officers and Cadets, Trustees and the Prospectus and Internal Regulations of the American Literary, Scientific, and Military Academy*, 25.

23. Baker, "Partridge Connection," 111; Edgar Fahs Smith, *Samuel Latham Mitchill: A Father in American Chemistry*, 11.

24. Ellis, *Norwich University*, 3; Walter Lowrie and Matthew St. Clair Clarke, eds., *American State Papers: Documents, Legislative and Executive, of the Congress of the United States*, Part 5, 1:834.

25. Baker, "Partridge Connection," 111; Robert Poirier, *By the Blood of Our Alumni: Norwich University Citizen Soldiers in the Army of the Potomac*, 7.

26. Alden Partridge, "Partridge Lectures," *American Journal of Education* 1 (1826): 395, 396.

27. Ibid., 397–98.

28. Baker, "Partridge Connection," 38, 40, 43.

29. Ibid., 49, 51.

30. The school with the longest record of the seventeen listed in table 3.1 was Jefferson Military College (Jefferson College and Military Institute) in Washington, Mississippi. The college was established in 1811 and transitioned to military format in 1829 under the plan provided by Captain Partridge. The school had short periods during which it was nonmilitary and was closed; however, it continued in operation until 1964.

31. Abraham Honeyman, ed., "Seven Generations of Lawyers," *New Jersey Law Journal* 44 (1921): 293–98; Ernest Ashton Smith, *Allegheny: A Century of Education, 1815–1915*, 59, 61.

32. Ellis, *Norwich University*, 399–401.

Chapter 4

1. Bradford Wineman, "Francis H. Smith: Architect of Antebellum Southern Military School and Education Reform," PhD diss., Texas A&M University, 2006, 122; Dean Baker, "The Partridge Connection: Alden Partridge and Southern Military Education,"

PhD, diss., University of North Carolina, 1958, 10, 96; Larry Manning, "The Contribution of Sylvanus Thayer and the United States Military Academy to Engineering Programs in Higher Education in the United States," PhD diss., Texas A&M University, 2003, 110; Peggy Kammen, "Contributions of West Point Graduates of the Pre-Thayer Era: 1802–1817," USMA Library/Digital Collections (1996), 7, http://digital-library. usma.edu/cdm/singleitem/collection/p16919c0111/id/17/rec/1.

2. Manning, "Contribution of Sylvanus Thayer," 103; École Polytechnique, "History and Heritage," accessed May 7, 2010, http://www.polytechnique.edu/home/ about-ecole-polytechnique/history-and-heritage/, copy in possession of the author.

3. Stephen Ambrose, *Duty, Honor, Country: A History of West Point*, 67, 68, 69.

4. Lance Bethos, *Carved from Granite: West Point Since 1902*, 9; Manning, "Contribution of Sylvanus Thayer," 103.

5. Manning, "Contribution of Sylvanus Thayer," 116; John H. B. Latrobe, *Reminiscences of West Point, from September, 1818, to March, 1882*, online copy posted as "West Point Reminiscences, from September, 1818, to March, 1882," annotated by Bill Thayer, accessed June 7, 2010, http://penelope.uchicago.edu/Thayer/E/Gazetteer/ Places/America/United_States/Army/USMA/LATREM*.html#Commandant_Bliss.

6. Lewis Sorley, *Honor Bright: History and Origins of the West Point Honor Code and System*, 17.

7. Manning, "Contribution of Sylvanus Thayer," 103.

8. Leigh Gignilliat, *Arms and the Boy: Military Training in Schools and Colleges*, 9; Manning, "Contribution of Sylvanus Thayer," 120–22.

9. Theodore J. Crackel, *West Point: A Bicentennial History*, 86; Latrobe, "Reminiscences," 10. The four charges leveled against Captain Bliss by cadets for acts between 14 October and 22 November 1818 and the charges that resulted in cadet dismissals are documented in Wilson Miles, Cary Fairfax, Thomas Ragland, and Nathaniel Hall Loring, *An Expose of Facts Concerning Recent Transactions, Relating to the Corps of Cadets of the United States Military Academy at West Point, New York*.

10. Crackel, *West Point*, 87.

11. Ibid.

12. Ibid., 87–88; Glenn Robins, *The Ministry and Civil War Legacy of Leonidas Polk*, 24, 61.

13. Larry Webb, "The Origin of Military Schools in the United States Founded in the Nineteenth Century," PhD diss., University of North Carolina, 1958, 260.

14. Robert Cowley and Thomas Guinzburg, *West Point: Two Centuries of Honor and Tradition*, 25, 34.

15. Manning, "Contribution of Sylanus Thayer," 95, 125.

16. Cowley and Guinzburg, *West Point*, 43. One of those pardoned by President Jackson was Cadet Robert T. P. Allen, class of 1834, who would later found Kentucky Military Institute. After the cadets burned what they considered an unsightly structure, Cadet Allen admitted to the act and was expelled, but he would not reveal his compatriots. William Sampson, "A History of the Kentucky Military Institute during the Nineteenth Century, 1845–1900," thesis, University of Louisville, 1954.

17. "Obituary of Colonel A. L. Roumford," *Germantown Independent-Gazette*, August 2, 1878, obtained from the Germantown Historical Society.

18. S. F. Hotchkin, *Ancient and Modern Germantown, Mount Airy, and Chestnut Hill*, 366–67.

19. George W. Cullum, *Biographical Register of the Officers and Graduates of the U.S. Military Academy*, 1:590.

20. Wineman, "Francis H. Smith," 159; Cullum, *Biographical Register*, 1:432.

21. Wineman, "Francis H. Smith," 159; Ezra J. Warner, *Generals in Gray: Lives of the Confederate Commanders*, 157–58; Christopher Olsen, *The American Civil War: A Hands-on History*, 1. Forrshey's school was not the same Texas Military Institute that was founded in 1893 as the West Texas Military Academy (WTMA).

22. The Citadel, "Citadel History," accessed May 1, 2010, http://www.citadel.edu/citadel-history/.

Chapter 5

1. William Couper, *One Hundred Years at VMI*, 1:34, 258; Dean Baker, "The Partridge Connection: Alden Partridge and the Southern Military Education," PhD diss., University of North Carolina, 1958, 102; Francis Smith, *The Virginia Military Institute: Its Building and Rebuilding*, 54–55.

2. Couper, *One Hundred Years*, 1:33; Smith, *Virginia Military Institute*, 23.

3. Francis Smith, *West Point Fifty Years Ago*, 7; Bradford Wineman, "Francis H. Smith: Architect of Antebellum Southern Military School and Education Reform," PhD diss., Texas A&M University, 2006, 26.

4. Wineman, "Francis H. Smith," 11.

5. Ibid., 45–46.

6. Ibid., 109.

7. Jennifer Green, *Military Education and the Emerging Middle Class in the Old South*, 40.

8. Wineman, "Francis H. Smith," 109.

9. John B. Strange, class of 1842, helped to transition Norfolk Academy in Virginia to the military format in 1843 as the assistant principal and then served as principal in 1847 (William Brown, "The Norfolk Academy of Virginia," *North South Trader's Trader's Civil War* 20, no. 5 [1993]: 33–36). He later started Albemarle Military Institute, Charlottesville, Virginia, in 1856. James L. Bryan, class of 1843, was the founder of Petersburg Classical and Military Academy in Virginia, which opened in 1848 (Wineman, "Francis H. Smith," 119). After the war he established Cambridge Military Academy in Maryland. Abingdon Military Academy in Virginia had a series of VMI men as principals, starting with B. F. Ficklin in 1849. John Henry Pitts, class of 1844, transitioned the Rumford Academy in Virginia to the Rumford Military Academy in 1849. Thomas Benton, class of 1850, was coprincipal at the Arkansas Military Institute when it opened in 1850. Valentine Saunders, class of 1842, chose the 1853–54 timeframe to open his Baton Rouge Military Institute in Louisiana, although it closed shortly thereafter due to a yellow fever epidemic (ibid.). James J. Phillips, class of 1853, established the Chuckatuck Military Academy in Chuckatuck, Virginia, in 1854. Titus V. Williams, class of 1859, established the Jeffersonville Military Academy in Tazewell County, Virginia (ibid., 118). Charles E. Lightfoot, class of 1854, was key in organizing the Culpeper Military Institute in Virginia in 1857. John W. Lewis, class of 1859, helped transition St. John's College, Arkansas, in 1859 as commandant. Thomas Harris, class of 1851, helped transition Elizabeth City Academy in North Carolina to military format in 1851. Charles Derby, class of 1848, was first superintendent of West Alabama Military Institute in 1851 (Charles Walker, *Biographical Sketches of the*

Graduates and Élèves *of the Virginia Military Institute Who Fell during the War Between the States*, 169–73, accessed October 16, 2014, https://archive.org/details/memorial virginia00walk). William Elisha Arnold, class of 1853, was the first superintendent of Lafayette Military Institute in Missouri in 1859.

Lynchburg Military College in Virginia was established in 1855, and Capt. James E. Blankenship, VMI class of 1852, was among the faculty as mathematics professor and a member of the military department, as may have been other VMI alumni. Blankenship would go on to help convert Randolph-Macon College in Virginia to a military format after the Civil War began. Also on the faculty with Captain Blankenship was James Thomas Murfee, VMI class of 1853, who taught natural sciences. Murfee is an excellent example of how far and long VMI's influence in the military school movement went. Prior to coming to Lynchburg, he taught at Madison College in Pennsylvania. Madison was not a military school, but it required uniforms for drill and had a demerit system (James Hadden, *A History of Uniontown: The County Seat of Fayette County, Pennsylvania*, 507). During the Civil War Murfee served as commandant for the University of Alabama, and after the war he converted Howard College to a military school as its president and later founded Marion Military Institute.

The Virginia Military Institute alumni contributed to an explosion of military schools in the South. Edward J. Magruder, class of 1855, ran the Rome Military Academy from 1855 until the start of the war in 1861 (George Battey, *The History of Rome and Floyd County, State of Georgia, United States of America*). Valentine M. Johnson, class of 1860, went to West Florida Seminary as commandant of cadets shortly before the war. He was joined there by James Cross, class of 1856, who became professor of mathematics and tactics. Cross had been principal of Winchester Military School in Virginia (established in 1856) during its early years (Virginia Military Institute, "James Lucius Cross," Virginia Military Institute Archives, Historical Rosters Database, accessed March 9, 2010, http://archivesweb.vmi.edu/rosters/record. php?ID=454). William Keiter, VMI class of 1859, started Shelbyville Military Academy in Tennessee in 1860. Frederick Bass, class of 1851, as first professor of military tactics and later president of Marshall College in Texas, transitioned the school to military format about 1858. James W. Keeble, class of 1857, was professor of tactics and principal of Sharon College in Mississippi when it converted to a military school in 1860. James W. Hairston, class of 1858, founded Hairston School in Mississippi in 1859. James H. Waddell, class of 1855, was the first commandant in 1860 for the Cappahoosie Military Academy in Virginia. James V. Hall, class of 1852, established the Craddockville Military Academy, in Accomack County, Virginia, about 1853.

Joseph H. Hebard and Edward C. Edmonds, both of the class of 1858, were hired at the Hampton Academy in Virginia as it changed to the Hampton Military Academy in the late 1850s (William Brown, "The Hampton Military Academy of Virginia." *North South Trader's Civil War* 24, no. 2 (1997): 148–58). Edwards left shortly thereafter to help found Danville Military Academy (Institute) in Danville, Virginia, in 1859 (Walker, *Biographical Sketches*, 185). William Mahone, class of 1847, and Thomas R. Thorston, class of 1852, were both employed at the Rappahannock Academy in Virginia and helped transform it to the Rappahannock Academy and Military Institute.

In addition to these twenty-six, an unidentified VMI graduate may have transitioned Elizabeth City Academy in North Carolina to a military school in 1851 (William Powell, ed., *Dictionary of North Carolina Biography*, Vol. 5, P–S [Chapel Hill: University of North Carolina Press, 1994], 120).

Besides Baton Rouge Military Institute, the other military school that research revealed closed due to illness was the Rectory School. In 1870 press coverage of a malaria outbreak at the school helped decrease enrollment from thirty-five to eight cadets by 1873, and the school closed. In 1849 and 1857 the Citadel closed for short periods because of yellow fever in Charleston. VMI closed for short periods in 1903 and 1910 due to typhoid. John Marshall High School, Richmond, and its corps of cadets stopped functioning while the school served as a hospital during the 1918 Spanish flu epidemic. The hardest hit by the Spanish flu of the military schools was likely Kearney Military Academy in Nebraska. That school had eighty-seven cases, of which five cadets died.

10. Richard Irby, *History of Randolph-Macon College, Virginia: The Oldest Incorporated Methodist College in America*, 151.

11. Ibid., 155.

12. William H. Buckley, *The Citadel and the South Carolina Corps of Cadets*, 2, 10; John Kraus, "The Civilian Military Colleges in the Twentieth Century: Factors Influencing Their Survival," PhD diss., University of Iowa, 1978, 256.

13. The Citadel, "Citadel History," accessed March 3, 2010, http://www.citadel.edu/citadel-history/; Green, *Military Education*, 184, 187.

14. B. Sharpe, "Gone but Not Forgotten: North Carolina's Educational Past: North Carolina Polytechnic Academy," accessed May 3, 2010, http://www.lib.unc.edu/ncc/gbnf, copy in possession of the author; John Wyeth, *History of La Grange Military Academy and the Cadet Corps, 1857–1862: La Grange College, 1830–1857*, 10. Another Citadel graduate who may have founded a military school was Evander Law, class of 1854, who may have established the Tuskegee Military High School in Alabama in 1860. The school closed upon the enlistment of older students in 1861. However, other sources state that Park High School for Boys functioned under that title from 1857 to 1891 and then became Alabama Military Institute. The other military school in Tuskegee was Alabama Scientific and Military Institute, which was known under several titles and opened in 1846.

15. John Thomas, *The History of the South Carolina Military Academy*, 48; Jerry West, *The Bloody South Carolina Election of 1876: Wade Hampton III, the Red Shirt Campaign for Governor, and the End of Reconstruction*, 179–80.

16. Francis Simkins, *A History of the South*, 3rd ed.

17. Christoper.J. Olsen, *The American Civil War: A Hands-on History*, 3–18; C. Vann Woodward, *The Burden of Southern History*, 62.

18. James B. Sellers, *History of the University of Alabama, 1818–1902*, 1:259; Wyeth, *History of La Grange Military Academy*, 9–10.

19. Lester A. Webb, "The Origin of Military Schools in the United States Founded in the Nineteenth Century," PhD diss., University of North Carolina, 1958, 260. The term *normal school* was a common designation for teachers colleges.

20. Clarence L. Mohr and C. R. Wilson, eds., *The New Encyclopedia of Southern Culture*, 17:81–83.

21. Green, *Military Education*, 39–42, 71.

Chapter 6

1. Hugh Miller, ed., *The Scroll of the Phi Delta Theta* 25:529; Charles Walker, *Biographical Sketches of the Graduates and* Élèves *of the Virginia Military Institute Who Fell during the War Between the States*, 314–315, accessed October 16, 2014, https://archive.org/details/memorialvirginia00walk.

2. Rossiter Johnson and John Brown, *The Twentieth Century Biographical Dictionary of Notable Americans*, 376.

3. James D. Stephens, *Reflections: A Portrait-Biography of the Kentucky Military Institute, 1845–1971*, 44; Rod Andrew Jr., "North Carolina Military Institute," NC Pedia, accessed January 21, 2016, http://ncpedia.org/north-carolina-military-institute; Walter Clark, ed., *Histories of the Several Regiments and Battalions from North Carolina in the Great War, 1861–65*, 5:645–46; E. B. Munson, *North Carolina Civil War Obituaries, Regiments 1 through 46: A Collection of Tributes to the War Dead and Veterans*, 23. North Carolina Military Institute was reopened by Col. John P. Thomas, Citadel class of 1851, in 1873 and remained open nine years.

4. Dick Brown, "St. John's College," The Encyclopedia of Arkansas History and Culture, 376, updated August 25, 2009, accessed October 12, 2014, http://encyclopediao farkansas.net/encyclopedia/entry-detail.aspx?entryID=3584; Eastern Digital Resources, "Cobb's Legion Infantry, Company B, Bowden Volunteers," accessed July 15, 2011, http://www.researchonline.net/gacw/rosters/cobbib.htm; Earl Coates and Michael McAfee, *Don Troiani's Regiments & Uniforms of the Civil War*, 267; John Wyeth, *History of La Grange Military Academy and the Cadet Corps, 1857–1862: La Grange College, 1830–1857*, 104–98.

5. Jerry West, *The Bloody South Carolina Election of 1876: Wade Hampton III, the Red Shirt Campaign for Governor, and the End of Reconstruction*, 179; Robert Wynstra, *The Rashness of That Hour: Politics, Gettysburg, and the Downfall of Confederate Brigadier General Alfred Iverson*, 294.

6 Michael Wallace, "The Use of the Virginia Military Institute Corps of Cadets as a Military Unit Before and During the War Between the States," master's thesis, US Army Command and General Staff College, 1999, 130; Gary Baker, *Cadets in Gray: The Story of the Cadets of the South Carolina Military Academy and the Cadet Rangers in the Civil War*, 49.

7. Wyeth, *History of La Grange Military Academy*, 159. Louisiana Seminary of Learning and Military Academy grew to seventy-three cadets before the war.

8. "The Hillsborough Military Academy (a.k.a. the North Carolina Military Academy)," accessed January 15, 2016, http://freepages.history.rootsweb.ancestry.com/~orangecountync/places/hma/hma.html; Clark, *Histories of the Several Regiments*, 5:638.

9. Baker, *Cadets in Gray*, 12–26.

10. Walter McTernan and Andrew Kulberg, "U.S. Marines Face Citadel Cadets at Tulifinny Crossroads," *Leatherneck Magazine*, March 2013, 44–48.

11. Baker, *Cadets in Gray*, 145; McTernan and Kulberg, "U.S. Marines Face Citadel Cadets," 46, 47.

12. McTernan and Kulberg, "U.S. Marines Face Citadel Cadets," 47. Seeing the manner of discipline displayed by the cadet formation, a veteran remarked, "Them Charleston people is the damndest politest officers to their men I ever struck up with in the army."

13. Ibid; Baker, *Cadets in Gray*, 145; McTernan and Kulberg, "U.S. Marines Face Citadel Cadets," 47.

14. Ibid., 48; Baker, *Cadets in Gray*, 145–47.

15. Clark E. Center, "The Burning of the University of Alabama," *Alabama Heritage* 16 (Spring 1990): 30–45.

16. James Conrad, *The Young Lions: Confederate Cadets at War*, 113.

17. Center, "Burning of the University of Alabama," 34; "The Historical Art of John Paul Strain: Alabama Corps of Cadets Call to Battle," accessed December 28, 2015, http://www.johnpaulstrain.com/art/alabama-corps-of-cadets.htm; *Conrad, Young Lions, 146.*

18. Center, "Burning of the University of Alabama," 38.

19. Conrad, *Young Lions*, 134; Dale Cox, *The Battle of Natural Bridge, Florida: The Confederate Defense of Tallahassee*, 33; David Cole and Robert Graetz, "The Garnet and Gray: West Florida Seminary in the Civil War," *The United Daughters of the Confederacy Magazine*, January 1989, 2–6. The 1861 enrollment figure of 250 may include female students, who were not cadets. In the final year of the war, enrollment was thirty-four male cadets and twenty-four females.

20. Cole and Graetz, "Garnet and Gray," 2–6; Cox, *Battle of Natural Bridge*, 101. The Florida Legislature in 1863 named the school the Florida Military and Collegiate Institute.

21. John C. Inscoe, "The Civil War in Georgia," *A New Georgia Encyclopedia Companion*, 125; Bowling Yates, *History of the Georgia Military Institute, Marietta, Georgia, Including the Confederate Military Service of the Cadet Battalion*, 1–2, 5, 7, 26–29.

22. Yates, *History of the Georgia Military Institute*, 8–9, 10, 30.

23. Ibid., 11–13.

24. Ibid., 10, 14–17, 30.

25. Conrad, *Young Lions*, 39–44.

26. Ibid., 54–58.

27. Ibid., 82–83.

28. Ibid., 93–94.

29. William C. Davis, *The Battle of New Market*, 122.

30. Charles R. Knight, *Valley Thunder: The Battle of New Market and the Opening of the Shenandoah Valley Campaign, May 1864*, 174; Davis, *Battle of New Market*, 122–24, 135.

31. Davis, *Battle of New Market.*, 136.

32. Ibid., 138–39, 198.

33. Colonel Patton was the great-grandfather of General George Patton, who would attend VMI a year before transferring to West Point and who would gain fame as a World War II commander.

34. Davis, *Battle of New Market*, 122; Conrad, *Young Lions*, 99–100.

35. Virginia Military Institute, "William Henry Gillespie," Virginia Military Institute Archives, Online Historical Rosters Database, accessed December 12, 2011, http://www9.vmi.edu/archiverosters/ArchiveRosters.asp?page=search.

36. Thomas Campbell, *Academy on the James: The Confederate Naval School*, 77–88.

37. Ibid., 93–104; J. Thomas Scharf, *History of the Confederate States Navy from Its Organization to the Surrender of Its Last Vessel*, 776.

38. Campbell, *Academy on the James*, 106–7.

39. Conrad, *Young Lions*, 87; Scharf, *Confederate States Navy*, 778–79.

40. Scharf, *Confederate States Navy*, 778.

41. Ibid., 778–79.

42. Jack Sweetman, *The United States Naval Academy: An Illustrated History*, 2nd ed., 59–60.

43. Harry Phillips, Letter dated May 19, 1861, in *Letters from Annapolis*, ed. Anne

Marie Drew, 60–64.

44. Sweetmen, *Naval Academy*, 63. The total number of midshipmen sent before graduation to help with an expanding navy (which would grow from 42 to 641 vessels) was 112.

45. Theodore J. Crackel, *West Point: A Bicentennial History*, 132, 133.

46. Robert Poirier, *By the Blood of Our Alumni: Norwich University Citizen Soldiers in the Army of the Potomac*, 153–54, 158.

47. Ibid.; Mary R. Cabot, *Annals of Brattleboro, 1681–1895*, 3:664; James Grant Wilson and John Fiske, eds. *Appleton's Cyclopedia of American Biography*, 1892, 2:185; Loeman Augustowski, "Echos of Honor: Peekskill Military Academy," *Cortlandt Magazine*, June 2001, article reprint, 2; Dione Longley and Buck Zaidel, *Heroes for All Time: Connecticut Civil War Soldiers Tell Their Stories*, 10.

48. John D. Hamilton, "The St. Albans Raid: The Confederate Raid on St. Albans, Vermont, October 19, 1864," *Bulletin of the American Society of Arms Collectors* 90 (October 2004): 48–49.

49. William Ellis, *Norwich University, 1819–1911: Her History, Her Graduates, Her Roll of Honor*, 418.

50. Ibid., 418–21.

51. Ibid., 421; Hamilton, "St. Albans Raid," 48–52.

52. Pennsylvania Military College, "Cadets during the Gettysburg Campaign," accessed January 7, 2016, http://pennsylvaniamilitarycollege.org/cadets-gettysburg-campaign/; John Jordan, ed., *Colonial and Revolutionary Families of Pennsylvania: Genealogical and Personal Memoirs*, 3:1359–60.

53. Pennsylvania Military College, "Cadets during the Gettysburg Campaign."

54. Bruce S. Allardice, "West Points of the Confederacy: Southern Military Schools and the Confederate Army," *Civil War History*, 43, no. 4 (December 22, 1997): 310–32; Marcus Cunliffe, *Soldiers and Civilians: The Martial Spirit in America, 1775–1965*, 353.

55. Clark, *Histories of the Several Regiments*, 5:637–46.

56. Allardice, "West Points of the Confederacy," 310–32; Tommy Young, *Character Makes the Man: Kentucky Military Institute, 1845–1971*, 17. Kentucky Military Institute would survive the Civil War. By 1913 the school was largely a secondary school with a junior college, and by 1925 it was strictly a secondary school. The school closed in 1971.

57. Jefferson College functioned as a secondary education school but was a civilian institute 1854–58.

58. Jennifer Green, *Military Education and the Emerging Middle Class in the Old South*, 261–64; Robert Cowley and Thomas Guinzburg, *West Point: Two Centuries of Honor and Tradition*, 90; Poirier, *By the Blood of Our Alumni*, 273.

59. Robert E. Krick, *Staff Officers in Gray: A Biographical Register of Staff Officers in the Army of Northern Virginia*, 6; Richard M. McMurry, *Two Great Rebel Armies: An Essay in Confederate Military History*, 99.

60. Bruce S. Allardice, *Confederate Colonels: A Biographical Register*, 8, 19–20. An additional three colonels received their military education in Europe.

61. Cunliffe, *Soldiers and Civilians*, 353.

62. Poirier, *By the Blood of Our Alumni*, 277; Cowley and Guinzburg, *West Point*, 90; Ezra J. Warner, *Generals in Blue: Lives of the Union Commanders*, xx.

63. Robert Collins, *General James G. Blunt: Tarnished Glory*, 16.

64. Weston Historical Society, "Jarvis Military Academy," accessed October 21, 2011, http://westonhistoricalsociety.org/sites.htm#jma; William P. Blake, *Town of Hamden, Connecticut*, 178, accessed September 7, 2011, https://archive.org/details/historytown hamd00blakgoog]; Clarence R. Moll, "A History of Pennsylvania Military College, 1821–1954," PhD diss., New York University, 1954, 39; Edward Miller, "VMI Men Who Wore Yankee Blue, 1861–1865," *VMI Alumni Review*, Spring 1996, 2–13; Warner, *Generals in Gray*, 54, 534.

Chapter 7

1. Alexander S. McKenzie, *Life of Stephen Decatur, a Commodore in the Navy of the United States*, 44–46; Benjamin Park, *The United States Naval Academy*, 21, 23; George R. Prowell, *History of Camden County, New Jersey*, 77–86, accessed June 26, 2016, https://archive.org/details/historyofcamdenc00prow; American Jewish Historical Society, "Biographical Note: Uriah P. Levy (1792–1862)," accessed August 19, 2004, http://findingaids.cjh.org/?pID=109192#a2.

2. Edward C. Marshall, *History of the United States Naval Academy: With Biographical Sketches, and Names of all Superintendents, Professors and Graduates*, 12–13.

3. Ibid., 14.

4. James R. Soley, *Report on the Foreign Systems of Naval Education*, 26–28, 101–2.

5. William P. Leeman, *The Long Road to Annapolis: The Founding of the Naval Academy and the Emerging American Republic*, 71; Brendan January, *The Aftermath of the Wars against the Barbary Pirates*, 27, 34–38.

6. Naval Historical Center, "Captain Josiah Tattnall, Confederate States Navy (1795–1871)," accessed July 13, 2005, https://www.ibiblio.org/hyperwar/OnlineLibrary/photos/pers-us/uspers-t/j-tattnl.htm.

7. Marshall, *History of the Naval Academy*, 14; Justin Winsor, ed., *The Memorial History of Boston, Including Suffolk County, Massachusetts, 1630–1880*, 3:348; Thomas Williams, *The American Spirit: The Story of Commodore William Phillip Bainbridge*, 332, 334.

8. Jennifer Speelman, "Nautical Schools and the Development of the United States Maritime Professionals, 1874–1941," master's thesis, Temple University, 2001, 2.

9. Ibid., 24; Thomas C. Armory, *The Life of Admiral Sir Isaac Coffin, Baronet: His English and American Ancestors*, 96–110; Robert Uhl, "Masters of the Merchant Marine," *American Heritage* 34, no. 3 (April–May 1983), accessed March 8, 2005, http://www.americanheritage.com/content/masters-merchant-marine.

10. Park, *United States Naval Academy*, 124–28.

11. Brant G. Filbert and Allen G. Kaufman, *Naval Law: Justice and Procedure in Sea Services*, 3rd ed., 8–10.

12. Murat Halstead, *Life and Achievements of Admiral Dewey: From Montpelier to Manila*, 101.

13. Ibid.

14. "Education in the Navy," *Southern Literary Messenger* 16, no. 9 (September 1850), 521; Leeman, *Long Road to Annapolis*, 9; Halstead, *Life and Achievements of Admiral Dewey*, 101; Park, *United States Naval Academy*, 157; William Ellis, *Norwich University, 1819–1911: Her History, Her Graduates, Her Roll of Honor*, 418.

15. Rossiter Johnson and John H. Brown, eds., *The Twentieth Century Biographical Dictionary of Notable Americans*, 494.

16. Park, *United States Naval Academy*, 157, 165, 167.

17. Jack Sweetman, *The U.S. Naval Academy: An Illustrated History*, 2nd ed., 31.

18. Ibid., 56–57.

19. Thomas Campbell, *Academy on the James: The Confederate Naval School*, 109, 169–94.

20. Admiral Farragut Academy is perhaps the best example of a successful secondary school among those listed. The original campus was in Pine Beach, New Jersey, and opened in 1933. That school was so successful and had such an excellent reputation that a second campus opened in 1945 in Saint Petersburg, Florida. Although the New Jersey campus closed in 1994, the Florida school remained open and had 348 cadets in 2009/2010.

21. John Hattendorf, "Stephen B. Luce: Intellectual Leader of the New Navy," in *Quarter and Bridge: Two Centuries of American Naval leaders*, ed. James Bradford, 203–18.

22. Ibid., 204–5.

23. T. H. Inkster, "McDougall's Whalebacks," *American Neptune*, 25, no. 3 (July 1965): 175; Hattendorf, "Stephen B. Luce," 203–18.

24. Ibid.

25. Stephen Luce, "The Manning of Our Navy and Mercantile Marine," *Record of the Naval Institute* 1 (November 13, 1873): 23.

26. Ibid.; Henery Barnard, *Military Schools and Courses of Instruction in the Science and Art of War in France, Prussia, Austria, Sweden, Switzerland, Sardinia, England, and the United States, Drawn from Recent Official Reports and Documents*.

27. Speelman, "Nautical Schools," 21.

28. Henry How, "Destruction of the Ocean Steamer *Arctic*, by Collision with the *Vesta* (1854)." In *Outrageous Seas: Shipwreck and Survival in the Waters off Newfoundland, 1583–1893*, ed. Rainer Baehre, 281–305; Timothy J. Runyan, ed., *Ships, Seafaring and Society: Essays in Maritime History*, 239–40.

29. J. D. B. De Bow, "Education of Seamen at the South," *De Bow's Review and Industrial Resources, Statistics, etc.* 26 (1859): 41; Runyan, *Ships, Seafaring and Society*, 246.

30. Jeffery L. Cruikshank and Cloe G. Kline, *In Peace and War: A History of the U.S. Merchant Marine Academy at Kings Point*, 32.

31. Speelman, "Nautical Schools," 1–2, 34.

32. U.S. Congress, *Causes of the Reduction of American Tonnage and the Decline of Navigation Interests: Being the Report of a Select Committee, Made to the House of Representatives of the United States on the 17th of February 1870*, H. Rep. 28, 41st Cong., 2nd sess., 200, 220; Joseph Nimmo, *Report to the Secretary of the Treasury in Relation to the Foreign Commerce of the United States and the Practical Workings of Our Relations of Maritime Reciprocity*, 22.

33. Irving H. King, *The Coast Guard Expands, 1865–1915: New Roles, New Frontiers*, 154; Burhoe, *United States Coast Guard Academy*, 3. The Coast Guard was formed in 1915 with the consolidation of the Revenue Cutter Service and US Life Saving Service.

34. Joanna R. Nicholls, "The United States Revenue Cutter Service: The History and Duties of an Important Branch of Our Navy," *Frank Leslie's Popular Monthly* 42, no. 4 (October 1896): 411–425, accessed June 22, 2016, https://babel.hathitrust.org/cgi/pt?id=inu.32000000494429;view=2up;seq=428; King, *Coast Guard Expands*, 6, 91.

35. Worth G. Ross, "Our Coast Guard: A Brief History of the United States Revenue Marine Service," *Harper's New Monthly Magazine* 73 (November 1886): 909–22; American Institute of Mining, Metallurgical, and Petroleum Engineers, "Obituary of Charles E.

Emery," *Transactions*, 1900, 29.

36. Burhoe, *United States Coast Guard Academy*, 7.

37. Luce, "Manning of Our Navy," 28.

38. Ibid., 34–37; Speelman, "Nautical Schools," 7, 13, 29, 30. The Morrill Land Grant Act and its impact on the establishment of the military colleges will be discussed in chapter 8.

39. Ibid., 31–32, 37, 40–41.

40. Ellen E Dickinson, "The Training School-ship *Minnesota*," in *How to Learn and Earn; or, Half-Hours in Some Helpful Schools*, 347–72; US Navy Department, *Report of the Secretary of the Navy: Being Part of the Messages and Documents Communicated to the Two Houses of Congress*, 45th Cong., 2nd sess., 1877.

41. Speelman, "Nautical Schools," 34.

42. Ibid., 27–28; US Department of Commerce, Office of Education, *Report of the Commissioner of Education* for the *Year 1875*, 42; San Francisco Chamber of Commerce, *Annual Report of the Chamber of Commerce of San Francisco*, 34, 49; William H. Stewart, *Admirals of the World: A Biographical Dictionary, 1500 to Present*, 142–43.

43. Speelman, "Nautical Schools," 34, 36; James H. Brown, ed., *Lamb's Biographical Dictionary of the United States*, 3:382–83.

44. Brown, *Lamb's Biographical Dictionary*, 40–41.

45. Ibid., 66–67.

46. William R. Cutter, *Historical Homes and Places and Genealogical and Personal Memoirs Relating to the Families of Middlesex County, Massachusetts*, 4:1691–92.

47. Speelman, "Nautical Schools," 66–67, 77–78, 91.

48. Ibid., 77–78, 104–6, 118–19, 127–28.

49. Ibid., 118–19.

50. Joint Board of Higher Curricula, *Second Annual Report of the Joint Board of Higher Curricula to the Governor of Washington*, 12, 14.

51. Speelman, "Nautical Schools," 103–7; US Congress, House Committee on Naval Affairs, *Sundry Legislation Affecting the Naval Establishment, 1921*, 77th Cong., 1st sess., 1005; Joint Board of Higher Curricula, *Second Annual Report*, 14.

52. Douglas Peterson, *The California Maritime Academy: A Brief History*, 3–7.

53. Speelman, "Nautical Schools," 131–32.

54. Ibid., 92.

55. Cruikshank and Kline, *In Peace and War*, 40–43, 63, 65–66, 69–70, 141–43.

56. Ibid., 43.

57. Ibid.,44.

58. René de la Pedraja Toman, *A Historical Dictionary of the U.S. Merchant Marine and Shipping Iindustry: Since the Introduction of Steam*, 582.

59. Cruikshank and Kline, *In Peace and War*, 46.

60. Ibid., 47, 59.

61. Ibid., 47, 70.

62. Ibid., 73.

63. Ibid, 73, 74.

64. Ibid., 74–75.

65. Ibid., 72, 121.

66. Ibid., 78–81.

67. Ibid., 117–19.

68. Ibid., 119–20.

69. Ibid., 132–34.

70. Ibid., 144, 281–82.

71. Independence Seaport Museum, "Steering a Course: A Short History of the Pennsylvania Nautical School and Pennsylvania Maritime Academy," accessed June 21, 2016, http://www.phillyseaport.org/pennsylvania-nautical-school/.

72. "Vandalism on Ship Denied: Pennsylvania Inquiry Discounts Sabotage Report on Cruise," *New York Times*, December 25, 1946; Cruikshank and Kline, *In Peace and War*, 120.

73. Maine Maritime Academy, "A Proud Heritage," paper in possession of the author.

74. Speelman, "Nautical Schools," 130; Joseph G. Dawson, "Texas Maritime Academy," *Handbook of Texas Online*, accessed March 22, 2006, http://www.tshaonline.org/handbook/online/articles/kct22.

75. Stephen J. Curley, *Aggies by the Sea: Texas A&M University at Galveston*, 20–25.

76. Ibid., 25, 34.

77. Ibid., 45. The 1969 opening of the Great Lakes Maritime Academy in Michigan as a civilian institution did not sway the remaining maritime academies from the Stephen B. Luce model of a maritime academy based on a military school format.

Chapter 8

1. Jennifer Green, *Military Education and the Emerging Middle Class in the Old South*, 194; William Ellis, *Norwich University, 1819–1911: Her History, Her Graduates, Her Roll of Honor*, 402.

2. Gary W. Gallagher and Alan T. Nolan, eds., *The Myth of the Lost Cause and Civil War History*, 1.

3. Rod Andrew Jr., *Long Gray Lines: The Southern Military School Tradition, 1839–1915*, 47.

4. James Morrison, *Rat Pants to Eagles and Tweeds: The Memoirs of a Soldier-Teacher*, 3.

5. Harry D. Temple, *The Bugle's Echo: The Chronology of Cadet Life at the Military College at Blacksburg, Virginia*, 1:66.

6. Ibid., 1:354.

7. Andrew, *Long Gray Lines*, 2.

8. Ibid., 49.

9. US Park Service, National Register of Historic Places, Fishburne Home School, August 21, 1984, accessed April 10, 2016, http://www.dhr.virginia.gov/registers/Cities/Waynesboro/136–0004%20-%20Fishburne%20Military%20School%20-%20Final%20Nomination.pdf.

10. James D. Blackwell, *The Poetical Works of James DeRuyter Blackwell*, 1:2–7; John T. Toler, "Bethel Military Academy, 1867–1911," *The Fauquier Historical Society* 18, no. 1 (Winter and Spring 1996): 1–13, http://www.rootsweb.ancestry.com/~vafauqui/bma1.htm (April 15, 2011). The University of Virginia in 1872 had 131 cadets (96 Virginians) organized into two companies. By 1883, four graduates were commissioned into the Virginia Militia annually, and in 1885 cadets were invited to the White House as guests of President Grover Cleveland. Annual scholarships to the University of Virginia, Virginia Military Institute, and two other Virginia colleges reflected the school's high standards. New cadets, called fish, entered the military prep school with high esprit.

11. U.S. Congress, "Morrill, Justin Smith (1810–1898)," accessed April 17, 2011, http://bioguide.congress.gov/scripts/biodisplay.pl?index=M000969; Ellis, *Norwich University*, 11.

12. Association of Public and Land Grant Universities, *The Land Grant Tradition*, 3, accessed June 25, 2016, http://www.aplu.org/library/the-land-grant-tradition/file.

13. Larry Webb, "The Origin of Military Schools in the United States Founded in the Nineteenth Century," PhD diss., University of North Carolina, 1958, 260.

14. After leaving Virginia Agricultural and Mechanical College (Virginia Tech), General Lane became a professor of the Agricultural and Mechanical College of Alabama (Auburn), which was a military school until 1905.

15. Temple, *Bugle's Echo*, 6:139.

16. Nancy Beck Young, "Texas Military Institute, Austin," *Handbook of Texas Online*, accessed January 23, 2016, http://www.tshaonline.org/handbook/online/articles/kbt17; Daniel Morley McKeithan, "James, John Garland," *Handbook of Texas Online*, accessed January 23, 2016, http://www.tshaonline.org/handbook/online/articles/fja18. There have been four schools titled Texas Military Institute. This one under discussion here was established in 1851, became Bastrop Military Institute in 1857, and became Texas Military Institute in 1868.

17. Temple, *Bugle's Echo*, 1:276.

18. Ibid., 3:2270.

19. John Adams, *Keepers of the Spirit: The Corps of Cadets at Texas A&M*, 35.

20. Ibid., 85.

21. Andrew, *Long Gray Lines*, 43; Edward M. Heyl, "Alabama Agricultural and Mechanical College," in *Annual Report to the Inspector-General of the Army*, 363–65.

22. Thomas Ruffin, Jo Jackson, and Mary Hebert, *Under Stately Oaks: History of LSU*, 3–17.

23. In 2013 North Georgia College and State University and Gainesville State College consolidated, and North Georgia became officially University of North Georgia. Part of the 1866 Morrill Land Grant Act was the detailing of active-duty army officers to Land Grant Colleges. See John Kraus, "The Civilian Military Colleges in the Twentieth Century: Factors Influencing Their Survival," PhD diss., University of Iowa, 1978, 277–79.

24. North Georgia Agricultural College, *Annual Catalogue, 1873–1880*, 10, 48; Georgia Department of Education, *Annual Report—Georgia Department of Education*, 20; North Georgia Agricultural College, *Annual Catalogue, 1873–1880*, 10, 48.

25. Kraus, "Civilian Military Colleges," 283, 294.

26. *Appleton's Annual Cyclopaedia and Register of Important Events of the Year 1886*, 384; Charles E. Jones, *Education in Georgia*, 51–53.

27. Edward Mayes, *History of Education in Mississippi*, 384; Jane Simpson, "Georgia Military College: A Brief History," video, 2008, retrieved February 16, 2011, http://vimeo.com/4627442.

28. Kraus, "Civilian Military Colleges," 90.

29. Ibid., 95; Mississippi Agricultural and Mechanical College, *Biennial Report of the Trustees, President, and Other Officers of the Mississippi Agricultural and Mechanical College*, 65, 101, 242.

30. Kraus, "Civilian Military Colleges," 98.

31. Ibid.; Michael B. Ballard, *Maroon and White: Mississippi State University. 1878–2003*, 60–61; Kraus, "Civilian Military Colleges," 98; Mayes, *History of Education in Mississippi*, 232.

32. Ballard, *Maroon and White*, 65–66.

33. Kraus, "Civilian Military.Colleges," 108.

34. Ibid., 62.

35. Ibid., 65.

36. Ibid., 72.

37. J. Baker, North Carolina State University Library, email message to the author, December 16, 2010; Alice E. Reagan, *North Carolina State University: A Narrative History*, 47. North Carolina College of Agriculture and Mechanic Arts, after a number of name changes, became North Carolina State University in 1965.

38. Michael Dennis, *Lessons in Progress: State Universities and Progressivism in the New South, 1880–1920*, 94.

39. Ibid.

40. Union veterans who started military schools on the West Coast include Wayne Scott Walker, former bugler with the 11th Indiana Cavalry. He had been head of United Brethren of Christ schools from 1876 to 1889 and then led the Pullman Military College in Pullman, Washington, from 1891 until it burned down in 1893. Ed Garretson, archivist, Whitman County Historical Society, email to the author, May 24, 2010.

41. J. H. Simberg, "Colonel Wright's Methods," *Printer's Ink*, April 23, 1902, 13–17; Samuel J. Rogal, *The American Pre-College Military School: A History and Comprehensive Catalog of Institutions*, 10–11, 178–79.

42. Simberg, "Colonel Wright's Methods," 14.

43. Rogal, *American Pre-College Military School*, 10; Simberg, "Colonel Wright's Methods," 13–17.

44. Hobart College, *Catalogue of Hobart College, Geneva, New York, 1910–1911*, 9:90.

45. William H. Powell, "Colonel Joseph Sumner Rogers, U.S.V.," in *Officers of the Volunteer Army and Navy Who Served in the Civil War*, 419, accessed March 4, 2010, http://www.all-biographies.com/soldiers/joseph_sumner_rogers.htm; J. C. Starbuck, "Michigan Military Academy at Orchard Lake," *Michigan History* 50 (September 1966), accessed March 4, 2010, http://community.ancestry.com/ViewUserContent. aspx?uid=00827709–0003–0000–0000–000000000000&pp=11&type=story, reprint with additional notes by Brian Bohnett in possession of author; Powell, "Colonel Joseph Sumner Rogers," 419.

46. Hyland C. Kirk, *A History of the New York State Teachers Association*, 149.

47. University of the State of New York, *University of the State of New York: Annual Report*, 302; *The Rochester Directory, Containing a General Directory of the Citizens, a Business Directory, and the City and County Register*, 21:282; W. P. Gaylord, *History of Floyd County, Iowa*, 2:759.

48. Blanche E. Little, "Military Methods in the Training of Boys" *School Journal* 67 (September 5, 1903): 204.

49. Mark A. Roeder, *A History of Culver and the Culver Military Academy*, 170–72.

50. William Hyde and Howard Louis Conrad, eds., *Encyclopedia of the History of St. Louis: A Compendium of History and Biography for Ready Reference*, 4:2256–58.

51. Rogal, *American Pre-College Military School*, 97–101. Ohio Military Academy ended operations in 1932. Ohio Military Institute, although founded in 1833, became a military school in 1890 and continued operation until at least 1958.

52. Walter Barlow Stevens, *Missouri, the Center State, 1821–1915*, 3:343; Little, "Military Methods," 204.

53. Robert W. Schramm, *Linsly School*, 17–25.

54. Mitchell C. Harrison, *New York State's Prominent and Progressive Men*, 85.

55. A. W. Dobyns, "The Military Feature at the Fanwood School," *American Annals of the Deaf*, 53 (1908): 16–21.

56. "Mutes, but Everyone is a Soldier," *Richmond Times*, December 23, 1900, 17, accessed April 2, 2010, http://chroniclingamerica.loc.gov/lccn/sn85034438/1900–12–23/ed-1/seq-17/; "A Military School for the Deaf, and Its Band of Deaf Musicians," *Volta Review* 13, no. 4 (September 1911): 201.

57. I took into account references that identified several other African American institutes of higher learning—Florida A&M, Georgia State Industrial College, and South Carolina State—as military schools for periods in their history. However, my research did not support that conclusion.

58. Donald F. Lindsey, *Indians at Hampton Institute, 1877–1923*, 3–4; L. C. Cooley, "The All-Around Training of the Hampton Cadet," *Southern Workman and Hampton School Record* 30, no. 1 (January 1901): 493–500; Donald Spivey, *Schooling for the New Slavery: Black Industrial Education, 1868–1915*, 27.

59. A. M. Bailey, "The Glorious Hampton Cadets," *Southern Workman and Hampton School Record* 21, no. 5 (May 1892): 77; Spivey, *Schooling for the New Slavery*, 32.

60. Marie B. Smith, "A History of St. Emma's Military Academy and St. Francis De Sales High School," PhD diss., Catholic University of America, 1949, 5, 13.

61. Charles Alexander, "St. Emma's Industrial and Agricultural College," *Alexander's Magazine* 4 (1907): 100–108.

62. T. L. Chenevert, "History" [St. Emma Military Academy], copy in possession of the author; Smith, "History of St. Emma's," 1–3.

63. Theodore J. Crackel, *West Point: A Bicentennial History*, 237–38.

64. Karen E. Petersen, "Harold Douglas Martin, Major, United States Army," Arlington National Cemetery Website, accessed October 3, 2014, http://www.arlingtoncemetery.net/hdmartin-002.htm; Peter Wallenstein, "The First Black Students at Virginia Tech, 1953–1963," *Diversity News* 4, no. 1 (Fall 1997): 3–5, accessed October 3, 2014, https://spec.lib.vt.edu/archives/blackhistory/timeline/blackstu.htm; Peter Wallenstein, "Not Fast, but First: The Desegregation of Virginia Tech," accessed October 3, 2014, http://www.vtmag.vt.edu/fa1197/feature1.html.

65. London Steverson, "Coast Guard Honors First Black Academy Graduate, April 3, 2012," accessed October 3, 2014, https://judgelondonsteverson.me/2012/04/03/coast-guard-honors-first-black-academy-graduate/; Jack Sweetman, *The U.S. Naval Academy: An Illustrated History*, 2nd ed., 207.

66. Michael Graczyl, "Texas A&M Cadets to Be Led by First Black Commander," *Dallas Morning News*, April 12, 2012, accessed October 12, 2014, http://www.dallasnews.com/news/state/headlines/20120412-texas-am-cadets-to-be-led-by-first-black-commander.ece; Alexander S. Macaulay, Jr., "Black, White, and Gray: The Desegregation of the Citadel, 1963–1973," in *Warm Ashes: Issues in Southern History at the Dawn of the Twenty-first Century*, ed. Winfred Moore, Kyle Sinisi, and David White, 301–19.

Chapter 9

1. Larry Manning, "The Contribution of Sylvanus Thayer and the United States Military Academy to Engineering Programs in Higher Education in the United States," PhD diss., Texas A&M University, 2003, 116; Frederick Rudolph, *The American College and University*, 17, 71–79, 144–49.

2. James Carper and Thomas Hunt, *The Praeger Handbook of Religion and Education in the United States*, 1:401–6.

3. William Ellis, *Norwich University, 1819–1911: Her History, Her Graduates, Her Roll of Honor*, 19, 80; John Thomas, *The History of the South Carolina Military Academy*, 39; Francis Smith, *The Virginia Military Institute: Its Building and Rebuilding*, 255–58.

4. John Kraus, "The Civilian Military Colleges in the Twentieth Century: Factors Influencing Their Survival," PhD diss., University of Iowa, 1978, 438, 439.

5. Allan L. Burleson, "A Paper Read at the Midwinter Fair, San Antonio, Texas, on Education Day," February 3, 1896, Archives of Texas Military Institute, San Antonio; Samuel J. Rogal, *The American Pre-College Military School: A History and Comprehensive Catalog of Institutions*, 27.

6. Thomas Whittaker, *Whittaker's Churchman's Almanac: The Protestant Episcopal Almanac and Parochial List*, xx.

7. William P. Blake, *History of the Town of Hamden, Connecticut, with an Account of the Centennial Celebration, June 15, 1886*, 173, accessed June 21, 2016, https://archive.org/details/historytownhamd00blakgoog.

8. Ibid., 174; A. Smyth, ed., "Military Organizations in Schools," *Ohio Journal of Education* 5 (1856): 149; Ernst Steiger, *Steiger's Educational Directory for 1878*, 3. After the Civil War the drillmaster was former cadet Ezra L. Stiles, who had served in the war as a corporal in Company A, 13th New York Cavalry.

9. Walter L. Prehn, "Episcopal Schools: History and Mission," 2011, manuscript in possession of the author, 19, 29.

10. Ibid., 39, 71–72.

11. Episcopal Church of St. Matthew, "History," accessed November 12, 2010, http://www.episcopalstmatthew.org/history.html; William R. Hamilton, "The Military Schools of the Pacific Coast," *Overland Monthly* 29 (May 1897): 465–81.

12. Hamilton, "Military Schools," 478.

13. Rogal, *American Pre-College Military School*, 208–9.

14. Peter Condon, "Knownothingism," *The Catholic Encyclopedia*, 8:2, accessed November 21, 2011, http://www.newadvent.org/cathen/08677a.htm.

15. Harold W. Hurst, *Alexandria on the Potomac: The Portrait of an Antebellum Community*, 54; Allan W. Robbins, "History of Saint John's Academy 1833–1895," *Alexandria History* 5 (1983): 25–31; James Grant Wilson and John Fiske, *Appleton's Cyclopedia of American Biography*, 1888, 2:462; Alexandria Library Special Collection, Alexandria City Records, 1853, Box 019II.

16. "Saint John's Academy Fiftieth Anniversary Speech," Saint John's Academy File, loose, Alexandria Library Special Collection, Alexandria City Records, 1883; Robbins, "History of Saint John's," 28.

17. Robbins, "History of Saint John's," 27.

18. Michael Kenney, *Catholic Culture in Alabama*, 348; *Spring Hill College, 1918–1919*, 11–12, 53–54; Richard Weaver, Spring Hill College, personal communication, email to author, March 9, 2011.

19. W. R. Harris, *The Catholic Church in Utah, 1776–1909*, 290–94.

20. Catholic Dioceses of Utah Archives, Correspondence, March 22, 1888—May 26, 1919.

21. Lucian L. Knight, *A Standard History of Georgia and Georgians*, 5:2499.

22. Paul Dans, "LaSalle Military Academy Prehistory Notes and Clason Point," manuscript, 1–3.

23. Jesse Grapes, "Message from the Headmaster," Benedictine High School, accessed May 9, 2011, http://www.benedictinecollegeprep.org/RelId/636273/ISvars/default/Message_from_the_Headmaster.htm.

24. Rogal, *American Pre-College Military School,* 150–51, 208; Leonard Hall Junior Naval Academy, "Proud History," accessed June 26, 2016, https://sites.google.com/site/lhjna2013/how-to-get-involved.

25. Leonard Hall Junior Naval Academy, "Proud History."

26. St. Catherine's Academy, "History," accessed February 7, 2010, http://www.stcatherinesacademy.org/about/history.aspx.

27. Suzanne S. Dietz and Amy L. Freiermuth, *Lewiston: Then and Now,* 42; Rogal, *American Pre-College Military School,* 181.

28. Kathleen Urbanic, Archivist, Sisters of St. Joseph, email to the author, October, 21, 2011.

29. Ibid., October 19, 2011.

30. Lester A. Webb, "The Origin of Military Schools in the United States Founded in the Nineteenth Century," PhD diss., University of North Carolina, 1958, 226; Margaret A. Hunter, "Education in Pennsylvania Promoted by the Presbyterian Church, 1726–1837," PhD diss., Temple University, 1937, 170.

31. Steiger, *Steiger's Educational Directory, 1878,* 62.

32. Theta Chi Fraternity, *History of Theta Chi, 1856–1927,* 74–85; William R. Hamilton, "The Military Schools of the Pacific Coast," *Overland Monthly* 29 (May 1897): 465–81.

33. US Department of the Interior, Office of Education, *Report of the Commissioner of Education for the Year 1874,* 25; Joseph E. Baker, ed., *Past and Present of Alameda County, California,* 2:136.

34. US War Department, *Annual Report of the Secretary of War for the Year 1892,* 1:4; Fort Plain Museum, "Clinton Liberal Institute," accessed June 7, 2010, http://www.montgomery.nygenweb.net/minden/ftplaininstitute.html.

35. Hamilton, "Military Schools of the Pacific Coast," 476–78.

36. Don Hedgpeth, *Proud Promise: The Story of Schreiner Institute/College,* 35, 46, 54, 56, 99.

37. Henry Crocker, *History of the Baptists in Vermont,* 503; War Department, *Annual Report, 1892,* 4:306.

38. Sean Flynt, Office of Communication, Samford University, personal communication by email to the author, February 9, 2010.

39. Ibid.

40. James Hadden, *A History of Uniontown: The Seat of Fayette County, Pennsylvania,* 505–8.

41. Ibid., 506–7; Dorothy T. Potter and Clifton W. Potter, *Lynchburg, 1757–2007,* Images of America.

42. Bastrop County [Texas] Museum, "A Survey of Texas Historical Markers in Bastrop County," research files of Bruce Allardice.

43. Kenneth Kesselus, *History of Bastrop County, Texas, 1846–65,* 73–84.

44. Ibid., 87; Daniel Morley McKeithan, "James, John Garland," *Handbook of Texas Online,* Texas State Historical Association, June 15, 2010, http://www.tshaonline.org/handbook/online/articles/fja18; Nancy Beck Young, "Texas Military Institute, Austin," *Handbook of Texas Online,* Texas State Historical Association, June 15, 2010, http://www.tshaonline.org/handbook/online/articles/kbt17.

45. Richard Irby, *History of Randolph-Macon College, Virginia: The Oldest Incorpo-rated Methodist College in America*, 153, 155.

46. Ibid., 146.

47. St. Charles College was established in 1832 and was closed when the Union Army occupied the town in 1861. The Methodist College patrons were Southern sym-pathizers, and their church became a Union prison for those of suspected loyalties. It was not until 1870 that the church returned to the Methodist Church and the college reopened. The return involved a state supreme court case.

48. Rogal, *American Pre-College Military School*, 190.

49. North Carolina Department of Cultural Resources, "Mount Pleasant Colle-giate Institute," 1962, accessed June 12, 2010, http://www.ncmarkers.com/Markers .aspx?MarkerId=L-65.

50. Ibid.; Webb, "Origin of Military Schools," 260.

51. Allan L. Burleson, "A Paper Read at the Midwinter Fair, San Antonio, Texas, on Education Day," February 3, 1896, Archives of Texas Military Institute, San Antonio; Rogal, *American Pre-College Military School*, 27; Kraus, "Civilian Military Colleges," 438–39.

Chapter 10

1. Harry D. Temple, *The Bugle's Echo: The Chronology of Cadet Life at the Military College at Blacksburg, Virginia*, 1:486, 489.

2. Ibid., 492.

3. Ibid.

4. Vermont, *Vermont Public Documents: Being Reports of State Officers, Departments and Institutions*, 3–4, accessed January 6, 2016, https://babel.hathitrust.org/cgi/pt?id =chi.095914113;view=2up;seq=1. Colonel Reeves was president of Norwich 1915–1917. In World War I he served with the 7th Division at Saint–Mihiel, where he was gassed. Later, he served with the 92nd Division and commanded the 137th Infantry Regiment of the 37th Division. He was key in founding the American Expeditionary Forces University in France.

5. There is one group of military schools that is not addressed in the text or ap-pendixes of this book. China operated the Western Military Academies in as many as twenty-one locations in the United States during the years 1902–1911. These schools had faculty of both Americans and Chinese, and the students were cadets who were obligated for military service back in China. The unusual story of these foreign schools on American soil is addressed in Carol Huang, "The Chinese Western Military Acad-emies in the United States, 1902–1911," in *Chartered Schools: Two Hundred Years of Independent Academies in the United States, 1727–1925*, edited by Nancy Beadie and Kim Tolley, 228–50.

6. Donna Peacock, *Parade Rest: Peacock Military Academy, 1894–1941*, Book One, 18–22. San Antonio, Texas, has likely had more military schools (eight) than any other city, including Chicago, which could boast of five in 2016. The first was Alamo Mili-tary Academy, which was operating as early as 1883 and lasted only a short period. Next was the West Texas Military Academy (WTMA), established in 1893. After in-corporating the upper school of San Antonio Academy, it was renamed Texas Mili-tary Institute and boasted 155 cadets in 2015. Peacock Military Academy became WTMA's biggest rival after it transitioned to a military school in 1900. The school was

very successful, but it closed in 1973. Faculty members of WTMA would start two military schools in the city: Rev. A. W. S. Garden, former rector, started Garden Military Academy in 1908, and Charles Lukin, a former principal, started Lukin Military Academy in 1917. Additionally, Lakeside Classical Institute functioned between 1905 and at least 1912; St. Mary's College, a male boarding school at the time, transitioned to military 1913 and operated as such until 1932. San Antonio Academy transitioned to military format in 1919 and was still in operation as a military grade school in 2016.

7. Roger Possner, *The Rise of Militarism in the Progressive Era, 1900–1914.*

8. Alvan C. Hadley, "The Association of Military Colleges and Schools of the United States (AMSCUS) and the Struggle for the Survival of the Military Preparatory Schools in America," PhD diss., University of Kentucky, 1999; Leigh Gignilliat, *Arms and the Boy: Military Training in Schools and Colleges,* 310.

9. Julius A. Penn, "Military Instruction at Civil Educational Institutions," *Infantry Journal* 6, no. 5 (1916): 679–703; Jane W. Dieffenbacker, *This Green and Pleasant Land: Fairfield, New York,* 196–210.

10. Dieffenbacker, *Green and Pleasant Land,* 196–210 There is some evidence, although not conclusive, that this school may have had America's first co-ed corps of cadets. With its small enrollment, the school was organized into two companies, and "girls students were also cadets and wore standard 'cadet gray' trimmed in black braid. Women teachers were uniformed as well with additional trimming to mark their rank. The female uniform cost about $13." Ibid., 204.

11. Clyde R. Terry, *Illinois Military School, Abingdon, Illinois,* 9, accessed November 19, 2008, http://www.archive.org/stream/illinoismilitary00abin#page/n1/mode/2up; Hadley, "Association of Military Colleges and Schools," 78.

12. Spencer E. Tucker, ed., *The Encyclopedia of the Spanish-American and Philippine-American Wars: A Political, Social, and Military History,* 1:22.

13. Rod Andrew, *Long Gray Lines: The Southern Military School Tradition, 1839–1915*; Mary Robson Farrand, "Hill Military Academy," Hill Military Academy Directory, n.d., accessed May 2006, http://gesswhoto.com/hill-academy.html.

14. Nancy Ford, *The Great War and America: Civil-Military Relations during World War I,* 17–19.

15. Ibid., 1, 9–13, 22–25, 77.

16. Chapultepec Inc., Almanac of Theodore Roosevelt, accessed September 21, 2014, http://www.theodore-roosevelt.com/treditorials.html.

17. Gignilliat, *Arms and the Boy,* 1–5, 115 Leigh R. Gignilliat was a graduate of the Virginia Military Institute, class of 1895, and superintendent of Culver Military Academy, 1910–1939. He was awarded the French Legion of Honour for service in World War I.; William Norton, ed., *Historic Gainesville and Hall County: Illustrated History,* 60–61.

18. US Congress, House Committee on Military Affairs, *Abolishment of Compulsory Military Training at Schools and Colleges,* Hearings on H.R. 8538, 69th Cong., 1st sess., 1926, 18, accessed December 22, 2009, http://babel.hathitrust.org/cgi/pt?view=image;size=100;id=mdp.39015076644916.

19. Friends of PMC, "Eddystone Disaster," Legacy of PMC, retrieved February 15, 2016, http://pennsylvaniamilitarycollege.org/category/legends/legacy-of-pmc/.

20. Louisiana State University functioned as a military school from 1860 to 1887.

21. Mark E. Grotelveschen, *The AEF Way of War: The American Army and Combat in World War I*, 11–12.

22. One impact of World War I was the influenza epidemic (Spanish flu) in the United States in 1918. At least one and likely many more military schools were hard hit. Of the eighty-seven cadets at Kearney Military Academy in Nebraska, five boys died. Kentucky Military Institute lost a cadet as well.

23. E. E. Calkins, "The Revolt of Youth," *The Rotarian*, February 1932, 52; Christopher Nack, "World War I Narratives and the American Peace Movement, 1920–1936," PhD diss., Florida State University, 2005, 2.

24. Robert W. Hamblin and Charles A. Peek, *A William Faulkner Encyclopedia*, 234; Nank, "World War I Narratives," 9.

25. Patti M. Peterson, "Student Organizations and the Anti-war Movement in America, 1900–1960," in *Peace Movements in America*, ed. Charles Chatfield, 137; Charles F. Howlett and Glen Zeitzer, *The American Peace Movement*, 37–38; Michael J. Cormack, *Ideology and Cinematography in Hollywood, 1930–39*.

26. Nank, "World War I Narratives," 109–11.

27. Calkins, "Revolt of Youth," 52.

28. "Committee on Militarism in Education, 1925–1940," DG009, Swarthmore College Peace Collection, accessed November 17, 2009, http://www.swarthmore.edu/library/peace/DG001–025/dg009cme.html.

29. US Congress, House Committee on Military Affairs, *Abolishment of Compulsory Military Training*, 17, 18.

30. Richard G. Davies, "Of Arms and Men: History of Culver Military Academy, 1894–1945," PhD diss., University of Indiana, 1983, 91, 92.

31. Rogers State University, "History: Oklahoma Military Academy," 2009, accessed January 9, 2010, http://www.rsu.edu/alumni/oklahoma-military-academy/. In 1947 the horses were replaced by tanks.

32. Davies, "Of Arms and Men," 91.

33. Ibid., 97.

34. Davies, "Of Arms and Men," 91. In 1920, 94 of 138 cadets, and in 1924, 126 out of 143 seniors, intended to go on to college.

35. John A. Adams, *Keepers of the Spirit: The Corps of Cadets at Texas A&M*, 106.

36. Ibid.; V. R. Cardozier, *Colleges and Universities in World War II*, 15.

37. William Trousdale, *Military High Schools in America*, 265; Robert S. McElvaine, *The Great Depression: America, 1929–1941*, 74–75.

38 . Samuel J. Rogal, *The American Pre-College Military School: A History and Comprehensive Catalog of Institutions*, 19–20, 67, 80–81; Morgan Park Academy, "History," accessed May 3, 2010, http://morganparkacademy.org.

39. Davies, "Of Arms and Men," 91; Rogal, *American Pre-College Military School*, 52, 68–69.

40. Rogal, *American Pre-College Military School*, 68.

41. Myron M. Stearns,"Where Are You Going to School?" *Boy's Life*, March 1934, 27.

42. Edwin V. O'Hara, "The School Question in Oregon," *Catholic World* 116 (October 1923): 442–49, 484.

43. Michael Imber and Tyll Van Geel, *Education Law*, 20; US Supreme Court, *Pierce v. Society of the Sisters of the Holy Names of Jesus and Mary and the Hill Military Academy*, 268 U.S. 510, Nos. 583, 584, 1925; Lyndsay M. Pinkus, *Moving beyond*

AYP: High School Performance Indicators, June 2009. Hill Military Academy was established by Dr. Joseph W. Hill after he left Bishop Scott Academy in 1901. The school enrolled 170 cadets in 1923–24. The school produced a number of American, British, and Canadian officers who served in the World Wars. Dr. Hill was assisted in administration of the school by his son James A. Hill, who was killed in North Africa as a captain in the US Army. The school added thirteen buildings at the beginning of the Depression, a gym in 1944, and a chapel after World War II. The school closed in 1962 with dropping enrollment.

44. Hadley, "Association of Military Colleges and Schools," 16.

45. Ibid.

46. Rogal, *American Pre-College Military School*, 151; Carnegie Foundation, *Carnegie Hero Fund Commission*.

47. Jerry Guinn, "A Short History of Roosevelt Military Academy," accessed April 4, 2008, http://www.rmaroughriders.org/history.htm; Terry, *Illinois Military School*.

48. Davies, "Of Arms and Men," 91; Missouri Military Academy, "History of Missouri Military Academy," accessed March 1, 2011, http://www.missourimilitaryacademy.org/about/history/. The Rectory School had a series of five fires between September 28 and October 7, 1868, when the barn and the attic of the house burned, and there were three fires in the schoolhouse. These fires killed three horses and nearly suffocated several cadets. A recently arrived cadet was arrested, and it was discovered that he intended to destroy the school. This was the only example I found of arson that closed a school. Hamden Historical Society, *Publications of the Hamden Historical Society, No. 1*, 30. For another case of arson with the objective of destroying an unsightly structure, see above, chapter 4, note 16.

49. "Three Students Die When Academy Burns: Nine Others Are Seriously Hurt at Gambier, Ohio," *New York Times*, February 24, 1908, accessed April 8, 2011, http://select.nytimes.com/gst/abstract.html?res=F20D13F93B5A12738DDDAC0A94DA4 05B868CF1D3&scp=1&sq=Three%20students%20die%20when%20academy%20 burns&st=cse (April 8, 2011).

50. Rogal, *American Pre-College Military School*, 67. Among the schools closed by fire in table 10.2 are Claverack College and Hudson River Institute. Opened in 1830 as Claverack Academy, the first school became military in 1854. The second was designated for female students in 1869, but by 1883 enrollment was 190 students, of whom approximately 30 were women. Between 1888 and 1890 Stephen Crane (author of *The Red Badge of Courage*) attended, rising to ranking cadet as the battalion adjutant over four cadet companies.

51. Henry Neil, "Collinwood School Fire," *The Encyclopedia of Cleveland History*, last modified March 27, 1998, accessed November 8, 2010, http://ech.case.edu/ech-cgi/article .pl?id=CSF; New York Military Academy, "New York Military Academy: A High Grade Preparatory School for all Colleges" (advertisement), *Cosmopolitan Magazine*, August 1912, 26, accessed September 2010, http://books.google.com/books?id=gz1XAAAAYAAJ&pg=RA2-PA198&dq=fire+proof+Military+school&hl=en#.

Chapter 11

1. Harry D. Temple, *The Bugle's Echo: The Chronology of Cadet Life at the Military College at Blacksburg, Virginia*, 1:viii.

2. Castle Heights added a junior college program in 1941 with R. C. Ford as dean. It is unclear how many years it functioned.

New Mexico Military Institute was founded as Goss Military Academy (Institute) with just 28 cadets in 1891 by James C. Lea and Colonel Robert Goss. In 1893 the school became public as the territorial school, the New Mexico Military Institute, with funding finally allocated in 1895. By 1896 the institute had two cadet companies and enrolled 105. In 1915 it expanded to include a junior college, and from 1948 to 1956 it was extended to four years of college.

3. V. R. Cardozier, *Colleges and Universities in World War II*, ix.

4. Laurence H. Suid, *Sailing on the Silver Screen: Hollywood and the U.S. Navy*, 31; Stephen Vaughn, *Ronald Reagan in Hollywood: Movies and Politics*, 97.

5. "The Major and the Minor," *Life Magazine*, August 17, 1942, 37–38; Wiley Lee Umphlett, *The Movies Go to College: Hollywood and the World of the College-Life Film*.

6. Dale Herbeck, "Football," in *The Columbia Companion to American History on Film: How the Movies Have Portrayed the American Past*, ed. Peter C. Rollins, 365–73; Jane Lockhart, "Looking at the Movies," *The Rotarian*, November 1951, 38–39.

7. P. J. Budahn, *What to Expect in the Military: A Practical Guide for Young People, Parents, and Counselors*, 201–2; Umphlett, *Movies Go to College*, 96.

8. *The Strange One*, Directed by Jack Garfein and Herb Gardener, Columbia Pictures, 1957.

9. Alvan C. Hadley, "The Association of Military Colleges and Schools of the United States (AMSCUS) and the Struggle for the Survival of the Military Preparatory Schools in America," PhD diss., University of Kentucky, 1999, 20–21.

10. Augusta Military Academy was established in Fort Defiance in 1879 as the Augusta Male Classical Academy (Roller School) by Charles Roller. Roller, a wounded veteran of the 1st Virginia Cavalry (CSA) changed the school to a military format in 1895 and was later assisted by his sons, one of whom was Charles S. Roller, VMI class of 1901. Enrollment in the late 1920s to the early 1950s ran between 164 and 265. In the early 1960s enrollment peaked at between 477 and 500. The academy closed in 1983.

11. US Congress, House Committee on Military Affairs, *To Increase the Efficiency of the Military Establishment of the United States*, Hearings, 64th Cong., 1st sess., 1916, 1, 451, accessed January 13, 2010, https://babel.hathitrust.org/cgi/pt?id=uc1.$b6542 40;view=2up;seq=6.

12. Hadley, "Association of Military Colleges and Schools," 20–21.

13. Ibid., 88, 267–69.

14. Ibid., 88.

15. Robert Cowley and Thomas Guinzburg, *West Point: Two Centuries of Honor and Tradition*, 170–71, 183–84.

16. Ibid., 170–73.

17. Ibid., 167, 183–84; Jack Sweetman, *The United States Naval Academy: An Illustrated History*, 2nd ed.

18. John Coulter, *Above and Beyond: The Story of Virginia Tech's Medal of Honor Recipients*, 106.

19. Barnes W. McCormick, Conrad F. Newberry, and Eric Jumper, *Aerospace Engineering Education during the First Century of Flight*, 363.

20. Louis E. Keefer, *The Army Specialized Training Program in World War II*, 1; Coulter, *Above and Beyond*, 106.

21. Hadley, "Association of Military Colleges and Schools," 111–12.

22. Robert Berlin, *U.S. Army World War II Corps Commanders: A Composite Biography*, 4–6. One of the West Point graduates, Gen. George Patton, also had attended Virginia Military Institute. Another also had attended Norwich for a year.

23. Gary Wade, *World War II Division Commanders*, CSI Report No. 7, 1–6. Additionally, one of the West Point graduates had attended Staunton Military Academy in Virginia (a secondary school). West Texas Military Academy is now known as Texas Military Institute. Biographies of six of the forty-five commanders were not detailed enough to determine the source of their education.

24. Walter R. Borneman, *The Admirals: Nimitz, Halsey, Leahy and King—The Five Star Admirals Who Won the War at Sea*, 11–26; Tom Bowerman, "World War II U.S. Navy Armed Guard and World War II U.S. Merchant Marine," accessed October 2, 2014, http://armed-guard.com/about-mm.html.

25. Douglas MacArthur made his comment on sports while superintendent of West Point shortly after World War I. Its accuracy was reflected in the leadership of both world wars. Douglas MacArthur, *Reminiscences*, 82.

The two nuclear bomb missions that ended World War II included four military school graduates in critical positions. Commanding the *Enola Gay*, which dropped the bomb on Hiroshima, was Col. Paul Tibbets, Western Military Academy class of 1933. The mission commander and weaponer was Capt. William S. Parsons, US Naval Academy class of 1922. Navigator on the second nuclear mission abord the *Bockscar* over Nagasaki was Maj. James Van Pelt, Virginia Tech class of 1940. Also on board *Bockscar* as weaponer was Cdr. Frederick Ashworth, US Naval Academy class of 1933.

26. St. Johns College was established in Annapolis in 1696 and functioned as a military college from 1889 to 1924.

The Korean War commander graduate of West Texas Military Academy was Gen. John B. Coulter, no relation to the author, WTMA class of 1912.

27. The Air Force Commands represented were Far East Forces, Fifth Air Force, Thirteenth Air Force, Twentieth Air Force, Far East Material Command, Far East Air Force Bomber Command (Provisional), Far East Air Force Cargo Command (Provisional).

The United States Air Force Academy was established in 1954, the last year of the Korean War; however, the first class was sworn in on July 11, 1955.

28. Samuel Rogal, *The American Pre-College Military School: A History and Comprehensive Catalog of Institutions*.

29. The Robinson Library, "Bernarr Macfadden," accessed May 10, 2009, http://www.robinsonlibrary.com/geography/recreation/physical/fitness/macfadden.htm; Tennessee State Library and Archives, "Ingram, Bowen, 1904–1980, Papers, 1856–1978," accessed June 9, 2011, http://share.tn.gov/tsla/history/manuscripts/findingaids/92–086.pdf. Among the publications Macfadden owned were the magazines *True Romances*, *True Ghost Stories*, and *True Detective*.

30. The Castle Heights junior college appeared not to have functioned long.

Also former cadets at Castle Heights Military Academy were Herbert S. Walters, Tennessee senator, and Dan Evans, founder of the Cracker Barrel Old Country Store chain of restaurants, and attendee Gen. Wesley Clark, US European NATO commander, Democratic Party nominee for President, and valedictorian of West Point.

31. Bob Dylan, "The Times They Are a-Changin'," released on the album of the same name, Warner Bros., 1964.

Chapter 12

1. Alvan C. Hadley, "The Association of Military Colleges and Schools of the United States (AMSCUS) and the Struggle for the Survival of the Military Preparatory Schools in America," PhD diss., University of Kentucky, 1999, 124, 148.

2. Wiley Lee Umphlett, *The Movies Go to College: Hollywood and the World of the College-Life Film*, 96.

3. Army and Navy Academy was established in 1910 by Thomas Davis with 13 cadets. It began in the Balboa Hotel as the San Diego Military Academy. The school expanded in new facilities on the harbor front, and in 1931–1932 it enrolled 315 cadets and included a junior college. In 1936, with the school under financial pressure, the San Diego property was sold, and the cadets moved north to Carlsbad. Through several name changes the school became the Army and Navy Academy. Enrollment in 2010 was 320.

4. John A. Coulter, *History of the Texas Military Institute Corps of Cadets*, 8.

5. Columbia Military Academy Alumni Association, "Columbia Military Academy History (1904–1979)," accessed May 27, 2011, http://cmaaa.com/history/history12. htm. Columbia Military Academy operated for seventy-four years and educated fifteen thousand cadets from thirty-eight states and eleven different countries.

6. Sarah Soloman, "Western Military Academy," *Illinois History: A Magazine for Young People*, December 1998, 15–16, Illinois Periodicals Online, accessed February 11, 2010, http://www.lib.niu.edu/1998/ihy981215.html.

7. Joseph Kraus, "The Civilian Military Colleges in the Twentieth Century: Factors Influencing Their Survival," PhD diss., University of Iowa, 1978, 429, 475. http:// www.worldcat.org/title/civilian-military-colleges-in-the-twentieth-century-factors-influencing-their-survival/oclc/27833308 (accessed Jan12, 2005).

8. Pennsylvania Military Academy was renamed Pennsylvania Military College in 1892.

9. John D. Kraus Jr., "The Civilian Military Colleges in the Twentieth Century: Factors Influencing Their Survival," PhD diss., University of Iowa, 1978, 178.

10. Ibid., 184.

11. Ibid., 186.

12. Ibid., 181.

13. Virginia Tech, "Module 09: The 1960s: Who Won? Student Protest and the Politics of Campus Dissent," Digital History Reader, accessed November 7, 2011, http:// www.dhr.history.vt.edu/modules/us/mod09_1960s/evidence.html; Steve MacGregor, "Crowd of Dissenters Rallies in Wash. D.C.," *Collegiate Times*, May 13, 1970, 3.

14. Virginia Tech, "Module 09: The 1960s," and "Evidence 1: Ed Miller, 'What caused demonstrations?' April 1970," in ibid., accessed February 8, 2009, http://www .dhr.history.vt.edu/modules/us/mod09_1960s/evidence_detail_01.html.

The author was the regimental S-2 in 1975–76 and had the responsibility for cadet recruiting. The cadets did their very best to recruit more cadets with love of the corps from their home high schools, but it was a very difficult task to find that level of commitment in the mid-1970s.

15. Tom Prunier, "Virginia Military School Tradition," *Virginia Living*, 2005 reprint, 5.

16. First Field Force, Second Field Force, 1st Cavalry Division (Air Mobile), 1st Infantry Division, 4th Infantry Division, 9th Infantry Division, 25th Infantry Division, 101st Airborne Division (Air Mobile).

17. Review of a sample of ten of the commanders of the I Marine Expeditionary Force, 1st Marine Division, and 3rd Marine Division reflected only one military school alumnus. It is interesting that this was very different than the results for the Korean War and the Gulf War.

18. US Army, "Medal of Honor Recipients," US Army Center of Military History, accessed October 16, 2014, http://www.history.army.mil/moh/index.html.

19. Peter Cookson and Caroline Persell, *Preparing for Power: America's Elite Boarding Schools*, 42.

20. Hadley, "Association of Military Colleges and Schools," 195–96.

21. Doris MacKenzie and Eugene Hebert, *Correctional Boot Camps: A Tough Intermediate Sanction*, 93; Dale Parent, *Correctional Boot Camps: Lessons from a Decade of Research*, 2, accessed September 25, 2014, https://www.ncjrs.gov/pdffiles1/nij/197018.pdf; Department of Justice, Office of Juvenile Justice and Delinquency Prevention, "Juvenile Justice Reform Initiatives in the States 1994–1996," accessed September 25, 2014, http://www.ojjdp.gov/pubs/reform/ch2_g.html.

The continued concern about military schools being equated to "boot camps" was alluded to by Santos Campos, principal of Riverside County Educational Academy, in an email to me on October 2014. In his discussion of actions taken to establish a military school culture at a new charter school, he noted, "Each year the culture improved . . . for the most part students conform or they are asked to leave we market our brand as an opportunity not a boot camp."

22. William Trousdale, *Military High Schools in America*, 193; Ronald Holmes, *How to Eradicate Hazing*, 1–8; Elizabeth J. Allan and Mary Madden, "Hazing in View: College Students at Risk," paper presented March 11, 2008, at the University of Maine, 2, accessed October 7, 2014, http://www.stophazing.org/wp-content/uploads/2014/06/hazing_in_view_web1.pdf.

23. Hank Nuwer, *The Hazing Reader*, xxv.

24. Oscar Booze was counted as one of the four hazing deaths in the text of this book.

25. Lance Betros, *Carved From Granite: West Point since 1902*, 38–39.

26. Rod Andrew, *Long Gray Lines: The Southern Military School Tradition, 1839–1915*, 72; Theodore Crackle, *West Point: A Bicentennial History*; Harry Temple, *The Bugle's Echo: The Chronology of Cadet Life at the Military College at Blacksburg, Virginia*, 1:284–86.

27. John A. Adams Jr. *Keepers of the Spirit: The Corps of Cadets at Texas A&M*, 171.

28. Trousdale, *Military High Schools in America*, 194.

29. Jane W. Dieffenbacker, *This Green and Pleasant Land: Fairfield, New York*, 204; Correspondence with Alex Herlihy, who has a photograph of female cadets at the Atlantic Air Academy, which operated from 1945 to 1949 or as late as 1953.

30. Amber Pearce, "*Faulkner v. Jones*: The Constitutionality of the Citadel's Single-Gender Admissions Policy," *New England Law Review*, Winter 1997.

31. Diane Diamond, "Good Cadets, Not Good Men: Gender Integration at the United States Military Academy at West Point and Gender Assimilation at Virginia Military Institute," PhD diss., Stony Brook University, 2005, 1.

32. Ibid., 1–2.

33. Ibid.

34. Julie M. Amstein, "*United States v. Virginia*: The Case of Coeducation at Virginia Military Institute," in *Educational Equity, Gender, and American Law: The Impact of the Law on the Lives of Women*, ed. Karen Maschke, 249–96.

35. Ronald Smothers, "Citadel Is Ordered to Admit a Woman to Its Cadet Corps," *New York Times* July 23, 1994, accessed October 1, 2014, http://www.nytimes .com/1994/07/23/us/citadel-is-ordered-to-admit-a-woman-to-its-cadet-corps.html.

36. Ibid.

37. Ibid.; Joan Biskupic, "Supreme Court Invalidates Exclusion of Women by VMI," *Washington Post*, June 27, 1996, accessed October 1, 2014, http://www.wash ingtonpost.com/wp-srv/local/longterm/library/vmi/court.htm.

38. Laura Brodie, *Breaking Out: VMI and the Coming of Women*, v, 87; Madelyn Rosenberg, "Changes Won't Reveal a Softer Side of VMI," *Roanoke Times*, August 20, 1997; Brodie, *Breaking Out*, 87; Ann D. Campbell and Francine D'Amico, "Lessons on Gender Integration from the Military Academies," in *Gender Camouflage: Women and the U.S. Military*, ed. Francis D'Amico and Laurie Weinstein, 67–80.

39. 1st Lt. Donaldson Tillar was West Point's single alumnus killed in the Gulf War. He had been a cadet at Virginia Tech prior to admission at West Point.

40. Gen. Binford Peay in 2003 became superintendent of Virginia Military Institute.

41. This trend of military school graduates was not reflected in the US Air Force's Gulf War leadership of the air component or its four major commands. Of those five commanders, only the commander of the 17th Air Division was a military school graduate: Brig. Gen. Patrick P. Caruana, US Air Force Academy class of 1963.

Chapter 13

1. Wayne K. Hoy and Cecil G. Miskel, *Educational Administration: Theory, Research, and Practice*, 8th ed., 293–95; Joseph Murphy and Catherine D. Shiffman, *Understanding and Assessing the Charter School Movement*; R. K. Blank, "Educational Effects of Magnet High Schools," in *Choice and Control in American Education*, ed. William H. Clune and John F. Witte, 2:77–110.

2. Hoy and Miskel, *Educational Administration*, 293; National Commission on Excellence in Education, "A Nation at Risk: Imperative for Educational Reform," 1983, 11, last modified October 7, 1999, http://www2.ed.gov/pubs/NatAtRisk/index.html.

3. Quote of President Reagan in L. H. Fitzharris, "An Historical Review of the National Assessment of Educational Progress from 1963 to 1993," PhD diss., University of South Carolina, 1993, 112. Murray-Wright Junior Naval Academy closed in 2007. The Magnet Military School was initially known as Toole Military School; it was later expanded to include high school.

4. Brooke Johnson, "From School Ground to Battle Ground: A Qualitative Study of a Military-Style Charter School," PhD diss., University of California, 2009, 27, 41–42.

5. "Meet Jerry Brown," 2010, accessed April 7, 2011, http://jerrybrown.org/about.

6. Associated Press, "State Board Approves Oakland Charter School," *Berkeley Daily Planet*, December 7, 2000, accessed February 9, 2011, http://www.berkeleydailyplanet .com/issue/2000–12–07/article/2540?headline=State-board-approves-Oakland-charter-school.

7. Democratic Leadership Council, "Idea of the Week: A Military Charter School," *New Democrat Daily*, December 15, 2000, 1.

8. California Senate, "California Senate Bill No. 537, California Cadet Corps," 2011, accessed January 10, 2012, http://www.leginfo.ca.gov/pub/11–12/bill/sen/ sb_0501–0550/sb_537_bill_20110713_amended_asm_v95.pdf.

9. Ann Therese Palmer, "Spit and Polish Comes to Chicago Schools," *Bloomberg Business Week*, June 25, 2001, accessed April 12, 2011, http://www.bloomberg.com/

news/articles/2001–06–24/spit-and-polish-comes-to-chicago-schools; Dawn Ruth, "Grading Paul Vallas," *New Orleans Magazine*, September 2010, http://www.my neworleans.com/New-Orleans-Magazine/September-2010/GRADING-PAUL-VAL LAS/; Paul Purpura, "New Orleans Military Maritime Academy Will Open in 2011 on Federal City Site," *New Orleans Times-Picayune*, October 20, 2010, accessed April 19, 2011, http://www.nola.com/military/index.ssf/2010/10/new_orleans_military_ maritime.html.

10. U.S. Department of Education, "Arne Duncan, U.S. Secretary of Education—Biography," 2010, accessed April 21, 2011, http://www.2.ed.gov/news/staff/bios/dun-can.html.

11. Bert L. Bershon, telephone communications with the author, November 4, 2009.

12. Ironically it was the president of the Florida Air Academy who blocked Bershon's attempt to get Sarasota Military Academy into the Association of Military Colleges and Schools of the United States. Florida Air Academy discontinued its military school requirements years later in June 2015.

13. Jerry Brown, "A Few Good Schools," *Education Next* 1, no. 2 (Summer, 2001): 1, accessed June 12, 2011, http://educationnext.org/a-few-good-schools/.

14. Ibid., 1, 2.

15. John D. Kraus, "The Civilian Military Colleges in the Twentieth Century: Factors Influencing Their Survival," PhD diss., University of Iowa, 1978, 438–39.

16. Hugh Price, "About Face!" *Educational Leadership*, May 2008, 28–34.

17. Western New York Maritime Charter School, "Mission Statement," accessed February 18, 2011, http://www.wnymcs9–12.com/domain/8; Public Safety Academy, "Our Mission Statement," accessed November 7, 2010, http://www.psasb.us/#!about-us/c1se; Philadelphia Military Academy, "About Us," accessed June 26, 2016, http://webgui.phila.k12.pa.us/schools/p/pma/about-us; Forestville Military Academy, "Our Mission as a Forestville Knight," accessed June 26, 2016, https://sites.google.com/a/pgcps.org/dixon-fma/home.

18. Michael Imber and Tyll Van Geel, *Educational Law*, 4th ed., 20; Brown, "A Few Good Schools," 1.

19. Southeast Academy High School, "Home," accessed June 26, 2016, http://www.southeastacademy.org/apps/pages/index.jsp?uREC_ID=342258&type=d&pREC_ID=749377; Public Safety Academy, "Our Mission Statement"; David Goodman, "Recruiting the Class of 2005," *Mother Jones*, January–February 2002, accessed April 7, 2010, http://www.motherjones.com/politics/2002/01/recruiting-class-2005.

20. Price, "About Face!" 28–34.

21. Payam Shahfari, "Uncle Sam Wants You Badly," Iranian.com Archives, May 1, 2007, accessed February 9, 2010, http://www.iranian.com/BTW/2007/May/Militarism/index.html (accessed Feb 9, 2010).

22. Johnson, "School Ground to Battle Ground," 16.; American Civil Liberties Union, *Soldiers of Misfortune: Abusive U.S. Military Recruitment and Failure to Protect Child Soldiers*, accessed January 4, 2011, https://www.aclu.org/files/pdfs/humanrights/crc_report_20080513.pdf.

23. Johnson, "School Ground to Battle Ground," 152.

24. Kenneth J. Saltman, "Education as Enforcement: Militarization and Corpo-ratization of Schools," *Race, Poverty, and the Environment*, Fall 2007, 2, 3, accessed November 8, 2011, http://reimaginerpe.org/node/1177; Johnson, "School Ground to Battle Ground," 60.

25. Johnson, "School Ground to Battle Ground," 25, 134, 137.

26. Ibid., 109; Saltman, "Education as Enforcement," 1.

27. Riva Enteen and Tommi A. Mecca, "JROTC Must Go Now," *San Francisco Bay Guardian*, May 14, 2008, 1, accessed November 7, 2011, http://jrotcmustgo.blogspot .com/2008/05/jrotc-must-go-now-by-riva-enteen-and.html.

28. Johnson, "School Ground to Battle Ground," 141, 151.

29. Ibid., 131.

30. ACLU, *Soldiers of Misfortune*, 2–4.

31. Ibid., 13, 15–17.

32. Adam Serwer, "No, Chicago Is Not More Dangerous Than Afghanistan," *Mother Jones*, August 15, 2012, accessed September 1, 2012, http://www.motherjones. com/mojo/2012/08/no-chicago-not-more-dangerous-afghanistan.

33. Hannah Fischer, *U.S. Military Casualty Statistics: Operation New Dawn, Operation Iraqi Freedom, and Operation Enduring Freedom*; U.S. Census Bureau, *Statistical Abstract of the United States: 2012*, 131st ed., accessed January 7, 2011, http://www .census.gov/library/publications/2011/compendia/statab/131ed.html.

34. Matthew B. Stannard, "Judge Tosses Laws Restricting Recruiters," *San Francisco Chronicle*, June 19, 2009, accessed July 4, 2016, http://www.sfgate.com/bayarea/ article/Judge-tosses-laws-restricting-recruiters-3295184.php.

35. Remi M. Hajjar, "The Public Military High School" *Armed Forces and Society*, October 2005, 49.

36. Ibid., 50; Author's personal observations of the Virginia Tech Corps of Cadets, 1997–2000, and the Texas Military Institute, 2000–2006; Johnson, "School Ground to Battle Ground," 60.

37. United Nations, Committee on the Rights of the Child, *Optional Protocol to the Convention of the Rights of the Child on Involvement of Children in Armed Conflict: Periodic Report*, January 22, 2010, 2, accessed January 19, 2011, http://www.state.gov/ documents/organization/135988.pdf.

38. Hajjar, "Public Military High School," 45; Chicago Public Schools, "Office of Military Academies and JROTC Fact Sheet," 2006, copy in possession of the author; Chicago Public Schools, "CPS Graduates on the Increase," 2007, accessed March 12, 2010, http://www.cps.edu/News/Press_releases/2007/Pages/04_24_2007_PR1.aspx.

39. Hajjar, "Public Military High School," 51; Robert L. Tarrant, "Leadership Development in Secondary Military Schools," PhD diss., Kansas State University, 1972.

40. Hajjar, "Public Military High School, 52.

41. Ibid., 56, 58.

42. Public Agenda, *Teaching Interrupted: Do Discipline Policies in Today's Public Schools Foster the Common Good?*, 2, 3, 27, accessed March 17, 2010, http://www .publicagenda.org/files/pdf/teaching_interrupted.pdf.

43. Detailed results published in John Coulter, "History of Military Schools of the United States: Origin, Rise, Decline, Resurgence, and Potential in Future Public Education," PhD diss., University of the Incarnate Word, 2013.

44. This figure of thirty-five does not include two public military schools and several charter military schools that had closed prior to 2014.

45. This does not include the unusual New Mexico Military Institute, which is state-supported but requires tuition and board and consists of both junior college and high school.

46. Gallup, "Confidence in Institutions," June 9–12, 2011, accessed July 4, 2016, http://www.gallup.com/poll/1597/Confidence-Institutions.aspx.

47. Paul T. Ringenbach, *Battling Tradition: Robert F. McDermott and Shaping the U.S. Air Force Academy*, 64. The supervisory positions that would later be taken by cadets as they progressed were manned by Air Force officers. The cadet wing commander was Beverly S. Parish, a Virginia Tech graduate.

48. All these institutions are state-supported except Norwich University and the all-female corps of cadets at Mary Baldwin College.

49. Mary Baldwin College, "History of VWIL," 2011, accessed January 17, 2012, http://www.mbc.edu/vwil/history.php.

50. The state-supported military junior colleges are Georgia Military College, New Mexico Military Institute, and Marion Military Institute in Alabama.

51. Echo Company, "The History of Echo Company: Kemper Military School," accessed May 5, 2011, http://www.echocompany.org/history/index.htm.

52. Samuel J. Rogal, *The American Pre-College Military School: A History and Comprehensive Catalog of Institutions*, 166.

53. Ibid., 153–54.

54. Mark Peterson, "Possibility is 'Very Strong' That Howe Military Academy Will Close," WNDU, May 19, 2014, accessed October 28, 2014, http://www.wndu.com/home/headlines/Possibility-is-very-strong-that-Howe-Military-Academy-will-close—259862791.html; Associated Press, "Howe Military Academy Faces Decision on Closing," *Northwest Indiana Post Tribune*, May 16, 2014, accessed October 28, 2014, http://wane.com/2014/05/16/howe-military-academy-faces-decision-on-closing/; T. Wireback, "Bailed Out by the Wrong Benefactor? Oak Ridge: Can the School Survive?" *Greensboro News-Record*, March 16, 2011, accessed March 27, 2011, http://www.greensboro.com/news/bailed-out-by-the-wrong-benefactor/article_e4476ab7–5980–52a8–8dbc-85697de35322.html.

55. This was Richmond's John Marshall High School's Corps of Cadets. Both New Mexico Military Institute and Oklahoma Military Academy were publicly supported, but the boarding nature and required parental contribution excluded them.

Chapter 14

1. Culver Military Academy (695 enrollment), a larger school, had a decrease of 52 percent, and Carlisle Military School (110 enrollment), a small military school, had a 54 percent decrease in enrollment.

2. San Francisco Chamber of Commerce, *Annual Report of the Chamber of Commerce of San Francisco*, 34.

3. The Puerto Rican military schools were Antilles Military Academy (established 1959), American Military Academy (1963), Bayamon Military Academy (1975), Caguas Military Academy (1975), Maita Luca Military Academy (1992), Christian Military Academy (2002), and North Point Military Academy (2003).

4. The four small schools that closed are Ontario Christian Military School, California, 2006–9; Low Country Military Academy, South Carolina, 2009–12; Wisconsin Air Academy, 2008–9; and Chamberlain-Hunt Military Academy, Mississippi, 1879–2014 (military since 1930).

5. The two secondary military schools that closed are East Side Cadet Academy, California, 2000–9, and Española Military Academy, Española, New Mexico, 2004–9.

6. *Anderson v. Laird*, US Court of Appeals for the District of Columbia Circuit, 466 F.2d 283 (D.C. Cir. 1972), accessed April 5, 2006, http://law.justia.com/cases/federal/appellate-courts/F2/466/283/424650/; Associated Press, "4th Circuit Won't Review Ruling Striking Down VMI Prayers," First Amendment Center, August 14, 2003; Anne Loveland, *Change and Conflict in the U.S. Army Chaplain Corps since 1945*, 192, 193. The US Naval Academy had a tradition of evening meal prayer similar to that of VMI.

7. Loveland, *Change and Conflict*, 183.

8. Ibid., 193.

9. Don Snider and Alexander Shine, *A Soldier's Morality, Religion, and Our Professional Ethic: Does the Army's Culture Facilitate Integration, Character Development, and Trust in the Professional?* 11, 13.

10. Zac Crippen, "The Sorry State of Religious Freedom at the Air Force Academy," *The Federalist*, March 14, 2014, accessed October 16, 2014, http://thefederalist.com/2014/03/14/the-sorry-state-of-religious-freedom-at-the-air-force-academy/.

11. ABC News, "U.S. Air Force Changes Its Enlistment Oath Policy to Allow Airmen to Omit 'So Help Me God,'" WJLA, September 18, 2014, accessed October 17, 2014, http://wjla.com/news/nation-world/u-s-air-force-changes-its-enlistment-oaths-policy-to-allow-airmen-to-omit-so-help-me-god—107271. The other services had made the phase optional earlier.

12. Todd Starnes, "Christian School's ROTC under Attack," Fox News Opinion, November 11, 2014, accessed November 12, 2014, http://www.foxnews.com/opinion/2014/11/11/christian-schools-rotc-under-attack/.

13. Snider and Shine, *Soldier's Morality*, 14; Tom Drake, Lt. Col., USA (Ret), Chaplain, conversations with the author 2012–13 in Camp Hill, Alabama.

14. Ivy Lamb, "Educating the Whole Person," *Contents: US Airlines*, November 2013, 68–133.

15. "Ranking College by Rhodes Scholarship," College Confidential, August 28, 2009, edited September 2012, accessed October 12, 2014, http://talk.collegeconfidential.com/college-search-selection/772208-ranking-college-by-rhodes-scholarship.html; *U.S. News & World Report Best Colleges*, 2013–2014; *The Alumni Factor: A Revolution in College Rankings*, 2013–14 ed.; Forbes, "America's Top Colleges," accessed November 12, 2014, http://www.forbes.com/sites/carolinehoward/2014/07/30/ranking-americas-top-colleges-2014/#2c7c6f2179f7.

Rhodes Scholarship awards began 1903, with the first American receiving a scholarship in 1904.

16. *Alumni Factor*, 2013–14.

17. Rachael Whitten, "Hokie Alums Prospering Post-college, Gallup Survey Shows," *Collegiate Times*, September 3, 2015, accessed January 1, 2016, http://www.collegiatetimes.com/news/hokie-alums-prospering-post-college-gallup-survey-shows/article_c86af3b4–528b-11e5-a4c9–87a88c52c046.html. Gallup Poll, *Great Jobs, Great Lives: The Gallup-Purdue Index Report—Virginia Tech*, 5, accessed January 3, 2016, http://www.vtnews.vt.edu/content/dam/vtnews_vt_edu/documents/2015–08-gallup.pdf, reports: "The Corps of Cadets, Virginia Tech's student military and citizen leader community, leads the way in well-being. More than one-quarter of cadets (28%) are thriving in all five elements, nearly three times the national average for college graduates."

18. Ralph Viggoda, "Leader of the Storm Gen. H. Norman Schwarzkopf, Familiar to the World as Commander of Operation Desert Storm, Has Been Familiar Much Longer to Some Who Share His Tie to the Philadelphia Area. He's a Graduate of Valley Forge Military Academy—The Place That He Says "Prepared Me For Life," Philly.com, February 26, 1991, accessed November 12, 2014, http://articles.philly.com/1991–02–26/news/25774609_1_commander-of-operation-desert-norman-schwarzkopf-cadets.

19. John Coulter, *History of the Texas Military Institute Corps of Cadets*, 11; Douglas MacArthur, "General Douglas MacArthur's Farewell Speech, Given to the Corps of Cadets at West Point, May 12, 1962, accessed November 12, 2014, http://www.national center.org/MacArthurFarewell.html.

Appendix B

1. Virginia Women's Institute for Leadership (VWIL) cadets who chose to attend ROTC instruction of choice at Virginia Military Institute.

2. Includes a high school program.

3. Ended association with Army JROTC in June 2016 but kept its military school requirements.

4. Closed August 2014.

5. In June 2015 the Florida Air Academy changed its name to Florida Preparatory Academy, and AJROTC was made optional. It appears that the school has ended its fifty-five years as a military school.

6. Offers a postgraduate year for additional college preparation. Ended association with Army JROTC prior to 2016 but kept its military school requirements.

7. Ended association with Army JROTC in June 2015 but kept its military school requirements.

8. The school focuses on police- and firefighter-style cadets.

9. Sometime prior to 2016 Philadelphia Military Academy at Leeds and Philadelphia Military Academy at Elverson merged into a single Philadelphia Military Academy.

10. The school has a naval theme, with focus on police and emergency response cadets. It has been adding grades annually.

11. The school has been adding grades annually.

Appendix C

1. Closed 1861–68 (1866?).

2. Closed 1863–65.

3. Both Marion Military Institute of Marion and Howard College of Birmingham claim as their origin the institute in Marion under the leadership of Col. James T. Murfee and named Howard College (1842–87). Howard College moved its operation to Birmingham, and its president remained in Marion and established Marion Military Academy.

4. Was operating as a military school in 1896.

5. Closed 1861–69, reopened 1869, closed 1874, reopened 1875.

6. Closed 1950–51.

7. Closed 1943–44?

8. Closed 1890–94.

9. Possibly open in 1910.

10. Civilian 1929–30.

11. Civilian 1903–8, 1909–18.

12. Academy of Richmond may have been a military school 1882–84.

13. Civilian 1923–26.

14. Closed 1861–65?

15. Civilian 1918, 1932.

16. Also operated as a military school from 1915 to 1932 and from 1951 to 1973.

17. Closed 1862–83.

18. May have been civilian in the 1950s.

19. Civilian 1892–1906.

20. Closed 1913–14.

21. In operation 1879.

22. In operation 1882.

23. In operation 1904.

24. Closed 1924–25.

25. Closed June 1861–April 1862 and closed April–October 1863.

26. Uniformed and armed as early as 1850.

27. May have been in operation 1860.

28. Closed 1864–67.

29. Burned 1855.

30. Closed 1861–?

31. In operation in 1915.

32. Closed 1942–44.

33. Closed 1861–65; may have reopened.

34. Civilian 1910–17.

35. Closed 1826–28; civilian 1854–58; remilitarized 1859; closed 1863–66.

36. Closed 1861?–67.

37. Closed 1856–61.

38. Closed 1896–99.

39. Closed 1862–66; may have closed in 1890s.

40. Closed 1943–52.

41. Remilitarized for a short time about 1932?

42. Possibly in operation in 1922.

43. Closed 1876–77.

44. May have closed in 1880 and reopened before 1890.

45. May have closed 1883 and reopened.

46. May have been in operation in 1936.

47. Closed 1861–62.

48. Closed 1861–66.

49. Closed 1861–73; reopened by Col. John P. Thomas, Citadel class of 1851.

50. Closed 1861–65.

51. Was known to have been in operation 1862–65.

52. Closed 1873–76.

53. Known to have been operational 1949–52.

54. Pennsylvania Military Preparatory School became a distinct preparatory department of Pennsylvania Military Academy (College) in 1916. In 1931 the faculty of the college and the prep school became separate. The school ended when the college took over its classroom and barracks space.

55. Closed 1914–19.

56. Closed 1865–82.

57. Closed 1861–66?

58. Closed 1930, but reopened by 1931.

59. Burned 1920–21.

60. May have reopened and been in operation in 1896.

61. Closed 1860–62.

62. Closed 1864–67; civilian 1867–68.

63. Closed 1861–65?

64. Operated for two years.

65. In operation 1920.

66. May have been in operation as late as 1917.

67. May have transitioned to military format as early as 1910.

68. Closed 1943–47.

69. May have transitioned back to civilian and then transitioned back to military in 1905.

70. May have been in operation in 1913.

71. Closed 1931–39.

72. Transitioned to female civilian, 1857–59; reopened as military, 1859.

73. Closed 1861–65.

74. Virginia Presbyterian School was operated as a nonmilitary school from 1933 to 1936.

75. Unclear if it was operating as a military school in 1838.

76. Became civilian for a few years about 1932.

77. Closed 1861–65.

78. Closed 1861–65.

79. Closed 1841–47.

80. Civilian 1907–10.

81. Civilian 1922–29.

References

ABC News, "U.S. Air Force Changes Its Enlistment Oath Policy to Allow Airmen to Omit 'So Help Me God.'" WJLA, September 18, 2014. http://wjla.com/news/nation-world/u-s-air-force-changes-its-enlistment-oaths-policy-to-allow-airmen-to-omit-so-help-me-god—107271.

Adams, John. *Keepers of the Spirit: The Corps of Cadets at Texas A&M*. College Station: Texas A&M University Press, 2001.

Adams, Oscar F. *Some Famous American Schools*. Boston, MA: Dana Estes & Co., 1903.

Aguirre, Adalberto, and Brooke Johnson. "Military Youth in Public Education: Observations from a Military-Style Charter School." *Social Justice* 15 (Fall 2005): 148–59.

Alachua County, Florida. "East Florida Seminary—Battalion." Alachua County Library District Historical Collection. 2002. http://heritage.acld.lib.fl.us/1051–1100/1082.html.

Alexander, Charles. "St. Emma's Industrial and Agricultural College." *Alexander's Magazine* 4 (1907): 100–108.

Alexandria Library Special Collection. Alexandria City Records. 1853. Box 019II.

Allan, Elizabeth J., and Mary Madden. *Hazing in View: College Students at Risk*. Paper presented March 11, 2008, at the University of Maine. Stophazing.org. http://www.stophazing.org/wp-content/uploads/2014/06/hazing_in_view_web1.pdf.

Allardice, Bruce S. *Confederate Colonels: A Biographical Register*. Columbia: University of Missouri Press, 2008.

———. Military schools research files. In possession of Bruce S. Allardice, Professor of History, South Suburban College, South Holland, IL. Reviewed 2010.

———. "West Points of the Confederacy: Southern Military Schools and the Confederate Army." *Civil War History* 43, no. 4 (December 22, 1997): 310–32.

All Biographies. "Colonel Joseph Rogers, U.S.V." 2011. http://www.all-biographies.com/soldiers/joseph_sumner_rogers.htm.Allen Academy. "Allen Academy History." 2010. http://www.allenacademy.org/about/history.

"Alton Western Military Academy HS 'Raiders.'" Illinois High School Glory Days. 2009. http://www.illinoishsglorydays.com/id528.html.

Altstetter, Mabel, and Gladys Watson. "Western Military Institute: 1847–1861." *Filson Club History Quarterly* 10 (April 1936): 100–115.

The Alumni Factor: A Revolution in College Rankings, 2013–14 ed. Avondale Estates, GA: The Alumni Factor, 2013.

Alvesson, Mats. *Understanding Organizational Culture.* Thousand Oaks, CA: SAGE Publications, 2002.

Ambrose, Stephen. *Duty, Honor, Country: A History of West Point.* Baltimore: The Johns Hopkins University Press, 1999.

American Civil Liberties Union. *Soldiers of Misfortune: Abusive U.S. Military Recruitment and Failure to Protect Child Soldiers.* New York: ACLU, 2008. https://www.aclu.org/files/pdfs/humanrights/crc_report_20080513.pdf.

American Institute of Mining, Metallurgical, and Petroleum Engineers. "Obituary of Charles E. Emery." *Transactions,* 1900, xxviii.

American Jewish Historical Society. "Biographical Note: Uriah P. Levy (1792–1862)." http://findingaids.cjh.org/?pID=109192#a2.

American Literary, Scientific, and Military Academy. *Catalogue of the Officers and Cadets, Trustees and the Prospectus and Internal Regulations of the American Literary, Scientific, and Military Academy.* Middletown: Starr & Niles, 1826.

Ames, G. W. *Gaston Griffin, a Country Banker.* Port Jervis, NY: Gazette Publishing Co., 1900.

Amstein, Julie M. "*United States v. Virginia*: The Case of Coeducation at Virginia Military Institute." In *Educational Equity, Gender, and American Law: The Impact of the Law on the Lives of Women,* edited by Karen Maschke, 249. New York: Garland Publishing, 1997.

Anderson v. Laird, US Court of Appeals for the District of Columbia Circuit, 466 F.2d 283 (D.C. Cir. 1972). http://law.justia.com/cases/federal/appellate-courts/F2/466/283/424650/.

Andrew, Rod, Jr. *Long Gray Lines: The Southern Military School Tradition, 1839–1915.* Chapel Hill, NC: University of North Carolina Press, 2001.

———. "North Carolina Military Institute." NCPedia. http://ncpedia.org/north-carolina-military-institute.

Andrews, John R. "Valuable to the Citizen as to the Soldier: Republicanism and Militarism in Southern Military Schools, 1839–1915." PhD diss., University of Georgia, Athens, 1997.

Appleton's Annual Cyclopaedia and Register of Important Events of the Year 1886. New York: D. Appleton and Company, 1887.

Armory, Thomas C. *The Life of Admiral Sir Isaac Coffin, Baronet: His English and American Ancestors.* Boston: Cupples, Upham & Company, 1886.

Army and Navy Academy. "History and Traditions." http://www.armyandnavy academy.org/about-ana-military-academy/school-history-traditions/.

Associated Press. "Charter Military School to Open in St. Paul." *Berkeley Daily Planet,* August 30, 2004.

———. "4th Circuit Won't Review Ruling Striking Down VMI Prayers." First Amendment Center, August 14, 2003.

———. "Howe Military Academy Faces Decision on Closing." *Northwest Indiana Post Tribune,* May 16, 2014. http://wane.com/2014/05/16/howe-military-academy-faces-decision-on-closing/.

———. "State Board Approves Oakland Charter School." *Berkeley Daily Planet,* Dec. 7, 2000. http://www.berkeleydailyplanet.com/issue/2000–12–07/article/2540?headline=State-board-approves-Oakland-charter-school.

Association of Military Colleges and Schools of the United States (AMCSUS). *Association of Military Colleges and Schools of the United States Constitution.* Springfield, VA: AMCSUS, 2000.

Association of Public and Land-Grant Universities. *The Land Grant Tradition.* Washington, DC: Association of Public and Land-Grant Universities, 2012. http://www.aplu.org/library/the-land-grant-tradition/file.

Augusta Military Academy Alumni Association. History of Augusta Military Academy. 1999. http://amaalumni.org/history_1.htm.

Augustowski, Loeman. "Echos of Honor: Peekskill Military Academy." *Cortlandt Magazine,* June 2001.

Bailey, A. M. "The Glorious Hampton Cadets." *Southern Workman and Hampton School Record* 21, no. 5 (May 1892): 77.

Baker, Dean. "The Partridge Connection: Alden Partridge and the Southern Military Education." PhD diss., University of North Carolina, 1986.

Baker, Gary. *Cadets in Gray: The Story of the Cadets of the South Carolina Military Academy and the Cadet Rangers in the Civil War.* Columbia: Palmetto Bookworks, 1989.

Baker, Joseph E., ed. *Past and Present of Alameda County, California,* vol. 2. Chicago: S. J. Clarke Company, 1914.

Ballard, Michael B. *Maroon and White: Mississippi State University, 1878–2003.* Jackson: University Press of Mississippi, 2008.

Barbes, Robert, and John McGrain, eds. "Timothy Hall." *HistoryTrails* 11, no. 3 (Spring 1977).

Barnard, Henry. *Military Schools and Courses of Instruction in the Science and Art of War in France, Prussia, Austria, Sweden, Switzerland, Sardinia, England, and the United States, Drawn from Recent Official Reports and Documents.* 1872. Reprint, The West Point Military Library; New York: Greenwood Press, 1969.

Bastrop County [Texas] Museum. "A Survey of Texas Historical Markers in Bastrop County." 2011. Research files of Bruce Allardice.

Bastrop Military Academy. *Annual Catalogue of Bastrop Military Academy.* Galveston, TX: Texas Christian Advocate, 1858.

Battey, George. *The History of Rome and Floyd County, State of Georgia, United States of America.* Floyd County: Webb, 1922.

Bazzel, R. J. "Organizational Citizenship Behavior and Student Achievement As Related to School District Size in the Public Schools of Salem County, New Jersey." PhD diss., Wilmington College, Wilmington, OH, 2007.

Beadie, Nancy. *Education and the Creation of Capital in the Early American Republic.* New York: Cambridge University Press, 2010.

Benjamin, Park. *The United States Naval Academy.* New York: G. P. Putnam's Sons, 1900.

Berlin, Robert. *U.S. Army World War II Corps Commanders: A Composite Biography.* Leavenworth: US Army Command and General Staff College, 1989.

Betros, Lance. *Carved from Granite: West Point since 1902.* College Station: Texas A&M University Press, 2012.

Betts, C. "Betts Family and Their Academy." *Stamford Historian* 1, no. 2 (1957): 163–68.

"Betts Academy Burned: Destruction of Old Institution of Stamford, Conn." *The Summary* 36, no. 4 (January 25, 1908): 4.

Biskupic, Joan. "Supreme Court Invalidates Exclusion of Women by VMI." *Washington Post*, June 27, 1996. http://www.washingtonpost.com/wp-srv/local/long term/library/vmi/court.htm.

Blackwell, James D. *The Poetical Works of James DeRuyter Blackwell*, vol. 1. New York: E. J. Hale and Son, 1879.

Blake, William P. *History of the Town of Hamden, Connecticut, with an Account of the Centennial Celebration, June 15, 1886.* New Haven: Price, Lee & Co., 1888. https://archive.org/details/historytownhamd00blakgoog.

Blank, R. K. "Educational Effects of Magnet High Schools." In *Choice and Control in American Education*, vol. 2, edited by William H. Clune and John F. Witte, 77–110. Bristol, PA: Falmer Press, 1990.

Borneman, Walter R. *The Admirals: Nimitz, Halsey, Leahy and King—The Five Star Admirals Who Won the War at Sea.* New York: Little, Brown and Company, 2012.

Bowen, B. *The Great Writings in Management and Organizational Behavior.* New York: Random House, 1987.

Bowerman, Tom. "World War II U.S. Navy Armed Guard and World War II U.S. Merchant Marine." http://armed-guard.com/about-mm.html.

Boyington, Edward. *West Point and Its Military Importance during the American Revolution, and the Origin and Progress of the United States Military Academy.* New York: D. Van Nostrand, 1871.

Bradberry, Bill. "Black Menagerie: Dreams of Future and Sins of Past. *Niagara Falls Report.* October 16, 2001. http://www.niagarafallsreporter.com/menag erie22.html.

Bradford, Bob. "The Roller School: A Loving History of Augusta Military Academy." Augusta Military Academy *Recall,* 1917.

Brodie, Laura. *Breaking Out: VMI and the Coming of Women.* New York: Random House, 2001.

Brown, Dick. "St. Johns' College," The Encyclopedia of Arkansas History and Culture. 2009. http://encyclopediaofarkansas.net/encyclopedia/entry-detail .aspx?entryID=3584.

Brown, James H., ed. *Lamb's Biographical Dictionary of the United States*, vol. 3. Boston: Lee and Shepard Publishers, 1900.

Brown, Jerry. "A Few Good Schools." *Education Next.* 1, no. 2 (Summer 2001): 1. http://educationnext.org/a-few-good-schools/.

Brown, William. "The Hampton Military Academy of Virginia." *North South Trader's Civil War* 24, no. 2 (1997): 148–58.

———. "The Norfolk Academy of Virginia." *North South Trader's Civil War* 20, no. 5 (1993): 33–36.

Buckley, William H. *The Citadel and the South Carolina Corps of Cadets.* Charleston: Arcadia Publishing, 2004.

Budahn, P. J. *What to Expect in the Military: A Practical Guide for Young People, Parents, and Counselors.* Westport, CT: Greenwood Publishing Group, 2000.

Bunker Hill Military Academy. Bunker Hill, IL: Bunker Hill Military Academy, 1913.

"Bunker Hill Military Academy." Illinois High School Glory Days. 2009. http://www.illinoishsglorydays.com/id811.html.

Burhoe, J Scott. *The United States Coast Guard Academy: A Brief History.* New London: US Coast Guard Academy, 2011. https://www.uscg.mil/history/uscghist/uscga_history_final.pdf.

Burke, James. "Military Culture." In *Encyclopedia of Violence, Peace, and Conflict,* vol. 3, edited by Lester Kurtz and Jennifer Turpin, 447–62. San Diego: Academic Press, 1999.

Burleson, Allan L. "A Paper Read at the Midwinter Fair, San Antonio, Texas, on Education Day," February 3, 1896. Archives of Texas Military Institute, San Antonio.

Cabot, Mary R. *Annals of Brattleboro, 1681–1895,* vol. 3. Brattleboro, VT: E. L. Hildreth & Co., 1922.

Cahn, C. "Town and Academy in Suffield: An Appreciation of Our Shared History." Email from Headmaster, Suffield Academy, Suffield, CT, 2009.

California Senate. "California Senate Bill No. 537, California Cadet Corps," 2011. http://www.leginfo.ca.gov/pub/11–12/bill/sen/sb_0501–0550/sb_537_bill_20110713_amended_asm_v95.pdf.

Calkins, E. E. "The Revolt of Youth" *The Rotarian,* February 1932, 52.

Camden Military Academy. "History of the School." 2010. http://www.camdenmilitary.com/history.php.

Campbell, Ann D., and Francine D'Amico. "Lessons on Gender Integration from the Military Academies." In *Gender Camouflage: Women and the U.S. Military,* edited by Francine J. D'Amico and Laurie L. Weinstein, 67–80. New York: New York University Press, 1999.

Campbell, Thomas. *Academy on the James: The Confederate Naval School.* Shippensburg, PA: Burd Street Press, 1998.

Cardozier, V. R. *Colleges and Universities in World War II.* Westport, CT: Greenwood Publishing Group, 1993.

Carnegie Foundation. *Carnegie Hero Fund Commission.* Pittsburg: Carnegie Foundation, 1919.

Carper, James, and Thomas Hunt. *The Praeger Handbook of Religion and Education in the United States,* vol. 1. Westport, CT: Praeger Publishers, 2009.

Cartmell, Thomas K. *Shenandoah Valley Pioneers and Their Descendants: A History of Frederick Country, Virginia.* Winchester, VA: Rudy Press Corp, 1909.

Caswell, Render R. "The History of Bowdon College" Master's thesis, University of Georgia, Athens, 1952.

Catholic Dioceses of Utah Archives. Correspondence, March 22, 1888–May 26, 1919.

Center, Clark E. "The Burning of the University of Alabama." *Alabama Heritage* 16 (Spring 1990): 30–45.

Chapman, B. *Brief History of Lebanon*. Lebanon, NH: Lebanon Historical Society, 1993.

Chapultepec Inc. Almanac of Theodore Roosevelt. http://theodore-roosevelt.com/treditorials.html.

Chenevert, T. L. "History" [St. Emma Military Academy]. Copy in possession of the author.

Chicago Military Academy. "Chicago Military Academy at Bronzeville." http://www.chicagomilitaryacademy.org/.

Chicago Public Schools. "Chicago Public Schools Website." http://www.cps.edu/Schools/High_schools/Pages/Highschools.aspx.

———. "CPS Graduates on the Increase." 2007. http://www.cps.edu/News/Press_releases/2007/Pages/04_24_2007_PR1.aspx.———. "Office of Military Academies and JROTC Fact Sheet." 2006. Copy in possession of the author.

Churchman Magazine. *Church Almanac and Year Book*. New York: Potts & Publishers, 1890, 1891, 1897, 1920.

The Citadel. "Citadel History." http://www.citadel.edu/citadel-history.

City of Chicago. "Mayor Daley Announces $2.1 Million Grant to Create First Naval Academy." Mayor's Office Press Release, July 19, 2004. Copy in possession of the author.

———. "Office of Military Academies and JROTC Fact Sheet," November 2005. Copy in possession of the author.

City of Oakland. "State Board Votes Unanimously for Oakland Military Charter School." Office of the Major, Press Release, December 6, 2000.

Clark, W. *The Burbank Military Academy*. 2009. http://wesclark.com/burbank/bma.html.

Clark, Walter, ed. *Histories of the Several Regiments and Battalions from North Carolina in the Great War, 1861–65*, vol. 5. Goldsboro: Nash Brothers Book and Job Printers, 1901.

Coates, Earl, and Michael McAfee. *Don Troiani's Regiments & Uniforms of the Civil War*. Mechanicsburg, PA: Stackpole Books, 2002.

Coffin, Selden J. *The Men of Lafayette, 1826–1893: Lafayette College, Its History, Its Men, Their Record*. Easton, PA: George West, 1891.

Cole, David, and Robert Graetz. "The Garnet and Gray: West Florida Seminary in the Civil War." *The United Daughters of the Confederacy Magazine*, January 1989, 2–6.

Collins, Robert. *General James G. Blunt: Tarnished Glory*. Gretna, LA: Pelican Publishing Company, 2005.

Columbia Military Academy Alumni Association. "Columbia Military Academy History (1904–1979)." http://cmaaa.com/history/history12.htm.

Committee on Diocesan Schools. *Journal of the Thirty-Fifth Annual Convention of the Protestant Episcopal Church*. Holly Springs, MS: Diocese of Mississippi, 1861.

"Committee on Militarism in Education, 1925–1940." DG009, Swarthmore College Peace Collection, Swarthmore College Library. https://www.swarthmore.edu/library/peace/DG001–025/dg009cme.html#bioghist.

Condon, Peter. "Knownothingism." *The Catholic Encyclopedia*, vol. 8. New York: Robert Appleton Company, 1910. http://www.newadvent.org/cathen/08677a.htm.

Conrad, Howard L., ed. *Encyclopedia of the History of Missouri: A Compendium of History and Biography for Ready Reference*, vol. 6. New York: Southern History Company, 1901.

Conrad, James L. *Rebel Reefers: The Organization and Midshipmen of the Confederate States Naval Academy*. Cambridge, MA: Da Capo Press, 2003.

———. *The Young Lions: Confederate Cadets at War*. Columbia: University of South Carolina Press, 1997.

Cookson, Peter, and Caroline Persell. *Preparing for Power: America's Elite Boarding Schools*. New York: Basic Books, 1985.

Cooley, L. C. "The All-Around Training of the Hampton Cadet." *The Southern Workman* 30, no. 1 (January 1901): 493–500.

Coon, Charles L. *North Carolina Schools and Academies, 1790–1840*. Raleigh, NC: Edwards & Beoughton Printing, 1915.

Cooper, Forest L. "Molders of Men: The Tupelo Military Institute Introduced a New Form of Education to North Mississippi in the Early 1900s." *Mississippi Magazine*, January–February 2007.

Cormack, Michael J. *Ideology and Cinematography in Hollywood, 1930–39*. New York: St. Martin's Press, 1994.

"Cosmopolitan Educational Guide." *Cosmopolitan Magazine*, July 1922.

Coulter, John A. *Above and Beyond: The Story of Virginia Tech's Medal of Honor Recipients*. Blacksburg: Virginia Tech Corps of Cadets Alumni, 2000.

———. *Bugle Notes: Cadet Handbook*. San Antonio: Texas Military Institute, 2006.

———. "History of Military Schools of the United States: Origin, Rise, Decline, Resurgence, and Potential in Future Public Education," PhD diss., University of the Incarnate Word, 2013.

———. *History of the Texas Military Institute Corps of Cadets*. San Antonio: Texas Military Institute, 2002.

———. "Military Colleges of the South." Austin Peay University. Clarksville, TN, 1984. Paper in possession of author.

Couper, William. *One Hundred Years at VMI*, vol. 1. Richmond, VA: Garrett & Massie, 1929.

Cowley, Robert, and Thomas Guinzburg. *West Point: Two Centuries of Honor and Tradition*. New York: Warner Books, 2002.

Cox, Dale. *The Battle of Natural Bridge, Florida: The Confederate Defense of Tallahassee*. Fort Smith: Dale Cox Publisher, 2007.

Crackel, Theodore J. *West Point: A Bicentennial History*. Lawrence: University Press of Kansas, 2002.

Creswell, John W. *Educational Research: Planning, Conducting, and Evaluating Quantitative and Qualitative Research*. Upper Saddle River, NJ: Pearson Merrill Prentice Hall, 2005.

Crippen, Zac. "The Sorry State of Religious Freedom at the Air Force Academy." *The Federalist*, March 14, 2014. http://thefederalist.com/2014/03/14/the-sorry-state-of-religious-freedom-at-the-air-force-academy/

Crocker, Henry. *History of the Baptists in Vermont.* Bellows Falls, VT: P. H. Goble Press, 1913.

Cruikshank, Jeffery L., & Cloe G. Kline. *In Peace and War: A History of the U.S. Merchant Marine Academy at Kings Point.* Hoboken: John Wiley & Sons, 2008.

Cullum, George W. *Biographical Register of the Officers and Graduates of the U.S. Military Academy,* vol. 1. New York: Houghton, Mifflin and Company, 1891.

Culver Educational Foundation, "Culver Academies." http://www.culver.org/.

Cummings, Thomas G., ed. *Handbook of Organization Development.* Thousand Oaks, CA: SAGE Publications, 2008.

Cunliffe, Marcus. *Soldiers and Civilians: The Martial Spirit in America, 1775–1965.* New York: The Free Press, 1973.

Curley, Stephen J. *Aggies by the Sea: Texas A&M University at Galveston.* College Station: Texas A&M University Press, 2005.

———. *The Ship That Would Not Die: USS Queens, SS Excambion, and USTS Texas Clipper.* College Station: Texas A&M University Press, 2011.

Cutter, William R. *Historical Homes and Places and Genealogical and Personal Memoirs Relating to the Families of Middlesex County, Massachusetts,* vol. 4. Boston: Goodspeed's Book Shop, 1919.

Daley R. M. "Boot School." *Education Matters,* Fall 2001, 4.

Dalin, Per; Hans-Gunter Rolff, and Bab Kleekamp. *Changing the School Culture.* New York: Cassell, 1993.

Dans, Peter. *LaSalle Military Academy: The Life and Death of a Catholic Military School.* Bedford, MA: Reynolds Dewalt, 2013.

Davies, Richard G. "Of Arms and Men: History of Culver Military Academy, 1894–1945," PhD diss., University of Indiana, 1983.

Davis, B. "Unit History." Manuscript addressing Louisiana State University, 1999. In possession of author.

Davis, Kristy. "Stratford Military Academy." *Cherry Hill (NJ) Courier Post,* February 25, 2009. Reprinted in "Stratford Military Academy," Borough of Stratford, http://www.stratfordnj.org/military-academy.html.

Davis, William C. *The Battle of New Market.* Baton Rouge: Louisiana State University Press, 1983.

Dawson, Joseph G. "Texas Maritime Academy." *Handbook of Texas Online.* http://www.tshaonline.org/handbook/online/articles/kct22.

Deal, Terrence E., and Kent D. Peterson. *Shaping School Culture: The Heart of Leadership.* San Francisco: Jossey-Bass Publishers, 1999.

De Bow, J. D. B. "Education of Seamen at the South." *De Bow's Review and Industrial Resources, Statistics, etc.,* 26 (1859).

De la Pedraja Toman, René. *A Historical Dictionary of the U.S. Merchant Marine and Shipping Iindustry: Since the Introduction of Steam.* Westport, CT: Greenwood, 1994.

DeLassus, D. "Military Programs: Incomplete Game Data." College Football Data Warehouse. http://cfbdatawarehouse.com/data/incomplete_data/incomplete_records_military.php.

DelVecchio, Valentine. *Cadet Gray: Your Guide to Military Schools, Military Colleges and Cadet Programs,* 2nd ed. Santa Barbara, CA: Reference Desk Books, 1997.

Democratic Leadership Council. "Idea of the Week: A Military Charter School." *New Democrat Daily*, December 15, 2000, 1.

Dennis, Michael. *Lessons in Progress: State Universities and Progressivism in the New South, 1880–1920*. Urbana: University of Illinois Press, 2001.

Diamond, Diane. "Good Cadets, Not Good Men: Gender Integration at the United States Military Academy at West Point and Gender Assimilation at Virginia Military Institute." PhD diss., Stony Brook University, 2005.

Dickinson, Ellen E. "The Training School-ship *Minnesota*." In *How to Learn and Earn; or, Half-Hours in Some Helpful Schools*, 346. Boston: D. Lothrop & Co., 1884.

Dieffenbacker, Jane W. *This Green and Pleasant Land: Fairfield, New York*. Fairfield: Steffen Publishing, 1996.

Dietz, Suzanne S., and Amy L. Freiermuth. *Lewiston: Then and Now*. Charleston: Arcadia Publishing, 2010.

Dobyns, A. W. "The Military Feature at the Fanwood School." *American Annals of the Deaf* 53 (1908).

Dood, William G. "Early Education in Tallahassee and the West Florida Seminary, Now Florida State University." *Florida Historical Quarterly* 27, no. 2 (October 1948): 157–80.

Dooley, Edwin L., Jr. "Gilt Buttons and the Collegiate Way: Francis H. Smith as Antebellum Schoolmaster." *Virginia Cavalcade* 36 (Summer 1986): 30–39.

Dunivin, Karen. "Military Culture: Change and Continuity." *Armed Forces & Society* 20, no. 4 (Summer 1994): 531–47.

Durston, Harry C. "History of the Manlius School." Manlius Pebble Hill School. 1966. http://www.mph.net/about/history_man.cfm.

Eastern Digital Resources. "Cobb's Legion Infantry, Company B, Bowden Volunteers." http://www.researchonline.net/gacw/rosters/cobbib.htm.

Echo Company. "The History of Echo Company, Kemper Military School." http://www.echocompany.org/history/index.htm.

École Polytechnique. "History and Heritage." http://www.polytechnique.edu/home/about-ecole-polytechnique/history-and-heritage/. Copy in possession of the author.

"Editorial: Game Changer." *Greensboro News & Record*, March 31, 2011.

Educational Supplement—Part III. *New York Times*, August 27, 2004.

"Education in the Navy." *Southern Literary Messenger* 16, no. 9 (September 1850): 521–29.

Elliot, N., ed. *Patterson's American Education*, vols. 59, 60. Mount Prospect, IL: Educational Directories Inc., 1962, 1963.

Ellis, William. *Norwich University, 1819–1911: Her History, Her Graduates, Her Roll of Honor*. Burlington, VT: Capital City Press, 1911.

Engs, Robert F. *Educating the Disfranchised and Disinherited: Samuel Chapman Armstrong and the Hampton Institute, 1839–1893*. Knoxville: University of Tennessee Press, 1999.

Enteen, Riva, and Tommi A. Mecca. "JROTC Must Go Now." *San Francisco Bay Guardian*, May 14, 2008, 1. http://jrotcmustgo.blogspot.com/2008/05/jrotc-must-go-now-by-riva-enteen-and.html.

Episcopal Church, Diocese of Connecticut. *Journal of the Annual Convention of the Protestant Episcopal Church, Diocese of Connecticut*, 1841–1850. New Haven, CT: Stanley & Chapin Printer, 1841–1850.

Episcopal Church, Diocese of Mississippi. *Convention, Episcopal Church, Diocese of Mississippi, Council—1855*. Natchez, MS: Natchez Daily Courier Book and Job, 1855.

Episcopal Church of St. Matthew. "History." http://www.episcopalstmatthew.org/history.html.

Evans, C., ed. *American College and Public School Directory*. Saint Louis, MO: C. H. Evans & Co., 1890 and 1893.

———. *Educational Year-Book and University Catalogue*. Saint Louis, MO: C. H. Evans & Co., 1883.

Farley, Douglas, and Ann Marie Linnabery. "DeVeaux School." Niagara County Historical Society's Bicentennial Moments. http://www.niagara2008.com/history149.html.

Favro, Tony. "American Mayors Welcome Military Schools into Poorer Neighborhoods." CityMayors, Education 23, June 3, 2008. http://www.citymayors.com/education/us-military-schools.html.

Filbert, Brant G., and Allen G. Kaufman. *Naval Law: Justice and Procedure in Sea Services*, 3rd ed. Annapolis, MD: US Naval Institute Press, 1998.

Finn, Chester, and Bruno Manno. *Charter Schools in Action: Renewing Public Education*. Princeton: Princeton University Press, 2000.

Firestone, William, and Karen Louis. "Schools as Cultures." In *Handbook of Research on Educational Administration*, 2nd ed., edited by Joseph Murphy and Karen Louis, 297–323. San Francisco: Jossey-Bass, 1999.

Fischer, Hannah. *U.S. Military Casualty Statistics: Operation New Dawn, Operation Iraqi Freedom, and Operation Enduring Freedom*. Washington, DC: Congressional Research Service, 2010.

Fitzharris, L. H. "An Historical Review of the National Assessment of Educational Progress from 1963 to 1993." PhD diss., University of South Carolina, 1993, 112..

Fleischman, Steve, and Jessica B. Heppen. "Improving Low Performance High Schools: Searching for Evidence of Promise." *The Future of Children* 19, no. 1 (Spring 2009): 105–33.

Florida Military Academy Alumni. "Cadet Alumni Pages." 2009. Paper in possession of the author.

Forbes. "America's Top Colleges." 2014. http://www.forbes.com/sites/caroline-howard/2014/07/30/ranking-americas-top-colleges-2014/#2c7c6f2179f7.

Ford, Nancy. *The Great War and America: Civil-Military Relations during World War I*. Westport, CT: Praeger Security International, 2008.

Forestville Military Academy, "Mission Statement." Accessed June 26, 2016. https://sites.google.com/a/pgcps.org/dixon-fma/home.

Fork Union Military Academy. "A Little about Fork Union Military Academy." http://www.forkunion.com/military-school-info/about-fork-union-military-academy.html.

Forsberg, C. "Western Military Academy." 2005. Paper in possession of the author.

Forshey, Caleb G. "Military Education in Civil Institution." *Southwestern Historical Quarterly* 48, no. 23 (October 1944). (Original address delivered July 4, 1855.)

FrancisEmma Inc. "History and Culture." http://www.francisemma.org/default .asp?Parent=19&Child=19.

Franklin, John H. *The Militant South*. Cambridge: Harvard University Press, 1956. Reprint, Urbana and Chicago: University of Illinois Press, 2002.

Franklin, Robert H. "History of Education in Gloucester County, VA." In *Fifteenth Annual Report of the Superintendent of Public Instruction for the Year Ending July 31, 1885*, 76–85. Richmond: State of Virginia, 1885.

Frederick Military Academy Alumni. "Historical Timeline of Significant Events." 1998. Paper in possession of the author.

Fredriksen, John. "Williams, Jonathan." In *The Encyclopedia of the War of 1812: A Political, Social, and Military History*, vol. 1, edited by Tucker Spencer, 783–84. Santa Barbara: ABC-CLIO, 2012.

Friends of PMC. "Eddystone Disaster." Legacy of PMC. http://pennsylvaniamili tarycollege.org/category/legends/legacy-of-pmc/.

Gallagher, Gary W., and Alan T. Nolan, eds. *The Myth of the Lost Cause and Civil War History*. Bloomington: Indiana University Press, 2000.

Gallup. "Confidence in Institutions [June 9–12, 2011]." http://www.gallup.com/ poll/1597/Confidence-Institutions.aspx.

———. *Great Jobs, Great Lives: The Gallup-Purdue Index Report—Virginia Tech*. Washington, DC: Gallup, 2015. http://vtnews.vt.edu/content/dam/vtnews_ vt_edu/documents/2015–08-gallup.pdf.

Ganfield, Jerry. "Minnesota Academy through Pillsbury Military Academy, 1877– 1957," Owatonna, MN, Steele County Historical Society, typescript, 2001. Copy in possession of the author.

Gay, Lorrie R., Geoff Mills, and Peter W. Airasian. *Educational Research: Competencies for Analysis and Applications*. Saddle River, NJ: Pearson Education, 2003.

Gaylord, W. P. *History of Floyd County, Iowa*, vol. 2. Chicago: Inter-State Publishing Company, 1882.

Georgia Department of Education. *Annual Report—Georgia Department of Education*. Atlanta: Franklin Printing and Publishing Company, 1906.

Gignilliat, Leigh. *Arms and the Boy: Military Training in Schools and Colleges*. Indianapolis: Bobbs-Merrill, 1916.

Goodman, David. "Recruiting the Class of 2005." *Mother Jones*, January–February, 2002. http://motherjones.com/politics/2002/01/recruiting-class-2005.

Graczyl, Michael. "Texas A&M Cadets to Be Led by First Black Commander." *Dallas Morning News*, April 12, 2012. http://www.dallasnews.com/news/state/ headlines/20120412-texas-am-cadets-to-be-led-by-first-black-commander.ece.

Grapes, Jesse. "Message from the Headmaster." Benedictine High School. http:// www.benedictinecollegeprep.org/RelId/636273/ISvars/default/Message_ from_the_Headmaster.htm.

Grayfow, G. *Boys and Men: Hundred Year History of Northwestern Military and Naval Academy*. Delafield, WI: St. John's Northwestern Military Academy, 1891.

Green, Jennifer. *Military Education and the Emerging Middle Class in the Old South*. Cambridge: Cambridge University Press, 2008.

Grotelveschen, Mark E. *The AEF Way of War: The American Army and Combat in World War I*. New York: Cambridge University Press, 2007.

Guinn, Jerry. "A Short History of Roosevelt Military Academy." http://www.rma roughriders.org/history.htm.

Hadden, James. *A History of Uniontown: The County Seat of Fayette County, Pennsylvania*. Uniontown: New Werner Co., 1913.

Hadley, Alvan. "The Association of Military Colleges and Schools of the United States (AMSCUS) and the Struggle for the Survival of the Military Preparatory Schools in America." PhD diss., University of Kentucky, 1999.

Hadley, J. "Greenbrier Military School: 161 Years Pride in Past—Confidence in the Future." 2008. Paper in possession of the author.

Hajjar, Remi. "The Public Military High School." *Armed Forces and Society*, October 2005, 44–62.

Halstead, Murat. *Life and Achievements of Admiral Dewey: From Montpelier to Manila*. Chicago: Our Possessions Publishing Co., 1890.

Hamblin, Robert W., and Charles A. Peek. *A William Faulkner Encyclopedia*. Westport, CT: Greenwood Press, 1999.

Hamden Historical Society. *Publications of the Hamden Historical Society, No. 1*. New Haven, CT: Ouinnipiack Press, 1938.

Hamilton, John D. "The St. Albans Raid: The Confederate Raid on St. Albans, Vermont, October 19, 1864." *Bulletin of the American Society of Arms Collectors* 90 (October 2004): 48–52.

Hamilton, William R. "The Military Schools of the Pacific Coast." *Overland Monthly* 29 (May 1897): 465–81.

———. "The Military Schools of the United States." *Outing Magazine* 20 (April–September 1892): 473–79.

———. "The Military Schools of the United States and Their Influence on the Nation." *Outing Magazine* 20 (April–September 1892): 330–36.

Harker History Committee. "Palo Alto Military Academy." 2005. Paper in possession of the author.

Harris, W. R. *The Catholic Church in Utah, 1776–1909*. Salt Lake City: Intermountain Catholic Church, 1909.

Harrison, Mitchell C. *New York State's Prominent and Progressive Men*. New York: New York Tribune, 1900.

Harvard-Westlake School. (2010). "History." http://www.hw.com/abouthw/tabid/2016/Default.aspx.

Haskins, Charles H., and William I. Hull. *A History of Education in Pennsylvania*. Washington, DC: Government Printing Office, 1902.

Hataway, Jimmie J. "History of the Department of Military Science at the University of Texas at Arlington." 1992. http://www.cadetcorps.org/PDFs/A%20 HISTORY%200F%20MS%20DEPT%20UTA.pdf.

Hattendorf, John. "Stephen B. Luce: Intellectual Leader of the New Navy." In *Quarter and Bridge: Two Centuries of American Naval Leaders*, edited by James Bradford. Annapolis: US Naval Institute Press, 1997.

Hays, Kim. *Practicing Virtues: Moral Traditions at Quaker and Military Boarding Schools.* Berkeley: University of California Press, 1994.

Hedgpeth, Don. *Proud Promise: The Story of Schreiner Institute/College.* Kerrville, TX: Press of the Guadalupe, 1998.

Herbeck, Dale. "Football." In *The Columbia Companion to American History on Film: How the Movies Have Portrayed the American Past,* edited by Peter C. Rollins. New York: Columbia University Press, 2003.

Heyl, Edward M. "Alabama Agricultural and Mechanical College." In *Annual Report to the Inspector-General of the Army,* 363–65. Washington, DC: Government Printing Office, 1890.

"Highland Military Academy." *Worcester Magazine* 4, no. 3 (1902): 83–93.

"The Hillsborough Military Academy (a.k.a. the North Carolina Military Academy)." http://freepages.history.rootsweb.ancestry.com/~orangecountync/places/hma/hma.html.

Hinde. E. R. "School Culture and Change: An Examination of the Effects of School Culture on the Process of Change." *Essays in Education* 12 (Winter 2004).

Hinkley, G. "Academy Was the West Point of the West." *Tradition: San Diego's Military Heritage,* November 1994. Reprint copy in possession of the author.

"The Historical Art of John Paul Strain: Alabama Corps of Cadets Call to Battle." http://www.johnpaulstrain.com/art/alabama-corps-of-cadets.htm.

Hobart College. *Catalogue of Hobart College, Geneva, New York, 1910–1911,* vol. 9. Geneva: Hobart College, 1911.

Holden, Edward. *The Centennial of the United States Military Academy at West Point, New York, 1802–1902.* Washington, DC: Government Printing Office, 1904.

Holmes, Ronald. *How to Eradicate Hazing.* Bloomington: AuthorHouse, 2013.

Holt, H. "The Beginnings of St. John's Military Academy: Commencement Exercise, 24 June." Unpublished manuscript, St. John's Military Academy Archives, Delafield, Wisconsin, 1885.

Honeyman, Abraham, ed. "Seven Generations of Lawyers." *New Jersey Law Journal* 44 (1921): 293–98.

Horvath, S. M. "Public Military Schools Catch On in Tough Districts." *Wall Street Journal,* August 28, 2002.

Hotchkin, S. F. *Ancient and Modern Germantown, Mount Airy, and Chestnut Hill.* Philadelphia; P. W. Ziegler & Co., 1889.

Hotek, D. "A Brief History of Lasallian in the Midwest District." Christian Brothers Conference, 2003. Paper in possession of the author.

How, Henry. "Destruction of the Ocean Steamer *Arctic,* by Collision with the *Vesta* (1854)." In *Outrageous Seas: Shipwreck and Survival in the Waters off Newfoundland, 1583–1893,* edited by Rainer Baehre. Montreal: Carleton Press, 1999.

Howe Military Academy. "Howe Military Academy." http://www.thehoweschool.org/wp/.

How to Learn and Earn. Boston: D Lothrop and Company, 1884.

Howlett, Charles F., and Glen Zeitzer. *The American Peace Movement.* Washington, DC: American Historical Association, 1985.

Hoy, Anita W., and Wayne K. Hoy. *Instructional Leadership: A Research-based Guide to Learning in Schools*, 3rd ed. Boston: Person Education Inc, 2009.

Hoy Wayne K., and Cecil G. Miskel. *Educational Administration: Theory, Research, and Practice*, 8th ed. Boston: McGraw Hill Company, 2008.

Huang, Carol. "The Chinese Western Military Academies in the United States, 1902–1911." In *Chartered Schools: Two Hundred Years of Independent Academies in the United States, 1727–1925*, edited by Nancy Beadie and Kim Tolley, 228–50. New York: Routledge/Falmer, 2002.

Hubbell, A. "New Orleans Military and Maritime Academy to Open in Algiers.." *New Orleans Times-Picayune*, May 6, 2011. http://www.nola.com/west-bank/index.ssf/2011/05/new_orleans_military_and_marit.html.

Hunter, Margaret A. "Education in Pennsylvania Promoted by the Presbyterian Church, 1726–1837," PhD diss., Temple University, 1937.

Hurst, Harold W. *Alexandria on the Potomac: The Portrait of an Antebellum Community*. Lanham, MD: University Press of America, 1991.

Hyde, William, and Howard Louis Conrad, eds. *Encyclopedia of the History of St. Louis: A Compendium of History and Biography for Ready Reference*, vol. 4. New York: The Southern History Company, 1899.

Imber, Michael, and Tyll Van Geel. *Education Law*. New York: Routledge, 2010.

Independence Seaport Museum. "Steering a Course: A Short History of the Pennsylvania Nautical School and Pennsylvania Maritime Academy." http://www.phillyseaport.org/pennsylvania-nautical-school/.

Inkster, T. H. "McDougall's Whalebacks." *American Neptune* 25, no. 3 (July 1965): 168–75.

Inscoe, John C. "The Civil War in Georgia." *A New Georgia Encyclopedia Companion*. Athens: University of Georgia Press, 2011.

Irby, Richard. *History of Randolph-Macon College, Virginia: The Oldest Incorporated Methodist College in America*. Richmond, VA: Whittet and Shepperson, 1894.

Ireland, R. E. "The Tar Heel Citadel." *Civil War Times* 40, no. 2 (May 2001): 38–44.

Izumi, L. "A Visit to the Oakland Military Institute." *Capital Ideas* 6 no. 41 (2001).

January, Brendan. *The Aftermath of the Wars against the Barbary Pirates.*" Minneapolis: Twenty-First Century Books, 2009.

Johnson, Brooke. "From School Ground to Battle Ground: A Qualitative Study of a Military-Style Charter School." PhD diss., University of California, Riverside, 2009.

Johnson, D. "High School at Attention." *Newsweek*, June, 21, 2002.

Johnson, Rossiter, and John H. Brown. *The Twentieth Century Biographical Dictionary of Notable Americans*. Boston: Biographic Society, 1904.

Joint Board of Higher Curricula. *Second Annual Report of the Joint Board of Higher Curricula to the Governor of Washington*. Olympia, WA: Frank Lamborn Public Printer, 1921.

Jones, Charles E. *Education in Georgia*. Washington, DC: Government Printing Office, 1889.

Jordan, John, ed. *Colonial and Revolutionary Families of Pennsylvania: Genealogical and Personal Memoirs*, vol. 3. New York, Lewis Publishing Co., 1911.

"Kamehameha Schools History." The Kamehameha Schools Archives. http://ka palama.ksbe.edu/archives/Timelines/Schools/KSHistory.php.

Kammen, Peggy. "Contributions of West Point Graduates of the Pre-Thayer Era: 1802–1817," 7. USMA Library/Digital Collections. 1996. http://digital-library.usma.edu/cdm/singleitem/collection/p16919c0111/id/17/rec/1

Kayler, K. "The Military Junior Colleges." PhD diss., Washington State University, 1989.

Keefer, Louis E. *The Army Specialized Training Program in World War II*. Jefferson, NC: McFarland, 2011.

Kenney, Michael. *Catholic Culture in Alabama*. New York: American Press, 1931.

Kerrick, H. *Military and Naval America*. Garden City, NY: Doubleday, Page & Co., 1918.

Kesselus, Kenneth. *History of Bastrop County, Texas, 1846–65*. Austin: Jenkins Publishing Co., 1987.

Khadaroo, Stacy T. "US High School Graduation Rate Climbs to 69.2 Percent." *Christian Science Monitor*, June 9, 2009. http://www.csmonitor.com/USA/2009/0609/p02s13-usgn.html.

Kiddle, H., and A. Schem, eds. *The Year-Book of Education for 1878*. New York: E. Steiger, 1878.

King, Irving H. *The Coast Guard Expands, 1865–1915: New Roles, New Frontiers*. Annapolis: Naval Institute Press, 1996.

Kirk, Hyland C. *A History of the New York State Teachers Association*. New York: E. L. Kellogg & Co., 1883.

Klein, R. "Marching On: The Chicago Military Academy." *Young Minds Magazine*, March–April 2003, 63.

Knight, Charles. *Valley Thunder: The Battle of New Market and the Opening of the Shenandoah Valley Campaign, May 1864*. New York: Savas Beatie, 2010.

Knight, Lucian L. *A Standard History of Georgia and Georgians*, vol. 5. Chicago: Lewis Publishing Company, 1917.Kraus, John D., Jr. "The Civilian Military College." *Military Review*, August 1976, 77–87.

——. "The Civilian Military Colleges in the Twentieth Century: Factors Influencing Their Survival." PhD diss., University of Iowa, 1978.

Krick, Robert E. *Staff Officers in Gray: A Biographical Register of Staff Officers in the Army of Northern Virginia*. Chapel Hill, NC: University of North Carolina Press, 2003.

Lagiarder, F. "Admiral Farragut Academy, Pine Beach NJ." http://www.largiader .com/farragut/.

Lamb, Ivy. "Educating the Whole Person." *Contents: US Airlines*, November 2013, 68–133.

Langins, Jannus. *Conserving the Enlightenment: French Military Engineering from Vauban to the Revolution*. Cambridge: Massachusetts Institute of Technology Press, 2004.

Latrobe, John H. B. *Reminiscences of West Point, from September, 1818, to March, 1882.* East Saginaw, MI: Evening News, Printers and Binders, 1887. Online copy, "West Point Reminiscences, from September, 1818, to March, 1882," annotated by Bill Thayer. http://penelope.uchicago.edu/Thayer/E/Gazetteer/Places/America/United_States/Army/USMA/LATREM*.html#Commandant_Bliss.

Leeman, William P. *The Long Road to Annapolis: The Founding of the Naval Academy and the Emerging American Republic.* Raleigh: University of North Carolina Press, 2010.

Leonard Hall Junior Naval Academy. "Proud History." https://sites.google.com/site/lhjna2013/how-to-get-involved.

Lindenberg, C. "Puget Sound Naval Academy." 2006. Paper in possession of the author.

Lindsey, Donald F. *Indians at Hampton Institute, 1877–1923.* Urbana: University of Illinois Press, 1995.

Linton Hall. "History." 2009. Paper in possession of the author.

Little, Blanche E. "Military Methods in the Training of Boys." *School Journal* 67 (September 5, 1903): 204–5.

Lockhart, Jane. "Looking at the Movies." *The Rotarian*, November 1951, 38–39.

Longley, Dione, and Buck Zaidel. *Heroes for All Time: Connecticut Civil War Soldiers Tell Their Stories.* Middletown, CT: Wesleyan University Press, 2015.

Lord, T. "Alden Partridge's Proposal for a National System of Education: A Model for the Morrill Land Grant Act." *History of Higher Education Annual* 18 (1998):11–22.

Loveland, Anne. *Change and Conflict in the U.S. Army Chaplain Corps since 1945.* Knoxville: University of Tennessee Press, 2014.

Lowrie, Walter, and Matthew St. Clair Clarke, eds. *American State Papers: Documents, Legislative and Executive, of the Congress of the United States,* Part 5, vol. 1. Washington, DC: Gales and Seaton, 1832.

Luce, Stephen. "The Manning of Our Navy and Mercantile Marine." *Record of the Naval Institute* 1 (November 13, 1873): 17–37.

Lunenburg, F. C., M. A. Sartori, and T. Bauske. "Classroom Climate, Teacher Control Behavior, and Student Self Control." Paper presented at the National Council of Professors of Education Administration, Jackson Hole, WY, 1999.

MacArthur, Douglas. "General Douglas MacArthur's Farewell Speech, Given to the Corps of Cadets at West Point, May 12, 1962," National Center for Public Policy Research, http://www.nationalcenter.org/MacArthurFarewell.html.

———. *Reminiscences.* New York: McGraw-Hill, 1964.

Macaulay, Alexander S. Jr. "Black, White, and Gray: The Desegregation of the Citadel, 1963–1973." In *Warm Ashes: Issues in Southern History at the Dawn of the Twenty-first Century,* edited by Winfred Moore, Kyle Sinisi, and David White, 301–19. Columbia: University of South Carolina Press, 2003.

MacGregor, Steve. "Crowd of Dissenters Rallies in Wash. D.C." *Collegiate Times,* May 13, 1970, 3.

MacKenzie, Doris, and Eugene Hebert. *Correctional Boot Camps: A Tough Intermediate Sanction.* Washington, DC: US Department of Justice, 1996.

Magill, M. "Clinton Liberal Institute, Fort Plain, N.Y." 2009. http://www.mont gomery.nygenweb.net/minden/ftplaininstitute.html.

Maine Maritime Academy. "A Proud Heritage." 2011. Paper in possession of the author.

"The Major and the Minor." *Life Magazine*, August 17, 1942, 37–38.

Managers American Teachers' Bureau. *American College and Public School Directory*, vols. 13 and 16. Saint Louis, MO: C. H. Evans & Co., 1890, 1893.

Mannering, M. "A Day at Historic Nazareth Hall." *The National Magazine*, vol. 40 (1914): 911–18.

Manning, Frederick J. "Morale, Cohesion, and Esprit de Corps." In *Handbook of Military Psychology*, edited by Reuven Gal and David Mangelsdorff, 453–54. Chichester, England: John Wiley & Sons, 1991.

Manning, Larry. "The Contribution of Sylvanus Thayer and the United States Military Academy to Engineering Programs in Higher Education in the United States." PhD diss., Texas A&M University, 2003.

Mansfield, Edward. "The United States Military Academy at West Point." *American Journal of Education* 30 (March 1863): 17–47.

Marist School. "Marist School History." 2013–14. http://www.marist.com/page/ About/Archives/School-History.

Marmor, J. "The Burette House: Enduring Landmark of Clifton-by-the-Sea." *Redondo Beach Historical Society Newsletter*, Winter 1991.

Marques, Pedro. "Oaklanders Blast Charter Schools Proposal." *San Francisco Bay Guardian*, May 3, 2000.

Marsh, Les. "William H. Keiter, C.S.A." 2009. Paper in possession of the author.

Marshall, Edward C. *History of the United States Naval Academy: With Biographical Sketches, and Names of all Superintendents, Professors and Graduates*. New York: D. Van Nostrand, 1862.

Mary Baldwin College. "History of VWIL." 2011. http://www.mbc.edu/vwil/his tory.php.

Maryland Military Academy. *Catalogue of the Officers and Cadets of the Maryland Military Academy*. Baltimore: John Murphy & Co., 1852.

Maslow, Abraham. "A Theory of Human Motivation." In *The Greatest Writings in Management and Organizational Behavior*, edited by Louis Boone and Donald Bowen. New York: Random House, 1987.

Massachusetts Association for the Blind. *Outlook for the Blind*, 3–4 (1909).

Matthews, Lloyd. "Ideals, Military." In *The Oxford Companion to American Military History*, edited by John Chambers and Fred Anderson. New York: Oxford University Press, 1999.

Mayes, Edward. *History of Education in Mississippi*. Washington DC: Government Printing Office, 1899.

McCausland, Walter. "Buffalo's First High School Had Its Own Defense Program: The Western Literary and Scientific Academy Was Once Housed in What Is Now the Pearl Block in Pearl Place." History of Buffalo. 1943. http://www .buffaloah.com/h/bflohs/index.html.

McClinton, C. "Scholar Practitioner Model." In *Encyclopedia of Distributed Learning*, edited by A. DiStefano, K. E. Rudestam, and R. Silverman, 393–96. Thousand Oaks, CA: Sage Publications, 2004.

McCormick, Barnes W., Conrad F. Newberry, and Eric Jumper. *Aerospace Engineering Education during the First Century of Flight.* Reston, VA: American Institute of Aeronautics and Astronautics, 2004.

McDonald, R. M. S. *Thomas Jefferson's Military Academy: Founding West Point.* Charlottesville, VA: University of Virginia Press, 2004.

McElvaine, Robert S. *The Great Depression: America, 1929–1941.* New York: Three Rivers Press, 2009.

McKate, Donald. *Tradition: A History of the Presidency of Clemson University.* Macon, GA: Mercer University Press, 1988.

McKeithan, Daniel Morley. "James, John Garland," *Handbook of Texas Online.* Texas State Historical Association, June 15, 2010. http://www.tshaonline.org/handbook/online/articles/fja18.

McKenzie, Alexander S. *Life of Stephen Decatur, a Commodore in the Navy of the United States* Boston: C. C. Little and J. Brown, 1846.

McMaster, Richard. "The Contribution of West Point to American Education." Master's thesis, University of Texas, 1951.

McMurry, Richard M. *Two Great Rebel Armies: An Essay in Confederate Military History.* Chapel Hill: University of North Carolina Press, 1989.

McTernan Walter, and Andrew Kulberg. "U.S. Marines Face Citadel Cadets at Tulifinny Crossroads." *Leatherneck Magazine*, March 2013, 44–48.

"Meet Jerry Brown." Brown for Governor Campaign. 2010. Accessed April 7, 2011. http://jerrybrown.org/about.

Miles, Wilson, Cary Fairfax, Thomas Ragland, and Nathaniel Hall Loring. *An Expose of Facts Concerning Recent Transactions, Relating to the Corps of Cadets of the United States Military Academy at West Point, New York.* Newburgh, NY: Uriah Lewis, 1819.

"A Military School for the Deaf, and Its Band of Deaf Musicians." *Volta Review* 13, no. 4 (September 1911).

Miller, A. "Delaware Military Academy: Charter School Navy Team Up for Education." *Community News: Greenville, Mill Creek, Hockessin, Brandywine,* November 21, 2002.

Miller, Edward. "VMI Men Who Wore Yankee Blue, 1861–1865," *VMI Alumni Review*, Spring 1996, 2–13.

Miller, Hugh, ed. *The Scroll of the Phi Delta Theta*, vol. 25. Indianapolis: Phi Delta Theta, 1901.

Mississippi Agricultural and Mechanical College. *Biennial Report of the Trustees, President and Other Officers of the Mississippi Agricultural and Mechanical College.* Nashville, TN: Press of Brandon Printing Company, 1909.

Missouri Military Academy. "History of Missouri Military Academy." http://www.missourimilitaryacademy.org/about/history/.

Missouri School Journal. "Missouri Military Academy Burned." *Missouri School Journal,* 13, no. 10 (October 1896): 652.

Mohr, Clarence L., and C. R. Wilson, eds. *The New Encyclopedia of Southern Culture*, vol. 17. Chapel Hill: University of North Carolina Press, 2011.

Moll, Clarence R. "A History of Pennsylvania Military College, 1821–1954." PhD diss., New York University, 1954.

Moore, B. *Bastrop County, 1691–1900*. Bastrop, TX: Bastrop County History Society, 1973.

Morgan, Tina. "The Making of Manlius Pebble Hill: Tale of Two Schools." Manlius Pebble Hill School, 2009. http://www.mph.net/pdf/Alumni/Tale%20°f%20Two%20Schools.pdf.

Morgan Park Academy. "History." http://morganparkacademy.org (accessed May 3, 2010).

Morrison, A. *The Beginning of Public Education in Virginia, 1776–1860: Study of Secondary Schools in Relation to the State Literary Fund*. Richmond, VA: State of Virginia, 1917.

Morrison, James. *Rat Pants to Eagles and Tweeds: The Memoirs of a Soldier-Teacher*. Kent, OH: Kent State University Press, 2004.

Morse, R. "A Very Different Mayor Brown." *San Francisco Chronicle*, July 25, 2001. http://www.sfgate.com/cgi-bin/article.cgi?f=/c/a/2001/07/25/MN21659.DTL.

Moss, J. "Private Schools of San Rafael Photo Album." Marin County Free Library. 2003. http://contentdm.marinlibrary.org/cdm/landingpage/collection/pssr.

Munson, E. B. *North Carolina Civil War Obituaries, Regiments 1 through 46: A Collection of Tributes to the War Dead and Veterans*. Jefferson, NC: McFarland & Company, 2015.

Murphy, Joseph, and Catherine D. Shiffman. *Understanding and Assessing the Charter School Movement*. New York: Teachers College Press, 2002.

"Mutes, but Everyone is a Soldier." *Richmond Times*, December 23, 1900, 17. http://chroniclingamerica.loc.gov/lccn/sn85034438/1900–12–23/ed-1/seq-17/.

Myer, David. "West Point and Jefferson's Constitutionalism." In *Thomas Jefferson's Military Academy: Founding West Point*, edited by Robert McDonald, 54–76. Charlottesville: University of Virginia Press, 2004.

Myers, Harry J. *American College and Private School Directory*, vol. 2. New York: Educational Aid Society, 1908.

Nagel, G. "The Partridge Military School in Harrisburg, 1845–1847." Camp Curtin Historical Society and Civil War Roundtable, 2005. Paper in possession of the author.

Nank, Christopher. "World War I Narratives and the American Peace Movement, 1920–1936." PhD diss., Florida State University, 2005.

National Commission on Excellence in Education. "A Nation at Risk: The Imperative for Educational Reform." 1983. http://www2.ed.gov/pubs/NatAtRisk/index.html.

Naval Historical Center. "Captain Josiah Tattnall, Confederate States Navy (1795–1871)." 2001. https://www.ibiblio.org/hyperwar/OnlineLibrary/photos/pers-us/uspers-t/j-tattnl.htm.

Nazareth Hall. "Click to Read Our Proud History." 2009. http://www.nazareth-hall.com/our-story/.

"The Nebraska Military Academy Burns." *New York Times*, October 18, 1908, 9.

Neil, Henry. "Collinwood School Fire." *The Encyclopedia of Cleveland History*. Last modified March 27, 1998. http://ech.case.edu/ech-cgi/article.pl?id=CSF.

New York Board of Education. *Thirty-Third Annual Report of the Board of Education for the City and County of New York for the Official Year Ending December 31, 1874.* New York: Cushing and Bardua, 1875.

New York Military Academy. "New York Military Academy: A High Grade Preparatory School for all Colleges" (advertisement). *Cosmopolitan Magazine*, August 1912, 26.http://books.google.com/books?id=gz1XAAAAYAAJ&pg=RA2-PA 198&dq=fire+proof+Military+school&hl=en#.

Nicholls, Joanna R. "The United States Revenue Cutter Service: The History and Duties of an Important Branch of Our Navy." *Frank Leslie's Popular Monthly* 42, no. 4 (October 1896): 411–25. https://babel.hathitrust.org/cgi/pt?id=inu .3200000494429;view=2up;seq=428.

Nielsen, M. "The Kearney Military Academy Serving the Community, 1892–1993." *Buffalo County Historical Society Magazine* 16, no. 1 (1993).

Nimmo, Joseph. *Report to the Secretary of the Treasury in Relation to the Foreign Commerce of the United States and the Practical Workings of Our Relations of Maritime Reciprocity.* Washington DC: Government Printing Office, 1871.

Norfolk Academy. "About NA: History." 2009. Paper in possession of the author.

North Carolina Department of Cultural Resources. (1939). "Bingham School." North Carolina Highway Historical Marker Program 1939. http://www.nc markers.com/Markers.aspx?MarkerId=G-36.

———. "Horner Military School. "North Carolina Highway Historical Marker Program 1939. http://www.ncmarkers.com/Markers.aspx?MarkerId=G-27.

———. "Mount Pleasant Collegiate Institute." 1962. http://www.ncmarkers.com/ Markers.aspx?MarkerId=L-65.

North Carolina Department of Public Instruction. *Report of the Superintendent of Public Instruction of North Carolina for Scholastic Years 1896–97 and 1897–1898.* Raleigh, NC: Guy V. Barnes Printer, 1898.

North Georgia Agricultural College. *Annual Catalogue, 1873–1880.* Athens: Banner Print, 1905.

Northwestern Military Academy. *Northwestern Military Academy Catalog.* Highland Park, IL: Northwestern Military Academy, 1888.

Norton, William, ed. *Historic Gainesville and Hall County: Illustrated History.* San Antonio: Lammert Publications, 2001.

Norwich University. *Cadet Handbook.* Norwich, VT: Norwich University, 2005.

Nuwer, Hank. *The Hazing Reader.* Bloomington: Indiana University Press, 2004.

Oakland Military Institute. "About." http://oakmil.org/about/.

"Obituary of Colonel A. L. Roumford." *Germantown Independent-Gazette*, August 2, 1878. Text obtained from the Germantown Historical Society.

O'Hara, Edwin V. "The School Question in Oregon." *Catholic World* 116 (October 1923): 483–90.

Olsen, Christopher. *The American Civil War: A Hands-on History.* New York: Hill and Wang, 2006.

Omen Technology Inc. "Western Military Academy." 2007. Paper in possession of the author.

"Onarga Military Academy High School 'Yellowjackets' & 'Spartans.'" Illinois High School Glory Days, 2009. http://www.illinoishsglorydays.com/id309.html.

Ouchi, William. *Theory z: How American Business Can Meet the Japanese Challenge.* Reading, PA: Addison-Wesley, 1981.

Pace, Edward D., and Thomas E. Shields, eds. "Schoolmaster and Priest." *Catholic Educational Review* 1 (January–May 1911): 375.

Palmer, Ann Therese. "Spit and Polish Comes to Chicago Schools." *Bloomberg Business Week*, June 25, 2001. http://www.bloomberg.com/news/articles/2001–06–24/spit-and-polish-comes-to-chicago-schools.

Parent, Dale. *Correctional Boot Camps: Lessons from a Decade of Research.* Washington, DC: US Department of Justice, 1997. https://www.ncjrs.gov/pdffiles1/nij/197018.pdf.

Park, Benjamin. *The United States Naval Academy.* New York: G. P. Putnam's Sons, 1900.

Partridge, Alden. "Partridge Lectures." *American Journal of Education* 1 (1826): 395–400.

Patterson, Homer L., ed. *Patterson's American Educational Directory.* Chicago, IL: American Educational Co., 1908, 1913, 1917, 1920, 1922, 1932, 1936, 1947, 1953.

Peacock, Donna. *Parade Rest: Peacock Military Academy, 1894–1941.* Book One. San Antonio: Peacock Military Alumni Association, 1990.

Pearce, Amber. "*Faulkner v. Jones:* The Constitutionality of the Citadel's Single-Gender Admissions Policy." *New England Law Review*, Winter 1997.

Penn, Julius A. "Military Instruction at Civil Educational Institutions" *Infantry Journal* 6, no. 5 (1910): 679–703.

Pennsylvania Military College. "Cadets during the Gettysburg Campaign." http://pennsylvaniamilitarycollege.org/cadets-gettysburg-campaign/

Petersen, Karen E. "Harold Douglas Martin, Major, United States Army." Arlington National Cemetery Website. http://www.arlingtoncemetery.net/hdmartin-002.htm.

Peterson, Douglas. *The California Maritime Academy: A Brief History.* Vallejo: California Maritime Academy, 2004.

Peterson, Mark. "Possibility is 'Very Strong' That Howe Military Academy Will Close." WNDU. May 19, 2014. http://www.wndu.com/home/headlines/Possibility-is-very-strong-that-Howe-Military-Academy-will-close—259862791.html.

Peterson, Patti M. "Student Organizations and the Anti-war Movement in America, 1900–1960." In *Peace Movements in America*, edited by Charles Chatfield, 116–32. New York: Schocken Books, 1985.

Pettus, Louise. "Asbury Coward: Soldier-Educator." Rootsweb, Ancestry.com. 2001. http://freepages.genealogy.rootsweb.ancestry.com/~york/5thSCV/ACoward.html.

Pezzaglia, Phillip. *Images of America: Rio Vista.* San Francisco: Arcadia Publishing, 2005.

Philadelphia Military Academy. "About Us." 2011. Accessed June 26, 2016. http://webgui.phila.k12.pa.us/schools/p/pma/about-us.

Phillips, Harry. Letter dated May 19, 1861, in *Letters from Annapolis,* edited by Anne Marie Drew, 60–64. Annapolis: Naval Institute Press, 1998.

Pinkus, Lyndsay M. *Moving beyond AYP: High School Performance Indicators.* Policy Brief. Washington, DC: Alliance for Excellent Education, 2009.

Poirier, Robert. *By the Blood of Our Alumni: Norwich University Citizen Soldiers in the Army of the Potomac.* Mason City, IA: Savas Publishing, 1999.

Porter-Gaud School. "Porter Military Academy." 2009. Paper in possession of the author.

Possner, Roger. *The Rise of Militarism in the Progressive Era, 1900–1914.* Jefferson, NC: McFarland and Company, 2009.

Potter, Dorothy T., and Clifton W. Potter. *Lynchburg, 1757–2007.* Images of America. Charleston: Arcadia Publishing, 2007.

Powell, William, ed. *Dictionary of North Carolina Biography,* vol. 5, P–S. Chapel Hill: University of North Carolina Press, 1994.

Powell, William H. "Colonel Joseph Sumner Rogers, U.S.V." In *Officers of the Volunteer Army and Navy Who Served in the Civil War.* Philadelphia: L. R. Hamersly & Co., 1893. http://www.all-biographies.com/soldiers/joseph_sumner_rogers.htm.

Prehn, Walter L. "Episcopal Schools: History and Mission." 2011. Manuscript in possession of the author.

Price, Hugh. "About Face!" *Educational Leadership,* May 2008, 28–34.

Prowell, George R. *The History of Camden County, New Jersey.* Camden, NJ: L. J. Richards & Co., 1886. https://archive.org/details/historyofcamdenc00prow.

Prunier, Tom. "Virginia Military School Tradition." *Virginia Living,* 2005 reprint.

Public Agenda. *Teaching Interrupted: Do Discipline Policies in Today's Public Schools Foster the Common Good?* New York: Public Agenda, 2004. http://www.publicagenda.org/files/teaching_interrupted.pdf.

Public Safety Academy. "Our Mission Statement." 2010. http://www.psasb.us/#!about-us/c1se.

Pugmire, T. "Military Charter Schools Promise Disciplined Education." Minnesota Public Radio News. September 7, 2004. http://news.minnesota.publicradio.org/features/2004/09/07_pugmiret_milcharters/.

Purpura, Paul. "New Orleans Military Maritime Academy Will Open in 2011 on Federal City Site." *New Orleans Times-Picayune,* October 20, 2010, http://www.nola.com/military/index.ssf/2010/10/new_orleans_military_maritime.html.

Pye, Beth. "Gordon State College." New Georgia Encyclopedia. 2004. http://www.georgiaencyclopedia.org/nge/Article.jsp?id=h-1442.

Ramsey, R. D. *Don't Teach the Canaries Not to Sing: Creating a School Culture That Boosts Achievement.* Thousand Oaks, CA: Corwin Press, 2008.

Randall, H. "The Senior Military Colleges: Analysis and Comparison of Senior Military Colleges and Military Junior Colleges with Respect to Seven Selected Characteristics." PhD diss., University of Southern California, Los Angeles, 1991.

"Ranking College by Rhodes Scholarship." College Confidential, August 28, 2009, edited September 2012. http://talk.collegeconfidential.com/college-search-selection/772208-ranking-college-by-rhodes-scholarship.html.

Ray, W. *Austin Colony Pioneers*. Austin, TX: Pemberton Press, 1970.

Reagan, Alice E. *North Carolina State University: A Narrative History*. Raleigh: North Carolina State University Foundation, 1987.

Reeves, Ira Louis. *Military Education in the United States*. Burlington, VT: Free Press Printing Co., 1914.

Reininger, Charles G. "Locus of Control and Performance at a Secondary Military Boarding School." PhD diss., University of Houston, 2004.

Reminiscences of West Point in the Olden Time. East Saginaw, MI: Evening News Printing and Binding House, 1886.

Ringenbach, Paul T. *Battling Tradition: Robert F. McDermott and Shaping the U.S. Air Force Academy*. Chicago: Imprint Publications, 2006.

Robbins, Allan W. "History of Saint John's Academy, 1833–1895." *Alexandria History* 5 (1983): 25–31.

Robbins, Stephen P., and Tim Judge. *Organizational Behavior*. Upper Saddle River, NJ: Pearson Prentice Hall, 2007.

Robins, Glenn. *The Ministry and Civil War Legacy of Leonidas Polk*. Macon, GA: Mercer University Press, 2006.

Robinson Library. "Bernarr Macfadden." http://www.robinsonlibrary.com/geogra phy/recreation/physical/fitness/macfadden.htm.

Robson, Mary Farrand. "Hill Military Academy." From Hill Military Academy Directory, n.d.http://gesswhoto.com/hill-academy.html.

The Rochester Directory, Containing a General Directory of the Citizens, a Business Directory, and the City and County Register, vol. 21. Rochester, NY: C. C. Drew Publisher, 1870.

Roeder, Mark A. *A History of Culver and the Culver Military Academy*. Lincoln, NE: iUniverse, 2004.

Rogal, Samuel J. *The American Pre-College Military School: A History and Comprehensive Catalog of Institutions*. Jefferson, NC: McFarland & Company, 2009.

Rogers State University, "History: Oklahoma Military Academy." 2009. http://www.rsu.edu/alumni/oklahoma-military-academy/.

Rosenberg, Madelyn. "Changes Won't Reveal a Softer Side of VMI." *Roanoke Times*, August 20, 1997.

Ross, Worth G. "Our Coast Guard: A Brief History of the United States Revenue Marine Service." *Harper's New Monthly Magazine* 73 (November 1886).

Royal Engineers Living History Group. "The Royal Military Academy, Woolwich." 2010. http://www.royalengineers.ca/RMA.html.

Rudolph, Frederick. *The American College and University*. Athens: University of Georgia Press, 1990.

Ruffin, Thomas, Jo Jackson, and Mary Hebert. *Under Stately Oaks: A Pictorial History of LSU*. Baton Rouge: Louisiana State University Press, 2006.

Runyan, Timothy J., ed. *Ships, Seafaring and Society: Essays in Maritime History*. Detroit: Wayne State University Press, 1987.

Russell, William, ed. "Partridge Lectures." *American Journal of Education* 1, no. 7 (July 1826): 395–401.

Ruth, Dawn. "Grading Paul Vallas." *New Orleans Magazine*, September, 2010.

"Sacred Heart Military Academy Here to Open Its Doors on Sept. 6." *Watertown Daily Times*, August 14, 1955. www.watertownhistory.org/articles/sacredheart.htm.

"Saint John's Academy Fiftieth Anniversary Speech." Saint John's Academy File, loose, Alexandria Library Special Collection, Alexandria City Records, 1883.

Saltman, Kenneth J. "Education as Enforcement: Militarization and Corporatization of Schools." *Race, Poverty, and the Environment*, Fall 2007, 28–30. http://reimaginerpe.org/node/1177.

Sampson, William. "A History of the Kentucky Military Institute during the Nineteenth Century, 1845–1900." thesis, University of Louisville, 1954.

San Francisco Chamber of Commerce. *Annual Report of the Chamber of Commerce of San Francisco.* San Francisco: Commercial Publishing Company, 1892.

Sargeant, Porter E. *A Handbook of the Best Private Schools of the United States and Canada.* Boston: Porter Sargeant, 1915, 1917, 1920, 1922.

Saxon, Gerald D. *The University of Texas at Arlington: 1895–1995.* Arlington: University of Texas at Arlington Press, 1995.

Scharf, J. Thomas. *History of the Confederate States Navy from Its Organization to the Surrender of Its Last Vessel.* New York: Rogers & Sherwood, 1887.

———. *History of Westchester County, New York, including Morrisania, Kings Bridge and West Farms, Which Have Been Annexed to New York City*, vol. 1, part 2. Philadelphia: L. E. Preston & Co., 1886.

Schein, Edgar. "Coming to a New Awareness of Organizational Culture." *Sloan Management Review* Winter 1984, 3–16.

———. *The Corporate Culture Survival Guide.* San Franciso: Jossey-Bass, 2009.

Schramm, Robert W. *Linsly School.* Charleston: Arcadia Publishing, 2003.

Sellers, James B. *History of the University of Alabama, 1818–1902*, vol. 1. Tuscaloosa: University of Alabama Press, 1953.

Serwer, Adam. "No, Chicago Is Not More Dangerous Than Afghanistan." *Mother Jones*, August 15, 2012. http://www.motherjones.com/mojo/2012/08/no-chicago-not-more-dangerous-afghanistan.

Shahfari, Payam. "Uncle Sam Wants You Badly." Iranian.com Archives. May 1, 2007. http://www.iranian.com/BTW/2007/May/Militarism/index.html (accessed Feb 9, 2010).

Shallat, Jonathan. *Structures in the Stream: Water, Science, and the Rise of the U.S. Army.* Austin: University of Texas Press, 1994.

Sharpe, B. "Gone but Not Forgotten: North Carolina's Educational Past: North Carolina Polytechnic Academy." North Carolina Collection, UNC Library, 1958. Copy in possession of the author.

Simberg, J. H. "Colonel Wright's Methods." *Printer's Ink,*" April 23, 1902, 13–17.

Simkins, Francis. *A History of the South*, 3rd ed. New York: Alfred A. Knopf, 1969.

Simpson, Jane. "Georgia Military College: A Brief History." Video, 2008. http://vimeo.com/4627442.

Smith, Edgar Fahs. *Samuel Latham Mitchill: A Father in American Chemistry*. New York: Columbia University Press, 1922.

Smith, Ernest Ashton. *Allegheny: A Century of Education, 1815–1915*. Meadville, PA: Allegheny College History Company, 1916.

Smith, Francis. *The Virginia Military Institute: Its Building and Rebuilding*. Lynchburg, VA: J. P. Bell Company, 1912.

——. *West Point Fifty Years Ago*. New York: D. Van Nostrand, 1879.

Smith, M. A. "St. Francis/St. Emma Military Academy Welcome Home." 2011. Paper in possession of the author.

Smith, Marie B. "A History of St. Emma's Military Academy and St. Francis De Sales High School." PhD diss., Catholic University of America, 1949.

Smothers, Ronald. "Citadel Is Ordered to Admit a Woman to Its Cadet Corps." *New York Times*, July 23, 1994. Top of Formhttp://www.nytimes.com/1994/07/23/us/citadel-is-ordered-to-admit-a-woman-to-its-cadet-corps.html.

Smull, John A., William P. Smull, Thomas B. Cochran, and W. Harry Baker, eds. "Pennsylvania Nautical School." *Smull's Legislative Hand Book and Manual of the State of Pennsylvania*. Harrisburg, PA: C. E. Aughinbaugh, 1912.

Smyth, A., ed. "Military Organizations in Schools." *Ohio Journal of Education* 5 (1856): 146–50.

Snider, Don, and Alexander Shine. *A Soldier's Morality, Religion, and Our Professional Ethic: Does the Army's Culture Facilitate Integration, Character Development, and Trust in the Professional?* Carlisle Barracks, PA: US Army War College Press, 2014.

Soley, James R. *Report on the Foreign Systems of Naval Education*. Washington, DC: Government Printing Office, 1880.

Soloman, Sarah. "Western Military Academy." *Illinois History: A Magazine for Young People*, December, 1998, 15–16. Illinois Periodicals Online. http://www.lib.niu.edu/1998/ihy981215.html.

Sorley, Lewis. *Honor Bright: History and Origins of the West Point Honor Code and System*. Boston: McGraw Hill, 2009.

South Carolina Department of Agriculture. *Handbook on South Carolina*. Charleston: Department of Agriculture, 1971.

South Carolina State Superintendent of Education. *Thirtieth Annual Report, 1898*. Reports and Resolutions of the General Assembly of the State of South Carolina, 1898, vol. 2. Columbia, SC: Bryan Printing, 1899. https://babel.hathitrust.org/cgi/pt?id=chi.096227585;view=1up;seq=9.

Southeast Academy High School. http://southeastacademy.org/.

Speelman, Jennifer. "Nautical Schools and the Development of the United States Maritime Professionals, 1874–1941." Master's thesis, Temple University, 2001.

Spivey, Donald. *Schooling for the New Slavery: Black Industrial Education, 1868–1915*. Westport, CT: Greenwood Publishing Group, 1978.

Spring Hill College, 1918–1919. Mobile, AL: Spring Hill College, 1919.

Stannard, Matthew B. "Judge Tosses Laws Restricting Recruiters." *San Francisco Chronicle*, June 19, 2009. http://www.sfgate.com/bayarea/article/Judge-tosses-laws-restrictingrecruiters-3295184.php.

Starbuck, James. "The Michigan Military Academy at Orchard Lake." *Michigan History Magazine* 50 (September 1966). Reprint with additional notes by Brian J. Bohnett, 9 pages, in possession of the author.

Starnes, Todd. "Christian School's ROTC under Attack." Fox News Opinion, November 11, 2014. http://www.foxnews.com/opinion/2014/11/11/christian-schools-rotc-under-attack/.

St. Catherine's Academy. "History." http://www.stcatherinesacademy.org/about/history.aspx.

Stearns, Alfred E., L. R. Gignilliat, and Milo H. Stuart. *Types of Schools for Boys.* Indianapolis: Bobbs-Merrill, 1917.

Stearns, Myron M. "Where Are You Going to School?" *Boy's Life*, March 1934, 27.

Steiger, Ernst. *Steiger's Educational Directory for 1878.* New York: E. Steiger, 1878.

Steiner, Bernard. *The History of Education in Connecticut.* Bureau of Education Circular No 2. Washington, DC: Government Printing Office, 1893.

Stephens, James D. *Reflections: A Portrait-Biography of the Kentucky Military Institute, 1845–1971.* Georgetown, KY: Kentucky Military Institute, 1991.

Stevens, Walter Barlow. *Missouri, the Center State, 1821–1915.* Vol. 3. Chicago: S. J. Clarke Publishing Company, 1915.

Steverson, London. "Coast Guard Honors First Black Academy Graduate." April 3, 2012. https://judgelondonsteverson.me/2012/04/03/coast-guard-honors-first-black-academy-graduate/.

Stewart, William H. *Admirals of the World: A Biographical Dictionary, 1500 to Present.* Jefferson, NC: McFarland & Company, 2009.

Stinnett, P. "New Leaders at Military School Include Students." *Oakland Tribune*, August 23, 2004.

St. John's College. "About St. John's College." 2008. http://www.stjohnscollege.edu/about/history.shtml.

St. John's Military Academy. *St John's Military Academy Catalog.* Delafield, WI: St. John's Military Academy, 1886.

St. Joseph Military Academy Alumni. "St. Joseph Military Academy: A Brief History." 2005. Paper in possession of the author.

The Strange One. Directed by Jack Garfein and Herb Gardener. Burbank, CA: Columbia Pictures, 1957.

Suid, Laurence H. *Sailing on the Silver Screen: Hollywood and the U.S. Navy.* Annapolis: Naval Institute Press, 1996.

Sweetman, Jack. *The U.S. Naval Academy: An Illustrated History,* 2nd ed. Annapolis: Naval Institute Press, 1995.

Sweetser, M. F. *King's Handbook of the United States.* Buffalo, NY: Moses King Corporation, 1891.

Tarrant, Robert L. "Leadership Development in Secondary Military Schools." PhD diss., Kansas State University, 1972.

Temple, Harry D. *The Bugle's Echo: The Chronology of Cadet Life at the Military College at Blacksburg, Virginia.* 6 vols. Blacksburg: Virginia Tech Corps of Cadets Alumni, 1996–2001.

———. "Cadet Life during the First Year of the Agricultural and Mechanical College at Blacksburg, Virginia, 1993." Manuscript in possession of the author.

———. *Donning of the Blue and Gray*. Blacksburg: Virginia Tech Corps of Cadets Alumni, 1992.

Tennessee State Library and Archives. "Ingram, Bowen, 1904–1980, Papers, 1856–1978." http://share.tn.gov/tsla/history/manuscripts/findingaids/92–086.pdf.

Terry, Clyde R. *Illinois Military School, Abingdon, Illinois*. Abingdon: Illinois Military Academy, 1934. http://www.archive.org/stream/illinoismilitary00abin#page/n1/mode/2up.

Tewsbury, Donald G. *The Founding of American Colleges and Universities before the Civil War*. New York: Teachers College, Columbia University, 1932.

Texas Military Institute. *Annual Catalogue of the Texas Military Institute*. Galveston: News Book and Job Office, 1859.

Theta Chi Fraternity. *History of Theta Chi, 1856–1927*. New York: Theta Chi Fraternity, 1927.

Thomas, Gage P. *Where to Educate, 1898–1898: A Guide to the Best Private Schools, Higher Institutions of Learning, etc., in the United States*. Boston: Boston and Company, 1899.

Thomas, John. *The History of the South Carolina Military Academy*. Charleston: Walker, Evans & Cogswell, 1893.

"Three Students Die When Academy Burns: Nine Others Are Seriously Hurt at Gambier, Ohio." *New York Times*, February 24, 1908. http://select.nytimes .com/gst/abstract.html?res=F20D13F93B5A12738DDDAC0A94DA405 B868CF1D3&scp=1&sq=Three%20students%20die%20when%20acad emy%20burns&st=cse.

Tierney, William G. Organizational Culture in Higher Education: Defining the Essentials. *Journal of Higher Education* 5 (1988): 2–21.

Toler, John T. "Bethel Military Academy, 1867–1911." *The Fauquier Historical Society* 18, no. 1 (Winter and Spring 1996): 1–13. http://www.rootsweb.ances try.com/~vafauqui/bma1.htm.

Torres, Kristina. "Despite Protests, DeKalb Proceeds with Marine School." *Atlanta Journal-Constitution*, March 23, 2009.

———. "Marine Corps School Would Give Kids 'a Niche.'" *Atlanta Journal-Consti tution*, April 6, 2009.

Traub, Peter E. "Military Hygiene: How Best to Enforce Its Study in our Military and Naval Schools and Promote Its Intelligent Practice in Our Army." *Journal of the Military Institute of the United States* 36, no. 133 (January–February 1905): 1–37.

Trejos, Nancy. "Chicago Public School Uses Military Methods to Make Model Students." *Washington Post*, January 2, 2002.

———. "Public Military Academy Brings Clash of Cultures." *Washington Post*, December 22, 2002. Reissue, Mindfully Green, http://www.mindfully.org/ Reform/2002/Military-Academy-Clash22dec02.htm.

Trousdale, William. *Military High Schools in America*. Walnut Creek, CA: Left Coast Press, 2007.

Tucker, Spencer E., ed. *The Encyclopedia of the Spanish-American and Philippine-American Wars: A Political, Social, and Military History*, vol. 1. Santa Barbara, CA: ABC-CLIO, 2009.

Turner, Bryan S., and Peter Hamilton. *Citizenship: Critical Concepts*, vol. 1. New York: Routledge, 1994.

Turner, Dorie. "Military Backed Public Schools on the Rise Despite Protests." *USA Today*, June 4, 2009. http://usatoday30.usatoday.com/news/education/2009–06–04-marine-schools_N.htm.

Uhl, Robert. "Masters of the Merchant Marine." *American Heritage* 34, no. 3 (April–May 1983). http://www.americanheritage.com/content/masters-mer-chant-marine

Umphlett, Wiley Lee. *The Movies Go to College: Hollywood and the World of the College-Life Film*. Cranbury, NJ: Associated University Presses, 1984.

UMS-Wright Preparatory School. "The UMS-Wright Tradition." 2006. http://www.ums-wright.org/page.aspx?pid=408.

United Kingdom Army. "The History of RMA Sandhurst." http://www.army.mod.uk/documents/general/history_of_rmas.pdf.

United Nations, Committee on the Rights of the Child. *Optional Protocol to the Convention of the Rights of the Child on Involvement of Children in Armed Conflict*. Periodic Report . . . , January 22, 2010. http://www.state.gov/documents/organization/135988.pdf.

University of Alabama. *Catalogue of the Officers and Students of the University of Alabama for the Academic Year 1899–1900*. Montgomery, AL: Brown Printing Co, 1900.

University of the State of New York. *University of the State of New York: Annual Report*. Albany: Jerome B. Parmenter State Printer, 1887.

University of St. Thomas. *Faculty Handbook*. Saint Paul, MN: University of St. Thomas, 2001.

US Army. *Junior ROTC and NDCC Units*. General Order No. 39, October 18, 1972.

———. *Junior ROTC and NDCC Units*. General Order No. 46, August 27, 1971.

———. *Junior ROTC Units at Military Schools*. General Order No. 49, August 5, 1969.

———. "Medal of Honor Recipients." US Army Center of Military History. http://www.history.army.mil/moh/index.html.

———. *The Soldier's Guide*. Field Manual 7–21.13. Washington, DC: Department of the Army, 2004.

US Census Bureau. *Statistical Abstract of the United States: 2012*, 131st ed. http://www.census.gov/library/publications/2011/compendia/statab/131ed.html.

US Congress. *Causes of the Reduction of American Tonnage and the Decline of Navigation Interests: Being the Report of a Select Committee, Made to the House of Representatives of the United States on the 17th of February 1870*. H. Rep. 28, 41st Cong., 2nd sess. Washington DC: Government Printing Office, 1870. https://archive.org/details/reductiontonnage00causrich.

———. *Goals 2000: Educate America Act*. H.R. 1804, 103rd Cong., 1994. http://www2.ed.gov/legislation/GOALS2000/TheAct/index.html.

———. *Index to the Executive Documents of the House of Representatives for the Second Session of the Forty-Sixth Congress, 1879–80*. 26 vols. Washington, DC: Government Printing Office, 1880.

———. "Morrill, Justin Smith (1810–1898)." Biographical Directory of the United States Congress. http://bioguide.congress.gov/scripts/biodisplay.pl?index=M000969.

———. No Child Left Behind Act of 2001, P.L. 107–110, 107th Congress, January 8, 2002. http://www2.ed.gov/policy/elsec/leg/esea02/107–110.pdf.

US Congress, House Committee on Military Affairs. *Abolishment of Compulsory Military Training at Schools and Colleges.* Hearings on H.R. 8538, 69th Cong., 1st sess., 1926. http://babel.hathitrust.org/cgi/pt?view=image;size=100;id=mdp.39015076644916.

———. *To Increase the Efficiency of the Military Establishment of the United States.* Hearings, 64th Cong., 1st sess., 1916. https://babel.hathitrust.org/cgi/pt?id=uc1.$b654240;view=2up;seq=6.

US Congress, House Committee on Naval Affairs. *Sundry Legislation Affecting the Naval Establishment, 1921.* Hearings, 77th Cong., 1st sess., 1922.https://archive.org/details/hearingsbeforec04affagoog.

US Department of the Army. *Senior Reserve Officers' Training Corps Program: Organization, Administration, and Training.* Army Regulation 145–1. Washington, DC: Department of the Army, 2011.

US Department of Education. "Arne Duncan, U.S. Secretary of Education—Biography." 2010. http://www2.ed.gov/news/staff/bios/duncan.html.

———. *High School Accountability and Assessment Systems.* High School Leadership Summit Issue Paper, 2002. ttps://www2.ed.gov/about/offices/list/ovae/pi/hsinit/papers/hsacct.pdf.

US Department of the Interior, Office of Education. Reports of the Commissioner of Education for the Years 1874, 1875 (vol. 2), 1883–1884, 1890–1891, 1895–1896 (2 vols.), and 1897–1898. Washington DC: Government Printing Office.

US Department of Justice, Office of Juvenile Justice and Delinquency Prevention. "Juvenile Justice Reform Initiatives in the States, 1994–1996." http://www.ojjdp.gov/pubs/reform/ch2_g.html.

US Navy Department. *Report of the Secretary of the Navy: Being Part of the Messages and Documents Communicated to the Two Houses of Congress.* 45th Cong., 2nd sess. Washington DC: Government Printing Office, 1877.

U.S. News & World Report Best Colleges, 2013–2014. New York: U.S. News & World Report, 2013.

US Park Service. "National Register of Historical Places Inventory—Nomination Form, Fishburne Home School." 21 August 1984. http://www.dhr.virginia.gov/registers/Cities/Waynesboro/136–0004%20-%20Fishburne%20Military%20School%20-%20Final%20Nomination.pdf.

US Supreme Court. *Pierce v. Society of the Sisters of the Holy Names of Jesus and Mary and the Hill Military Academy.* 268 U.S. 510. Nos. 583, 584, 1925. http://www.scribd.com/doc/269919/pierce-v-society-of-the-sisters.

US War Department. *Annual Report of the Secretary of War for the Year 1891,* vol. 1. Washington, DC: Government Printing Office, 1892.

———. *Annual Report of the Secretary of War for the Year 1891,* vol. 4. Washington, DC: Government Printing Office, 1892.

———. *Annual Report of the Secretary of War for the Year 1892*, vol. 4. Washington, DC: Government Printing Office, 1892.

———. *Annual Report of the Secretary of War for the Year 1893*, vol. 4. Washington, DC: Government Printing Office, 1893.

———. *Annual Report of the Secretary of War for the Year 1914*, vol. 1. Washington, DC: Government Printing Office, 1914.

"Vandalism on Ship Denied: Pennsylvania Inquiry Discounts Sabotage Report on Cruise." *New York Times*, December 25, 1946.

Vaughn, Stephen. *Ronald Reagan in Hollywood: Movies and Politics*. New York: Cambridge University Press, 1994.

Vermont. *Vermont Public Documents: Being Reports of State Officers, Departments, and Institutions*. Rutland, VT: Tuttle Company, 1918. https://babel.hathitrust .org/cgi/pt?id=chi.095914113;view=2up;seq=1.

Viggoda, Ralph. "Leader of the Storm Gen. H. Norman Schwarzkopf, Familiar to the World as Commander of Operation Desert Storm, Has Been Familiar Much Longer to Some Who Share His Tie to the Philadelphia Area. He's a Graduate of Valley Forge Military Academy—The Place That He Says 'Prepared Me For Life.'" Philly.com, February 26, 1991. http://articles.philly .com/1991–02–26/news/25774609_1_commander-of-operation-desert-norman-schwarzkopf-cadets.

Virginia, Department of Public Instruction. *Virginia School Report, 1885: Fifteenth Annual Report of the Superintendent of Public Instruction*. Richmond, VA: Public Printing Office, 1885.

Virginia Military Institute. "James Lucius Cross." Virginia Military Institute Archives, Historical Rosters Database. http://archivesweb.vmi.edu/rosters/ record.php?ID=454.

———. "William Henry Gillespie." Virginia Military Institute Archives, Historical Rosters Database. http://archivesweb.vmi.edu/rosters/record.php?ID=1072. Virginia Military Institute Archives, "New Market Ceremony History." http://www.vmi.edu/archives/civil-war-and-new-market/battle-of-new-mar ket/new-market-ceremony-history/.

Virginia Tech. "Module 09: The 1960s: Who Won? Student Protest and the Politics of Campus Dissent." Digital History Reader. http://www.dhr.history.vt.edu/ modules/us/mod09_1960s/.

The Visitor Center. "Tadeusz Kosciuszko." Lafayette Square. http://www.the-visitor-center.com/pages/Lafayette-Square/slides/lafayette-square-050.htm.

Wade, Gary. *World War II Division Commanders*. CSI Report No. 7. Leavenworth, KS: Combat Studies Institute, 1983.

Walker, Charles. *Biographical Sketches of the Graduates and Élèves of the Virginia Military Institute Who Fell during the War Between the States*. Philadelphia: J. B. Lippincott & Co., 1875. https://archive.org/details/memorialvirginia00walk.

Wallace, Michael. "The Use of the Virginia Military Institute Corps of Cadets as a Military Unit Before and During the War Between the States." Master's thesis, US Army Command and General Staff College, 1999.

Wallenstein, Peter. "The First Black Students at Virginia Tech, 1953–1963." *Diversity News* 4, no. 1 (Fall 1997): 3–5. https://spec.lib.vt.edu/archives/blackhistory/timeline/blackstu.htm.

———. "Not Fast, but First: The Desegregation of Virginia Tech." http://www.vtmag.vt.edu/fa1197/feature1.html.

Warner, Ezra J. *Generals in Blue: Lives of the Union Commanders.* Baton Rouge: Louisiana State University Press, 1992.

———. *Generals in Gray: Lives of the Confederate Commanders.* Baton Rouge: Louisiana State University Press, 1992.

Washington, George. "Sentiments on a Peace Establishment." In *Soldier-Statesmen of the Constitution*, by Robert Wright and Morris MacGregor, 193–99. Washington, DC: Government Printing Office, 2004.

Webb, Lester A. "The Origin of Military Schools in the United States Founded in the Nineteenth Century." PhD diss., University of North Carolina, 1958.

Wedekind, Jennifer. "The Childrens Crusade: Military Programs Move into Middle Schools to Fish for Future Soldiers." *In These Times*, June 3, 2005. http://www.inthesetimes.com/article/2136/.

Welcker, William T. *Military Lessons: Military Schools, Colleges, and Militia.* New York: Ivison, Blakeman, Taylor & Co., 1874.

Wells, R. "St. John's Military Academy Priceless Heritage." Paper in St. John's Military Academy Archives, Delafield, WI. 1960.

West, Jerry. *The Bloody South Carolina Election of 1876: Wade Hampton III, the Red Shirt Campaign for Governor, and the End of Reconstruction.* Jefferson, NC: McFarland, 2011.

WestEd. *Successful Magnet High Schools: Innovations in Education.* Report for the United States Department of Education, September 2008. http://www2.ed.gov/admins/comm/choice/magnet-hs/magneths.pdf.

Western New York Maritime Charter School. "The Mission." http://www.wnymcs9–12.com/domain/8.

Weston Historical Society. "Jarvis Military Academy." http://www.westonhistoricalsociety.org/jarvis-military-academy/.

"Where Are You Going to School?" *Boy's Life* 24, no. 3 (March 1934): 43.

Whitehead, M. G. *Collections of the Early County Historical Society*, vol. 2. Blakely, GA: Early County Historical Society, 1979.

Whittaker, Thomas. *Whittaker's Churchman's Almanac: The Protestant Episcopal Almanac and Parochial List.* New York: T. Whitaker, 1901.

Whitten, Rachael. "Hokie Alums Prospering Post-college, Gallup Survey Shows." *Collegiate Times*, September 3, 2015. http://www.collegiatetimes.com/news/hokie-alums-prospering-post-college-gallup-survey-shows/article_c86af3b4–528b-11e5-a4c9–87a88c52c046.html.

Widener University. "Vision & History." http://www.widener.edu/about/vision_history/.

Williams, M. "Lake Elsinore: Former Military Academy on the Lake Continues Decline." 2009. Paper in possession of the author.

Williams, Thomas. *The American Spirit: The Story of Commodore William Phillip Bainbridge*. Bloomington, IN: AuthorHouse, 2010.

Wilson, James Grant, and John Fiske. *Appleton's Cyclopaedia of American Biography*, 1888, vol. 2. New York: D. Appleton and Company, 1888.

——, eds. *Appleton's Cyclopaedia of American Biography*, 1892, vol. 2 New York: The Press Association Compilers, 1892.

Wineman, Bradford. "Francis H. Smith: Architect of Antebellum Southern Military School and Education Reform." PhD diss., Texas A&M University, 2006.

Winsor, Justin, ed. *The Memorial History of Boston, Including Suffolk County, Massachusetts, 1630–1880*, vol. 3. Boston: Ticknor & Company, 1881.

Wireback, T. "Bailed Out by the Wrong Benefactor? Oak Ridge: Can the School Survive?" *Greensboro News-Record*, March 26, 2011. http://www.greensboro.com/news/bailed-out-by-the-wrong-benefactor/article_e4476ab7–5980–52a8–8dbc-85697de35322.html.

Wong, E. "The Legacy of the University of Texas at Arlington Corps of Cadets Post-WWII Years Onwards." University of Texas at Arlington Army ROTC, 2005. Copy in possession of the author.

Woodward, C. Vann. *The Burden of Southern History*. Baton Rouge: Louisiana State University Press, 2008.

Woodward Academy. "Our History." 2009. https://www.woodward.edu/about/history.

Wyeth, John Allan. *History of La Grange Military Academy and the Cadet Corps, 1857–1862: La Grange College, 1830–1857*. New York: Brewer Press, 1907.

Wynstra, Robert. *The Rashness of That Hour: Politics, Gettysburg, and the Downfall of Confederate Brigadier General Alfred Iverson*. New York: Savas Beatie, 2010.

Yates, Bowling. *History of the Georgia Military Institute, Marietta, Georgia, Including the Confederate Military Service of the Cadet Battalion*. Marietta, GA: n.p., 1968.

Young, Nancy Beck. "Texas Military Institute, Austin," *Handbook of Texas Online*, Published by the Texas State Historical Association. http://www.tshaonline.org/handbook/online/articles/kbt17.

Young, Tommy. *Character Makes the Man: Kentucky Military Institute, 1845–1971*. Bloomington, IN, Trafford Publishing, 2013.

Yung, Roger. "What Did Lea Do? Lea's Army: Chinese Soldiers Trained on American Soil." http://www.homerleasite.com/Site/Current.html.

Zollinger, R. K. "A Case Study of Transformative Education in Action: Partnership Education for Sustainability." Doctorial diss., California Institute of Integral Studies, San Francisco, 2010.

Index

designations, xviii–xix; post-war trends, 113–14; school closures, 125, 166, 250; sinking of *Sultana,* 97; Union Army military school alumni, 83–85; Union Army occupation of Confederate schools, 370*n*47; Union Army veterans' impact on military school, 132–35

Clark, Gen. Wesley, 375*n*30

Clason Point Military Academy. *See* La Salle Military Academy (Oakland, New York)

class, socioeconomic, 37, 46, 58, 233, 238–39

Classical and Mathematical Academy, 160

Claverack Academy (College), 373*n*50

Clemson University (Clemson Agricultural College), 10–11, 129–30, 191, 197, 200

Cleveland, Grover, 175

Cleveland Junior Naval Academy, 224

Clinton, George, 24

Clinton Liberal Institute, 162

Clio (ship), 88

closures: administrative failures, 155, 157; arson, 373*n*48; availability of public schools, 187; Civil War enlistments, 59–64, 77–78, 357*n*14; desegregation, 142; due to Union Army occupation, 370*n*47; enrollment, 125, 156, 372–73*n*43, 381*n*55, 381*nn*4–5; financial problems, 156, 168, 246–47; fires, 187–89; Great Depression, 185–86, 190, 252; illness, 355–57*n*9; influence of pacifist movements, 158, 175–76; lack of space, 124; legal challenges, 186–87; low graduate success, 104; maritime military schools, 103–4; mergers, 125, 168; Vietnam era, 207–9, 252

CME (Committee on Militarism in Education), 181–82

CNN Money rankings, 260

coastal fortifications, 29–30

Coffin, Adm. Sir Isaac, 88

cohesion, 6–8

Colcock, Maj. Richard W., 48, 53, 54

Cold War, 160

colleges: admission of women, 217–20;

cadet corps participation requirements, 122, 123, 130, 132; Catholic, 155; cohesion and esprit de corps, 8; conflicts between military and civilian programs, 121, 126, 128, 131, 137, 163–64, 181; current numbers of, xv; enrollment, 243–45; growth, 239, 243–45; Lost Cause, 114–19; preparation for, 183–84; transition to military schools during Civil War, 57; Vietnam War, 209–13, *212,* 252. *See also* antimilitary attitudes; Morrill Land Grant Acts; *specific schools by name*

Collegiate Institute, 168

Collins, Robert, 84

Colonial education, 21

Columbia Military Academy, 208–9

USS *Columbus* (ship), 94

Committee on Militarism in Education (CME), 181–82

community colleges. *See* junior military colleges

Compulsory Education Act, State of Oregon (1922), 186–87

conduct, codes of: influence of Partridge, 36; maritime schools, 103; religious influence, 144; school culture, 3–4, 17, 193, 206–7, 235–37; Thayer Method, 43; US Air Force Academy, 242; at USMA, 44–45; VMI, 50

Cone, Adm. Hutchinson, 105

Confederate Conscription Act (1862), 62, 66, 69

Confederate Naval Academy, 74–76, 92

Congregational Church, 168, *169*

USS *Congress* (ship), 94

USS *Constitution* (ship), 76–77

Continental Congress, 23

Cornelison, Charles, 124

Couch, Maj. Gen. Darius N., 80

Coulter, John, 86

Coulter, Gen. John B., 375*n*26

counterculture, 206–13, 252

Courtenay, Edward H., 46

Cox, Samuel, 165

Crackel, Theodore, 44

Crane, Stephen, 373*n*50

Crawford, Richard Tyler, 127

Crawford, William, 33